127 Topics in Current Chemistry

Fortschritte der Chemischen Forschung

Managing Editor: F. L. Boschke

Organic Chemistry

With Contributions by
M. Asami, K. G. Heumann, H. J. Keller,
K. P. Meurer, T. Mukaiyama, Z. G. Soos,
F. Vögtle

With 138 Figures and 64 Tables

Springer-Verlag
Berlin Heidelberg GmbH
1985

This series presents critical reviews of the present position and future trends in modern chemical research. It is addressed to all research and industrial chemists who wish to keep abreast of advances in their subject.

As a rule, contributions are specially commissioned. The editors and publishers will, however, always be pleased to receive suggestions and supplementary information. Papers are accepted for "Topics in Current Chemistry" in English.

ISBN 978-3-662-15229-4   ISBN 978-3-540-39231-6 (eBook)
DOI 10.1007/978-3-540-39231-6

Library of Congress Cataloging in Publication Data. Main entry under title:
Organic chemistry.
(Topics in current chemistry = Fortschritte der chemischen Forschung; 127)
Bibliography: p. Includes index.
1. Chemistry, Organic — Addresses, essays, lectures.
I. Asami, M. II. Series: Topics in current chemistry; 127.
QD1.F58 vol. 127 [QD255] 540s [547] 84-26879

2152/3020-543210

Managing Editor:

Dr. *Friedrich L. Boschke*
Springer-Verlag, Postfach 105280, D-6900 Heidelberg 1

Editorial Board:

Table of Contents

Helical Molecules in Organic Chemistry

Kurt P. Meurer[1] and Fritz Vögtle[2]

1 BAYER AG, Wissenschaftliches Hauptlaboratorium 5090 Leverkusen, FRG
2 Institut für Organische Chemie und Biochemie der Universität Bonn, Gerhard-Domagk-Straße 1,
  D-5300 Bonn 1, FRG

Table of Contents

Kurt P. Meurer and Fritz Vögtle

1 Introduction

1.1 Screw Structures in Chemistry

Screw structures or helices (helix: Greek ἕλιξ = winding, convolution, spiral) are encountered in various variations in nature and technique. Propeller-shaped, helical structures play an important role in architecture, physics, astronomy and biology. Screw-shaped macromolecular skeletons of nucleic acids, proteins and polysaccharides are important structural elements in biochemistry. Their helix turns often are stabilized through hydrogen bonds, metal cations, disulfide linkages and hydrophobic interactions.

α-Amylose was the first helix postulated for a natural macromolecule. In solution it contains approx. six glucose units per turn and is stabilized through hydrogen bridges connecting the three hydroxy groups. Normally, the amylose "host molecule" contains on the average, one "guest" water molecule per glucose unit. The cavities of the helix are just large enough to take up approx. one I_2 molecule per helix turn [1a, 2a]. The enclosed iodine guest causes the characteristic intense blue color. The helical cavities of amylose, having a diameter of approx. 450–700 pm (6-helix and 7-helix, resp.) [2b] also enclose stilbenes and other compounds of appropriate size. The 6-helix forms the most stable amylose and can harbor benzene as well as aliphatic compounds in its cavities [2b].

One of the most important natural compounds with a helical structure is deoxyribonucleic acid (DNA). It was studied by Watson and Crick [3] based on the X-ray analysis of Franklin and Gossling [4]. Two right-handed polynucleotide chains are wound around a central axis. The sugar phosphate chains form an outer screw line, whereas the purine and pyrimidine bases are projecting inside. Their ring planes are

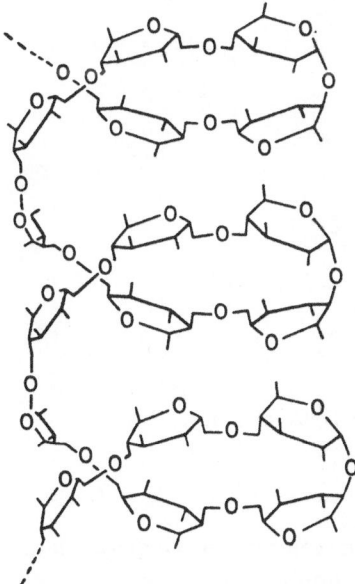

Fig. 1. Structure (simplified) of the α-amylose helix (in solution)

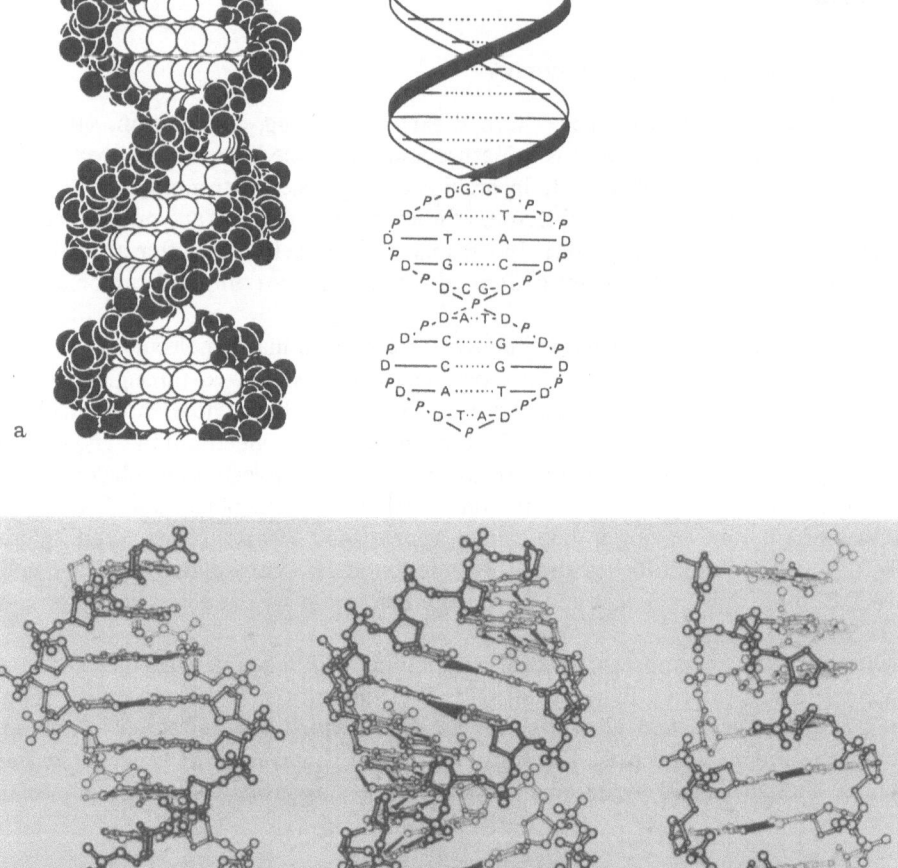

Fig. 2. a Classical DNA structure, **b** structure of A-, B- and Z-DNA (cf. Saenger, [5b])

parallel to each other and approx. perpendicular to the central axis. Hydrogen bonds between complementary bases allow the legendary "double-helix structure", which is stabilized also by hydrophobic interactions between facing bases. A complete screw winding contains ten base pairs (pitch of the helix: 3400 pm; see Table 1). The sequence of the purine or pyrimidine bases in one of the strands determines the second polynucleotide chain; both carry a complementary information. Through despiralization of the double helix catalysed by the Rep-protein called "helicase" [3] and formation of two daughter double helices, which are identical with the former, DNA is able to reduplicate identically [5a], the foundation of (biological) growth and heredity (Fig. 2). Today we know that DNA can build up several conformations, of strongly different structures. This flexibility allows it to fulfill many biological functions. The biologically active species, which is formed at high hydration grades, is B-DNA. If water is stripped from B-DNA, caused through higher ion concentrations, A-DNA is formed, which is only hydrated at the strongly polar phosphate groups.

Until recently, only right-handed double helices were known. A left-handed helix structure was observed for the first time in the synthetic hexameric DNA containing alternating cytosine and guanine bases [d(CGCGCG)]. Both polynucleotide strands are wound around themselves in a left-handed sense. The phosphate groups show a zick-zack-type pattern along the screw; this structure therefore was named Z-DNA (Fig. 2).

Table 1 compares the characteristic structural features of A-, B- and Z-DNA. Z-DNA segments seem to occur in chains with long alternating CG-sequences, as they are found in nature, within otherwise B-configurated DNA. Special regulatory functions (e.g. activation of DNA transcription by the CAP protein) were ascribed to these left-handed helices. Such structural changes can be caused by high salt concentrations, but also by alkylation of the guanine units.

Helical structures also are encountered in *proteins*. Three conformational types characterize their secondary structure: the α-structure (helix), the β-helix is a right-handed screw with 3.6 amino acids forming one turn — its pitch height is 540 pm.

Table 1. Comparison of the structures of A-, B- and Z-DNA

Helix Type	A	B	Z
Screw tendency	right	left	left
Corrugated conformation of sugar units	C-3'-endo	C-2'-endo	G: C-3'-endo C: C-2'-endo
Orientation around the glycosidic bond	anti	anti	G: syn C: anti
Orientation around the C-4'—C-5' bond	(+)-gauche	(+)-gauche	G: trans C: (+)-gauche
Rotation per base pair	32.7°	36°	60° per CG dimer
Number of bases per pitch of the helix	11	10	12
Pitch of the helix	28.2 Å (2820 pm)	34 Å (3400 pm)	44.5 Å (4450 pm)
Inclination of the base pairs against the helix axis[a]	19°	—6°	—7°

[a] Positive values: Inclination of the base pairs counterclockwise.

For an α-helix structure the formation of a maximum number of hydrogen bridges [7] is important. As a prerequisite to this, the configurations of the amino acid components should be uniform; this is one of the main reasons why mainly L-amino acids have been found in natural proteins. Right-handed helices result for energetic reasons.

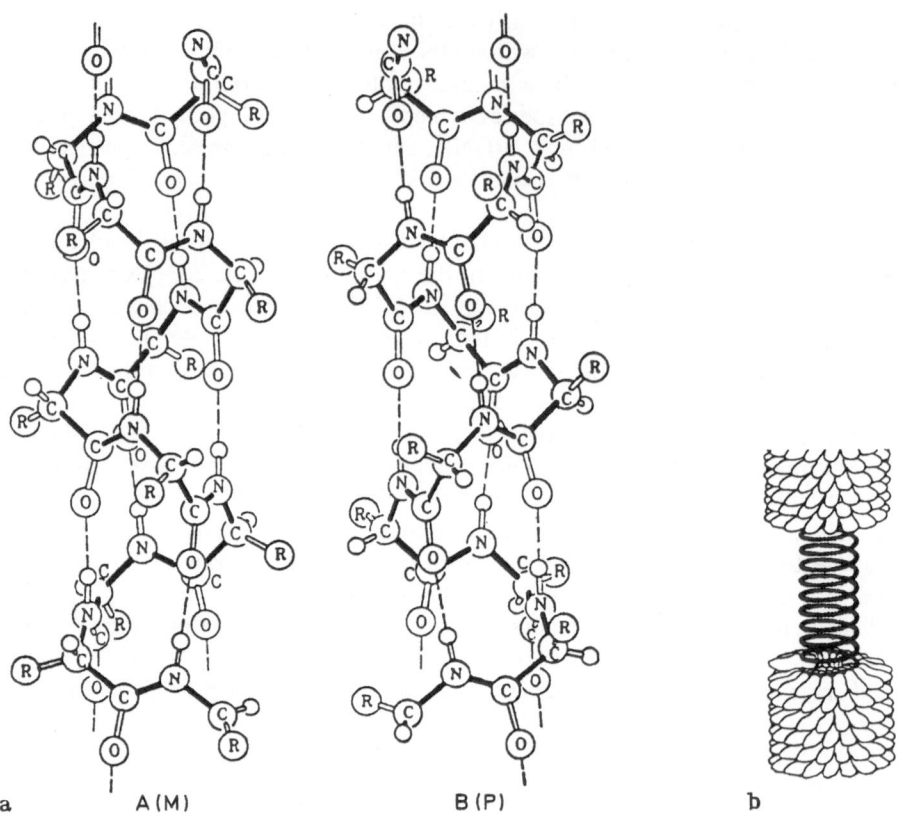

a   A (M)   B (P)   b

Fig. 3. a Left-handed (A) and right-handed (B) α-helix conformation in proteins. **b** Screw arrangement of proteins in the tobacco mosaic virus (TMV) [1b]

Table 2. Formation of the α-helix: Stabilizing, destabilizing and breaking off amino acids

Amino acids stabilizing the formation of the α-helix:			
Ala	Phe	Met	
Cys	Tyr	AspN	
Leu	Try	GluN	
Val	His		
Amino acids destabilizing the formation of the α-helix:			
Ser	Glu	Lys	Ile
Thr	Asp	Arg	Gly
Amino acids breaking off the formation of the α-helix:			
Pro			

Apart from the α-helix in proteins (Fig. 3, Table 2), a helical structure is also found for the conformation of the structural subunits of collagen and tropocollagen [8]: three screw-shaped polypeptide chains, constructed alternately mainly from glycine, proline and hydroxyproline, are wound around each other and form a rigid·cable. The single strands of this helix form a structure by far wider open than those known in the polypeptide helix series. Other molecules which are structurally similar to collagen are found in other areas where mechanical endurance is important. Myosin

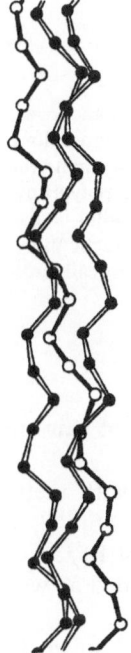

Fig. 4. Tropocollagen helix [8]

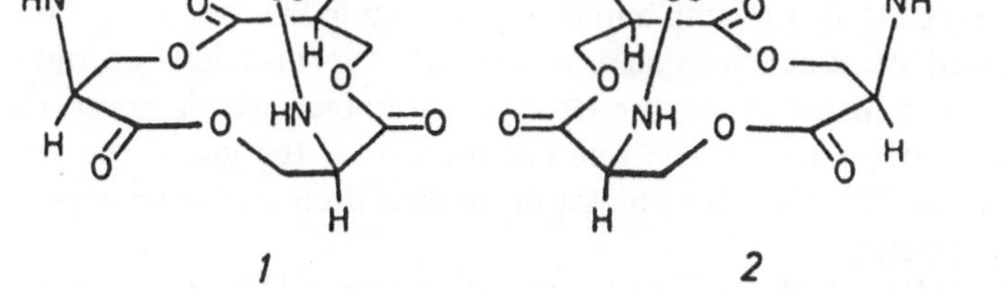

Fig. 5. Δ-cis-Fe(III)-enterobactin (*1*) and Λ-cis-Fe(III)-enantio-enterobactin (*2*)

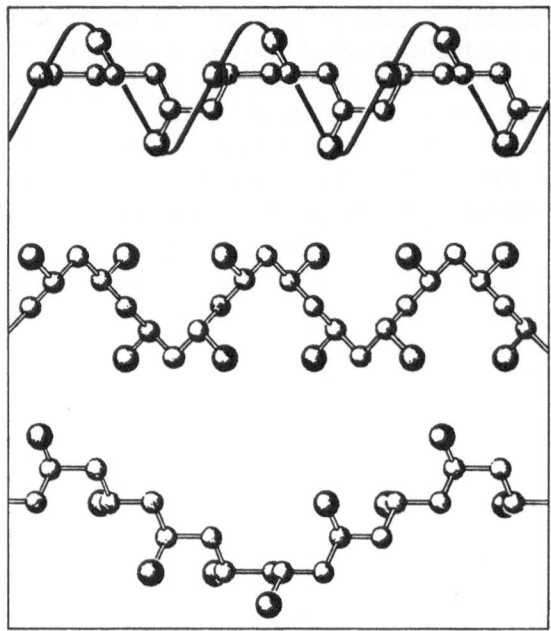

Fig. 6. Polypropylene: isotactic arrangement of methyl groups along the helix (described by a tape); the structure of syndiotactic polypropylene (drawing in the middle) is also regular; the chain in atactic polypropylene (below) is statistically irregular [1a, 12]

and tropomyosin of muscle and skin, the fibrin in blood clot and keratine of hair are examples [9].

Helical structures in nature also are formed through complexation with metal cations: Some microbes, including Escherichia coli, produce enterobactin [10], a siderophor, which complexes Fe(III) with a very high complex constant of 10^{52}.

Six ligand atoms coordinate the cation in such a way that a right-handed as well as a left-handed helix could exist; but only the right-handed helix is found.

In *macromolecular chemistry* helically wound carbon chains are well known. In isotactic polypropylene (Fig. 6) the methyl groups are arranged on a helix. Some other chiral polymers form helices, like polymethylmethacrylate, polytriphenylmethyl-methacrylate [11], poly-1,2-butadiene and poly-tert-butylethylene oxide [1a, 12].

Helical (+)-poly(triphenylmethyl-methacrylate) has been applied successfully as a chromatographic material for the separation of racemates [11]. An overview of the helix types exhibited by crystalline macromolecules [12] and some examples are given in Table 3.

Further helical structures in chemistry are formed, e.g. by *urea*, which in the course of the crystallization from suitable solvents builds up a hexagonal lattice. It contains long channels, surrounded by urea molecules, which either form a left- or right-handed helix. If urea is crystallized in the presence of a suitable racemic guest, the right-handed lattice preferably encloses the one enantiomer of the guest, the left-handed lattice the other one [13a]. This discrimination of enantiomers can be applied for the separation of racemates.

Bilatrienes, e.g. Biliverdin of the all-Z-all-syn-configuration [13b], are helically structureɑ in the solid state; two molecules in case of appropriate configuration form

Table 3. Helical arrangement of polymers [12]

Conformation	Spatial arrangement along the chain	Helical type[a]	Rotation angle	Examples of polymers
···ttt···		1_1	0/0	polyethylene
		2_1	0/0	polymethylene
		1_1	0/0	st-polyvinyl chloride
···tgtg···		13_1	16/16	polytetrafluoroethylene
		3_1	0/120	it-polypropylene ($R = CH_3$)
			0/120	it-polystyrene ($R = C_6H_5$)
			0/120	it-poly-5-methyl-1-heptene ($R = CH_2-CH_2-CH-CH_2-CH_3$)
				$\quad\qquad\qquad\qquad |$
				$\quad\qquad\qquad\qquad CH_3$
		7_2	−13/110	it-poly-4-methyl-1-pentene
		11_3	−16/104	it-poly-3-methylstyrene
		4_1	−24/96	it-poly-3-methyl-1-butene
			0/90	it-poly-o-methyl-styrene
			−45/95	it-polyacetaldehyde
···ggg···		9_5	103/103	polyoxymethylene
···ttgttg···		7_2	−12/12/120	polyethyleneglycol
		7_2	−36/0/106	polyglycine II
···tggtgg···		11_3	0/122/143	α-helix of polypeptides
···ttgg···		4_2	0/−120/−120	st-polypropylene

[a] Building blocks per turn.

a kind of a dimer helix [13c)] caused by intermolecular NH ... O bonds. From red algae a substance called carrangeenan has been isolated, which consists of two fractions, the ϰ-fraction (D-galactose, 3,6-anhydro-D-galactose and their sulfates) and the λ-fraction (galactopyranosyl-D-galactose, partly as the sulfate). In warm water it is well soluble, cooling yields a thixotropic gel. The stability of this gel is reached by carrageenan forming a double helix structure. It is therefore applied in food chemistry as a gel and binding material [14)].

1.2 Definition of a Helix in Chemistry

Helicity (screw sense) is a (geometric) structural property, by which the arrangement of atoms in molecules can be described as a combined rotation and translation process. Centro-chiral, planar-chiral molecules can be considered also as helical, having in mind their spatial environment (helicity around a point, a plane or an axis). In numerous cases, the understanding of chiral structures under the aspect of helicity is advantageous.

Cahn, Ingold and Prelog [15)] described and defined the helix as follows:

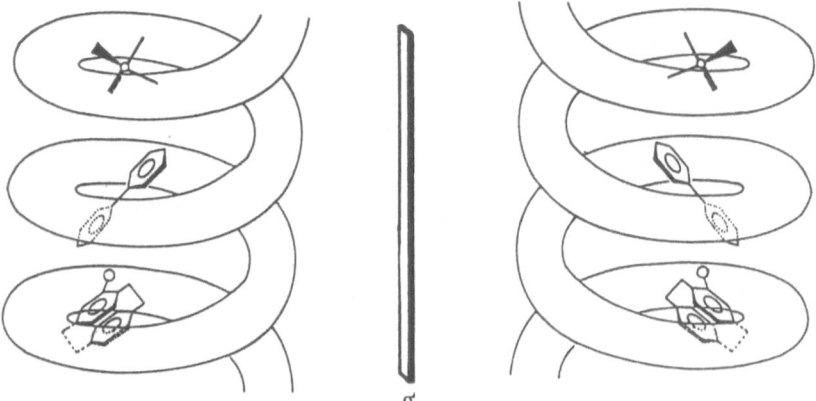

Fig. 7. Point, axis and plane in a helical surrounding

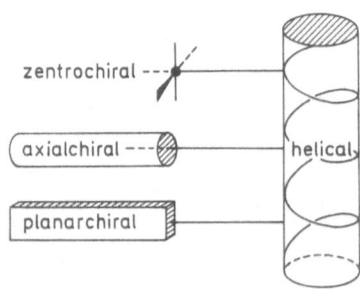

Fig. 8. Interdependence of centrochirality, axial chirality, planar chirality and helicity (schematic illustration)

"A helix is characterised by a helical sense, a helical axis, and a pitch (ratio of axially linear to angular properties). Thus, helicity is a special case of chirality, because it implies two additional properties. When a natural object or process has all the three properties, the helical model, just because it provides so well-filled-in a portrait, may be the easiest abstract model to recognise and use, even though it summarises more properties than may be relevant to the matter under discussion. Of course, when a chiral object has not all the helical properties, nor by a simple adaptation can be given them, the specialised helical model is inapplicable, and the object belongs to the general case of chirality."

A molecular helix is defined by an axis, by its screw sense and its pitch. If the radius is constant, there is a cylindrical helix, if it changes constantly, a conical helix exists. A palindromic helix is characterized by a constant pitch.

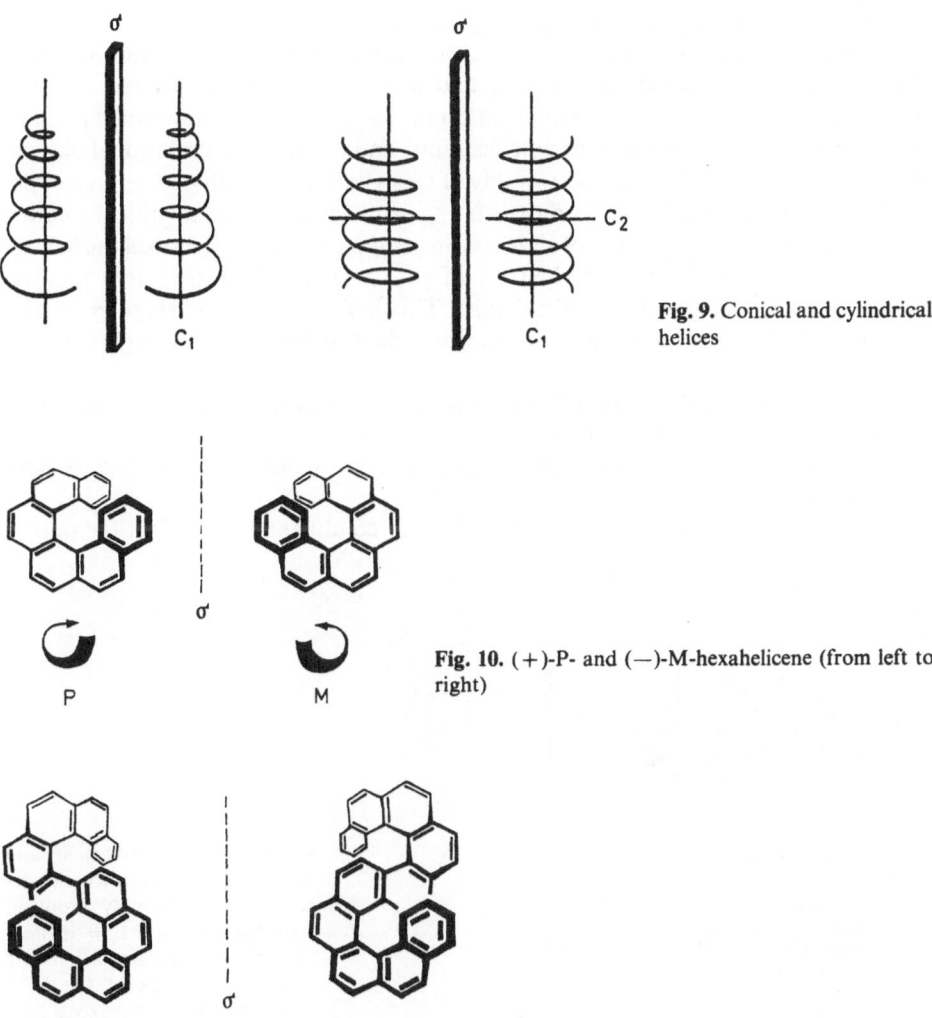

Fig. 9. Conical and cylindrical helices

Fig. 10. (+)-P- and (—)-M-hexahelicene (from left to right)

Fig. 11. M,P-decahelicene and P,M-decahelicene

11

A helix is considered to wind starting from the viewers eye to a point distant from the viewer; the screw sense is right-or left-handed according to the turns of the watch-fingers (Fig. 10). Turning the helix in space relative to the observer doesn't change the screw sense. A right-hand helix is symbolized by P (Plus), a left-handed helix by M (Minus). As Fig. 10 (left side) shows, the absolute configuration of (+)-rotating hexahelicene has been determined to be P, and (−)-hexahelicene to have the M configuration (see chapter 3).

If a molecule contains several helical segments, which may be of opposite screw sense (amphiverse helix, see also Fig. 14), these helical subunits should be described using M or P (Fig. 11).

In such cases, especially with large (biochemical) molecules, it is often useful to differentiate between the absolute helicity, which is the sum of all M- and P-helical domains, and the net helicity, defined by the difference of the volumes of left-handed and right-handed helix domains. The net helicity is of interest for chemical use with respect to the observed chiroptical properties.

Brewster [16] developed a "helical conductor model" for the description of helical arrays of atoms according to geometric and mathematical points of view. He also pointed out the difficulties of the application of the helical conductor model (helicity of atom chains, which function as a helical unit) and of the "conformational dissymmetry model", which bases on an assembly of small helical units (that may be of opposite helicity). Both models may lead to differing results with respect to optical rotation values and to the assignment of absolute configurations. Brewster described the helicity of nonlinear lines and defined the helix as a figure, which is characterized by a movement of a point along an axis (helical axis). This helical axis is archetypally characterized by a circular and a linear movement, which follows either a right- or left-handed screw sense.

A longer atom chain, composed of four or five tetrahedrally bound atoms, can be described by P-, M- or T-configurations.

In addition, step helices, amphiverse helices, radial helices and star helices have been differentiated [16]:

An impressive example of the use of the helical conductor model [19] of the optical

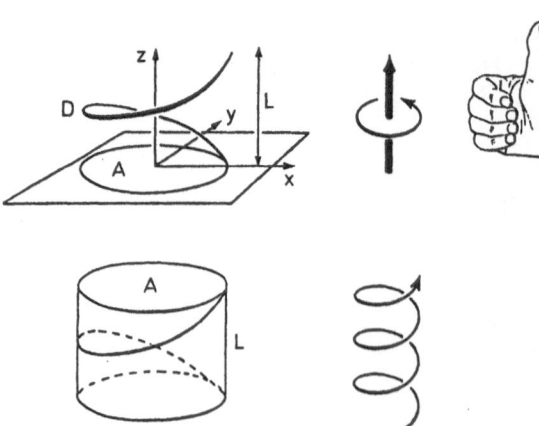

Fig. 12. Right-handed helicity and its description: Geometrical description of a simply crooked, right-handed helix: L = inclination of the helix parallel to the z-axis, A = area of the helix projection on to the x,y-plane; D: length of the helix line [16]

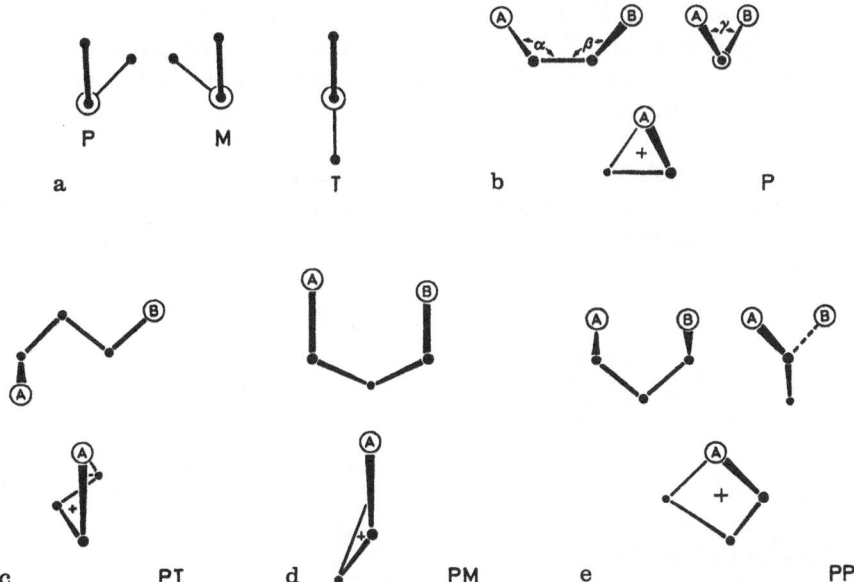

Fig. 13a–e. P-, M-, T-configurations. Absolute configurations of four- to five-membered atomic chains: **a, b** four-membered atomic chain with the absolute conformation P, M and T; **c** five-membered atomic chain with the absolute configuration PT; **d** five-membered atomic chain with the absolute configuration PM; **e** five-membered atomic chain with the absolute configuration PP [16]

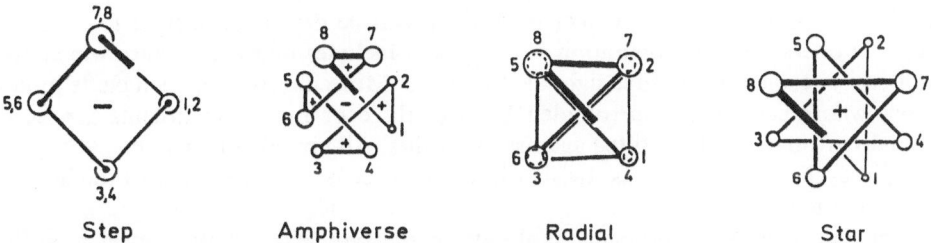

Fig. 14. Step helix, amphiverse helix, radial helix, star helix (cf. Brewster [16])

activity is the rather exact theoretical prognosis of the optical rotation of twistane (*3*) [20]. The high experimental rotation value of twistane [M_D] 570 [17], 590 [18] (in ethanol), was approached fairly good according to this model with a value of [M_D] 484 for the (+)-enantiomer, theoretically. According to the conformational dissymmetry model [21] for twistane only a theoretical rotation value of [M_D] 140 was evaluated, which is not very close to the experimental value [20]. A correlation of the circular dichroism using the octant rule for 2- and 4-twistanone (*4, 5*) was not very successful with respect to the rotational sense and to the absolute configuration of twistane.

The helical conductor model is useful for the prediction of optical rotations and configurations of aliphatic helices which do not contain chromophoric skeletons, and for twisted chains of atoms. In studies of the chiroptical properties of isotactic polymers produced from chiral alkenes, the conformational dissymmetry model was somewhat more successful [22, 23]. Brewster concludes in a survey given in 1974 [16]: "*Thus the story is not done.*" The difficulties to exactly describe helical molecular structures and to interprete and correlate their physical properties theoretically are continuing.

Recent studies of molecular chirality in which precise new definitions are used, such as "chiropticity" and "stereogenicity", specifying chirality and topicity, give rise to some optimism also in the field of helical structures [24, 25].

3   *4*   *5*   **Fig. 15.** Twistane (*3*), 2-twistanone (*4*), 4-twistanone (*5*)

1.3 Molecular Propellers

Propeller structures are well known from nature, science and, in various types, in technology, e.g. in the fields of movement techniques and energy production. Propellers consist of interrupted screw windings mostly of the same screw sense. Methyl groups substituted by arene or hetarene substitutents often act as lipophilic haptophores for drugs [26f]. Molecules with two- and three-bladed propellers have been reviewed in detail elsewhere. It seems sufficient here to give a short glance on the subject [26a–f]. Molecular propellers [26] have skeletons which can be described using a D_n- or C_n-($n > 1$) operation. This definition is broad and includes many molecules which are not easily recognized on first sight as propellers. Molecular propellers include molecules whose centers are surrounded by two, three or more substituents arranged radially, symmetrically or tilted against each other. All propeller blades, e.g. aromatic or spirocyclic rings, should be twisted in the same sense and thus lead to the helical conformation.

Here, we consider a propeller molecule of the type A_3Z whose central atom Z is substituted by three aryl groups. The C atoms bound to the central atom Z form a reference plane which cuts the three phenyl rings perpendicularly. The ortho substituents A, B and C can be arranged alternating above or below this plane. Thereby, we obtain eight (2×4) different d, l pairs. The one pair in which all substituents lie above the reference plane is shown in Fig. 16 (top). Turning the aryl rings, we generate four helical arrangements starting with both planar structures. If the central atom Z is replaced by one with central chirality, we obtain eight different isomers (Fig. 16, bottom). If the same symmetry operations are carried out with the three pairs omitted in Fig. 16, we obtain $4 \times 8 = 32$ isomers.

Trimesitylmethane (*12a*) [32, 34] is an example of an Ar_3ZX propeller possessing C_3 symmetry. The X-ray analysis of dimesityl(2,4,6-trimethoxyphenyl)methane (*12b*) proves the propeller conformation for the solid state [33]. For the interconverting exo- and endomethoxyl groups of *12* in a temperature range of 118–176 °C, $\Delta H^{\neq} = 74$ kJ/mol, $\Delta S^{\neq} = 9.8$ eu are found, whereas for *13* a lower entropy of activation

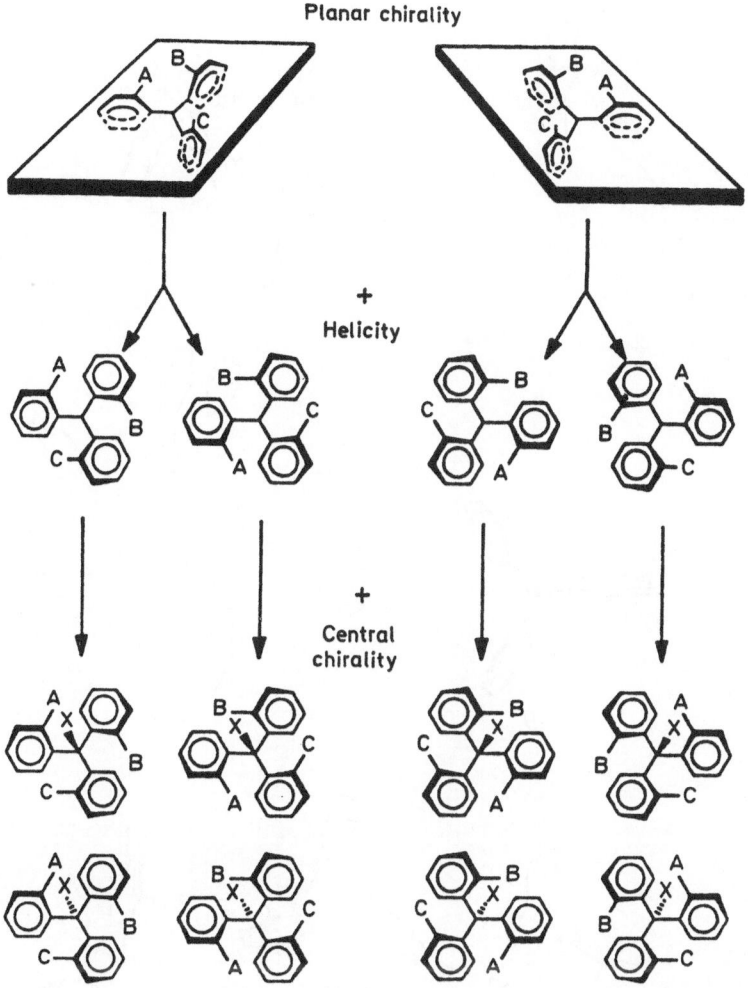

Planar chirality

+

Helicity

+

Central chirality

Fig. 16. Symmetry and isomerism of the propeller skeleton A_3Z (cf. Mislow et al. [26])

of $\Delta H^{\neq} = 33.5$ kJ/mol, $\Delta S^{\neq} = -10$ eu in a temperature range of T = -69 up to $+81$ °C is obtained [26]. The helical structure of some further triphenylmethane derivatives has been ascertained by NMR spectroscopy in solution and electron scattering in the gas phase. Using dynamic NMR spectroscopy, the ring isomerization of the arylboranes *15* [26] and *25–27* has been studied: Barriers between $\Delta G^{\neq} = 46$ kJ/mol and a maximum of 67.8 kJ/mol have been measured (Table 4). For the corresponding triarylcarbenium ions, dependent on the substituents, activation energies of 36.4 and 59.0 kJ/mole have been evaluated.

Wittig et al. [37a] separated biphenyl-α-phenyl-naphthylphosphine (*16*) into its enantiomers by using D-(+)-campher-10-sulfonic acid; the rotational strengths are: $[\alpha]_{436}^{20} = +9.7$, $[\alpha]_{436}^{20} = -8.5$, $[\alpha]_{578}^{24} = -5.2$ (c = 3.38, CH$_2$Cl$_2$). Musso et al. [37b] succeeded in determining the absolute configuration of the ter- and quaterphenyl

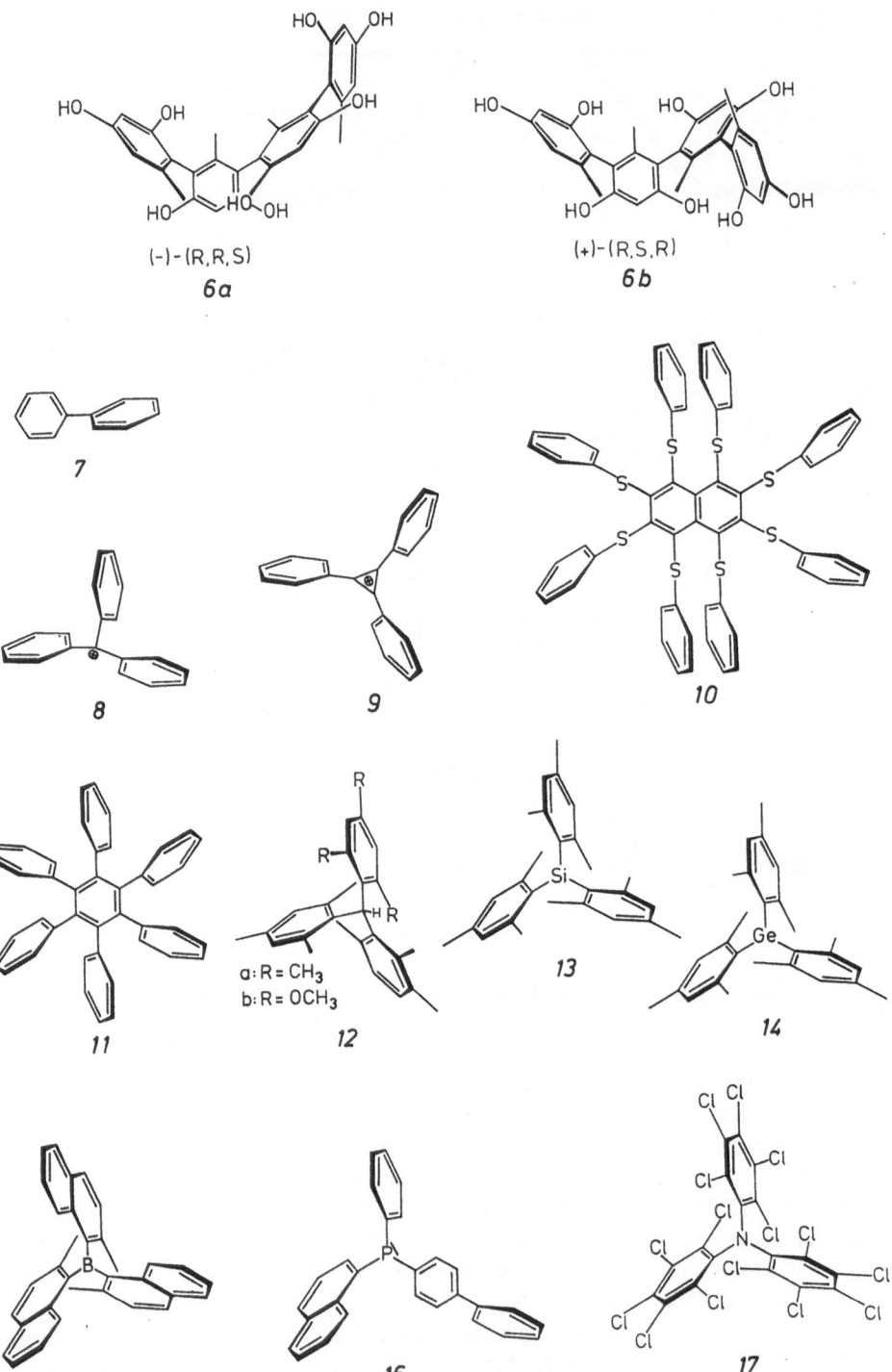

Fig. 17. Some examples of molecular propeller structures (*6–17*) [26–43]

Table 4. Ring Isomerization of Triarylcarbenium Ions *18–24* and of Triarylboranes *15, 25–27*

Triarylcarbenium ions	$\Delta G^{\neq}$ [kJ/mol]	Triarylboranes	$\Delta G^{\neq}$ [kJ/mol]	T [°C]
18	59.0	*15*	67.8 / 66.6 / 61.1	20 / 20 / 20
19	54.8	*25*	58.2	15
20	53.2			
21	48.6	*26*	64.5	20
22	51.5	*27*	46.0	-68
23	43.5			
24	36.4			

derivatives *6a, b* of orcin and measured their optical rotation; they were resolved into their enantiomers by chromatography on potato starch: $(-)$-(R,R,S)-*6a*$[-48.9]_D^{RT}$ (ethanol); $(+)$-(R,S,R)-*6b*$[+117.4]_D^{RT}$ (ethanol). The optical rotations of the resolved enantiomers did not allow any configurational assignments and point to a high negative contribution of the central (R)-configuration or molecular M-helicity. A barrier of racemization of $\Delta G^{\neq} = 91.7$ kJ/mol [26] has been found for trimesitylmethane *12a*. This is in accord with force field calculations (83.7 kJ/mol). The trimesitylsilane *13* and the corresponding trimesitylgermane *14*, both related to triphenylmethane, exhibit a lower barrier of $\Delta G_{-47} = 45.6$ kJ/mol [38] and of $\Delta G_{-80} = 38.5$ kJ/mol. resp. [26].

Perchlorotriphenylamine (*17*, D_3 symmetry) [39] has been ascertained by X-ray analysis to be a molecular propeller and was enantiomerically enriched using triacetyl cellulose: 0.4% ee for (+)-*17*, 1% ee for (−)-*17*. The racemization barrier (measured: $E_a = 116.2$ kJ/mol, $\Delta H^{\neq} = 113.9$ kJ/mol, $\Delta S^{\neq} = -45.2$ kJ/mol, $\Delta G^{\neq} = 131.4$ kJ/mol at T = 120 °C), which was well predicted by force field calculations (104.7–113 kJ/mol), demonstrates the optical stability of the antipodes.

This propeller compound has been completely separated into its enantiomers using (+)-poly(triphenylmethyl-methacrylate) (PTrMa, *116* [11,40]: $[\alpha]^{25}_{435} = +2385$, $[\alpha]^{25}_{546} = +1193$, $[\alpha]^{20}_{589} = +985$ (CCl$_4$); $[\alpha]^{25}_{435} = -2344$, $[\alpha]^{25}_{546} = -1200$, $[\alpha]^{25}_{589} = -967$ (CCl$_4$). The CD absorptions of the (+)- and (−)-enantiomer show Cotton effects at 270 (intensive), 300 (intensive) and 330 nm.

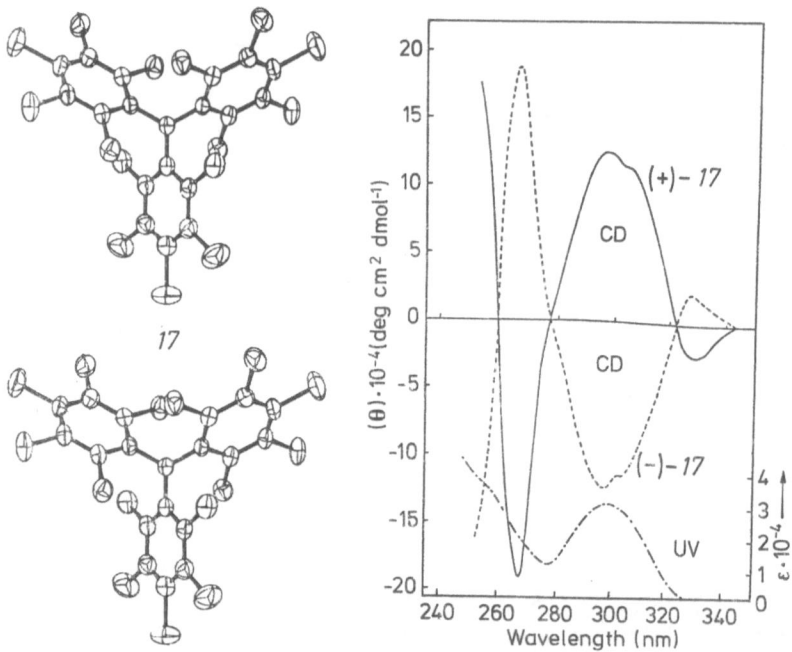

Fig. 18. X-ray structure of perchlorotriphenylamine *17* and CD- and UV curves of (+)- and (−)-*17* [38,40]

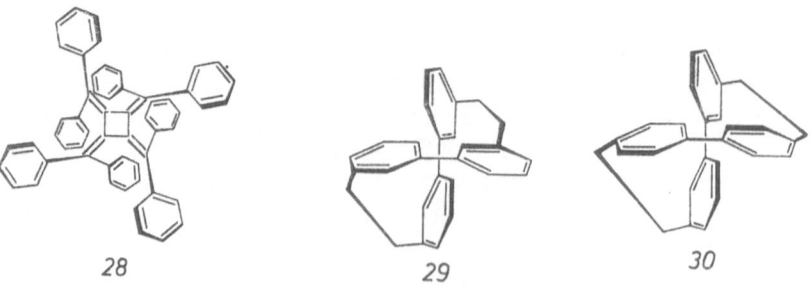

Fig. 19. Some additional propeller molecules (*28*–30) [41,42]

18

An attractive molecular propeller is octaphenyltetramethylencyclobutane (*28*, Fig. 19) [41]. To date, the synthesis has been carried out only for the heptaphenyl derivative, but the propeller structure can be recognized already from the high-field shifts of overlapping phenyl rings ($\delta = 6.85$ ppm).

The two-bladed propeller molecules *20* and *30* also are helical [42a, b]: $\delta_{H_i} = 5.97$ ppm, $E_a = 34$ kJ/mol for *20*, $\delta_{H_i} = 5.77$ ppm, $E_a = 53.1$ kJ/mol for *30*.

There are several more propeller-shaped molecular structures. Here, a selection had to be made to facilitate the understanding of the bridged and clamped propeller molecules introduced in Section 4.2.

2 Aliphatic and Alicyclic Helices

2.1 Atrop Isomers

The axial chirality of the atropisomeric oxalic acid derivatives *31a,b* [43] according to Brewster [19] also can be classified as helicity. (R)-$(+)$-N,N,N′,N′-Tetramethyl-dithiooxamide *31a* then constitutes one of the smallest molecular helices, the enantiomers of which have been separated semipreparatively using liquid chromatography

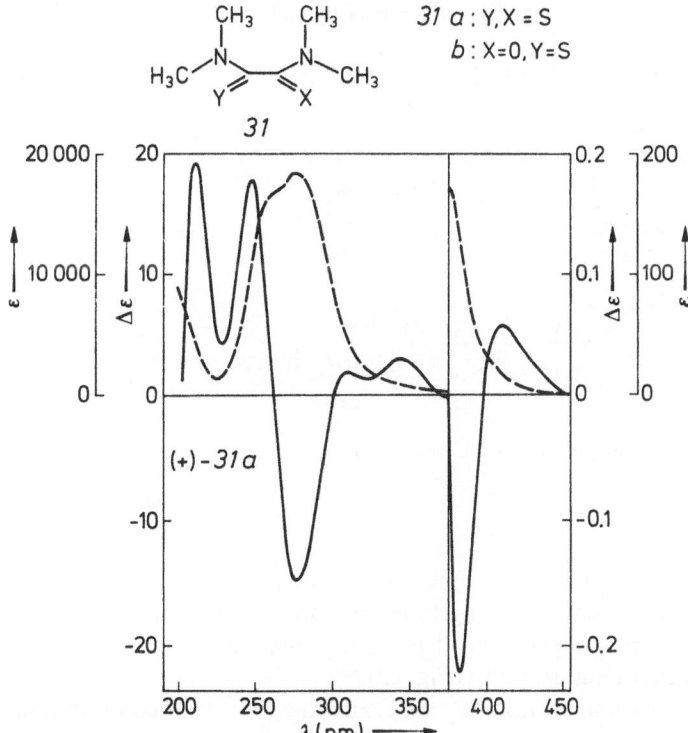

Fig. 20. *31a, b*; CD- and UV spectra of *31a* (in ethanol) [43]

on triacetyl cellulose. The CD spectra (Fig. 20) have been interpreted by qualitative MO theory [43]. For crystalline *31a* a torsion angle θ (SCCS) of 93.1° has been found [44], for *31b* an SCCO angle of 87.4° [45]. The racemization of *31a* in chloroform at 60.9° yields a barrier for the partial rotation around the $=C—C=$ single bond of $\Delta G^{\neq}$ = 107.2 kJ/mol. The nonplanarity of substituted dithiooxamides in solution had been detected by Carter and Sandström [45].

The CD spectrum of (+)-*31a* and the UV spectrum can be seen from Fig. 20. There are strong Cotton effects at 276 and 248 nm, which correlate with strong absorption bands at 273 and 259 nm of the UV spectrum. In the short wave range a strong Cotton effect is found at 211 nm. Of the Cotton effects at longer wave lengths (411, 381, 343 and 308 nm) the one at 343 nm corresponds to a weak UV maximum. The absolute configuration of $(+)_{365}$-*31a* has been assigned by CD spectroscopy and qualitative MO arguments to be (R). The enantiomeric purity of *31a* was 0.5 ± 0.1.

2.2 Spiro Isomers

Spirochiral molecules also can be understood as helical compounds. The polyspiro alkanones *32*, *33* are composed mainly of 2,3-connected tetrahydrofurane building blocks. They have been named "helixanes" because of their helical topology [47]. Their physical and chiroptical properties have been scarcely investigated. Only for the cyclopentyl[4]helixane *32* an X-ray analysis has been published, which gives insight into the helical geometry of this molecule (Fig. 21).

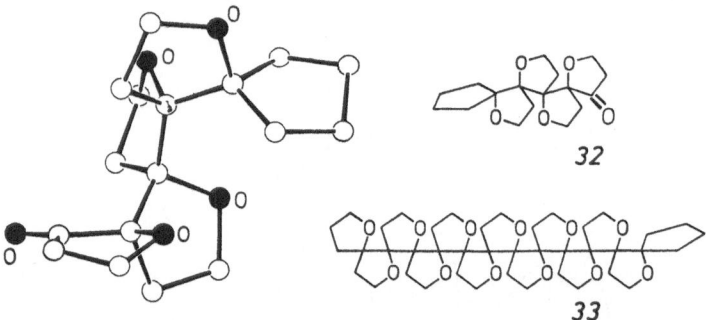

Fig. 21. Helixanes *32* and *33*; X-ray structure of *32* (left side)

2.3 Twist Isomers

The twisted boat form (twist boat) of the cyclohexane molecule is a simple example of a helical alicyclic hydrocarbon skeleton. It is a conformation of the cyclohexane skeleton with low conformational stability (Fig. 22) [48]

By bridging of suitable carbon atoms of cyclohexane, the twist boat conformation is fixed. Two ethano bridges in the 1,4- and 1,5-positions of cyclohexane as clamps lead to tricyclo[4.4.0.0^{3,8}]decane (twistane, *3a,b*). In addition to four equally sub-

Table 5. Structures and optical rotations of twistane derivatives *3a, b* and *34–50*

No.	Structure	Name	Optical rotation α (solvent)	Ref.
3a		(+)-twistane	$[+414]_D^{22}$, $[+434]_D^{22}$ (ethanol)	49, 50, 53, 54)
3b		(—)-twistane	$[-437]_D^{22}$, $[-440]_D^{22}$ (ethanol)	49, 50, 53, 54)
34		(+)-2,7-twistadiene	$[+444 \pm 11]_D^{22}$ (ethanol) $[+423 \pm 11]_D^{22}$ (trichloromethane)	55)
35		(—)-2,7-twistadiene	$[458 \pm 11]_D^{22}$ (ethanol) $[-423 \pm 11]_D^{22}$ (trichloromethane)	55)
36		2-twistene	$[+414]_D^{22}$ (ethanol)	54)
37		(+)-2,7-dioxa-4,9-twistadiene	$[+326 \pm 8]_D^{22}$ (trichloromethane)	55, 56)
38		(—)-2,7-dioxa-4,9-twistadiene	$[-321 \pm 7]_D^{22}$ (trichloromethane)	55, 56)
39		(+)-2,7-dioxatwistane	$[+222 \pm 8]_D^{22}$ (trichloromethane)	55, 56)
40		(—)-2,7-dioxatwistane	$[-217 \pm 6.5]_D^{22}$ (ethanol) $[-229 \pm 5.5]_D^{22}$ (trichloromethane)	55, 56)
41		2-azatwistane	$[-423.6]_D^{20}$ (ethanol)	57)

21

Table 5 (continued)

No.	Structure	Name	Optical rotation α (solvent)	Ref.
42		2-oxatwistbrendane	$[-142]_D^{15}$ (ethanol)	[58]
43		4,5-diazatwist-4-ene	$[+1165]_D^{RT}$ (n-hexane) $\theta_{220} = +8.2 \times 10^{-3}$ $\theta_{375} = -16.2 \times 10^{-3}$ (n-hexane)	[50, 52]
44		twistbrendane	$[-284]_{436}^{20}$ (abs. ethanol)	[61]
45		brexane	$[-94.3]$ (ethanol)	[60]
46			$[-233]_D^{RT}$ (trichloromethane)	[59]
47		C_2-bishomocubane	$[-44]_D^{RT}$ (trichloromethane)	[59]
48		D_3-trishomocubane	$[-165]_D^{RT}$ (trichloromethane)	[59]
49		C_2-bismethanotwistane	$[-293]_D^{RT}$ (trichloromethane)	[59]
50		9-nortwistbrendane	$[+99]_D^{RT}$ (n-hexane)	

stituted chiral centers, this molecule shows right- or left-handed helicity, which explains the high rotational values ($3a$: $[\alpha]_D^{22} = -437 \pm 12$, in ethanol).

According to Brewster [19], a net P-helicity (P,P,M,P,M) and a net M-helicity (M,M,P,M,P) can be assigned to P-twistane ($3b$) and M-twistane ($3a$), respectively. The first synthesis of optically active twistane has been reported by Nakazaki [49]. He postulated — as did shortly afterwards Tichy — an absolute configuration for (+)-twistane, which was corrected later by Tichy [50] as follows: (+)-$3b$ corresponds to ($1R,3R,6R,8R$), (−)-$3b$ corresponds to ($1S,3S,6S,8S$) [56]. But even today there are some uncertainties with respect to the absolute configuration in the twistane series [62]. Numerous aliphatic and heterocyclic molecules having helical twistane skeletons have been synthesized and characterized stereotopologically. Some of them are shown in Table 5 together with some chiroptical data.

Fig. 22. Ring inversion of the twist-boat conformation of cyclohexane

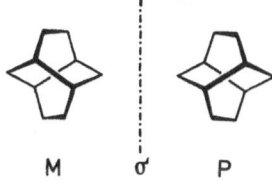

M σ P   **Fig. 23.** M- and P-helicity of the twistanes $3a, b$ [48]

The high rotational strengths of molecules like 3 and 34–50 are remarkable. The helicity of the skeleton may account mainly for this. Molecules exhibiting the same rotational sign show the same helicity too, e.g. the enantiomer rotating counter clockwise is correlated to minus (M) helicity [19].

The dioxa analogues 37–40 of the twistanes $3a,b$ and of 2,7-twistadienes 34, 35 show lower rotation strengths compared to the corresponding carbocyclic skeletons.

The rotational strengths of the twistbrendane 44, brexane 45 and of the cubane derivatives 46–48 also are smaller than those of twistane and 2,7-twistadiene. The circular dichroisms of the twistanones and twistanediones have been investigated [50, 62] (Table 6).

Paquette et al. [51] stated the linear dependence of the rotational values at the Na$_D$ line from the bridge lengths in twistane derivatives (Fig. 25).

The optical rotation values $[\alpha]_D$ ($3b$, 44, 45, 47–49) significantly decrease in the series twistane ($\equiv 3$), twistbrendane 44 and brexane 45. This can be interpreted in terms of helicity as follows: Whereas the twistane molecule 3 contains five 6-membered rings in the twist conformation of equal helicity, in the case of brendane 44 there are only three 6-membered and two less twisted 5-membered rings; brexane 45 only contains two 6-membered and three 5-membered rings.

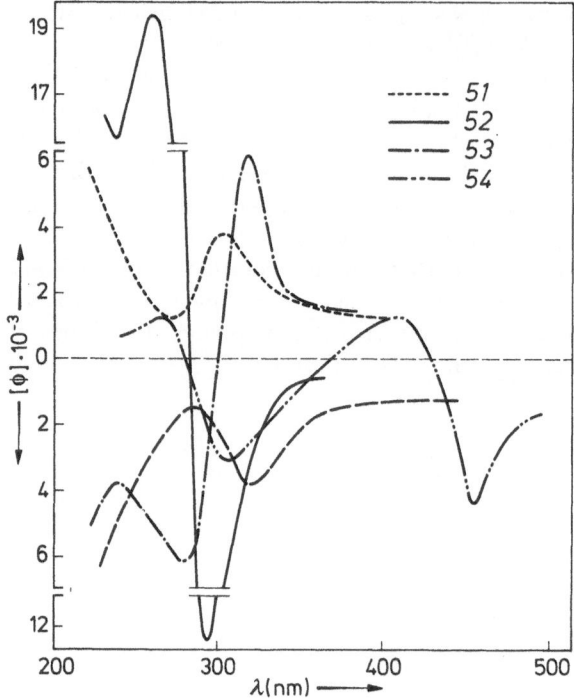

Fig. 24. CD curves of *51–54*

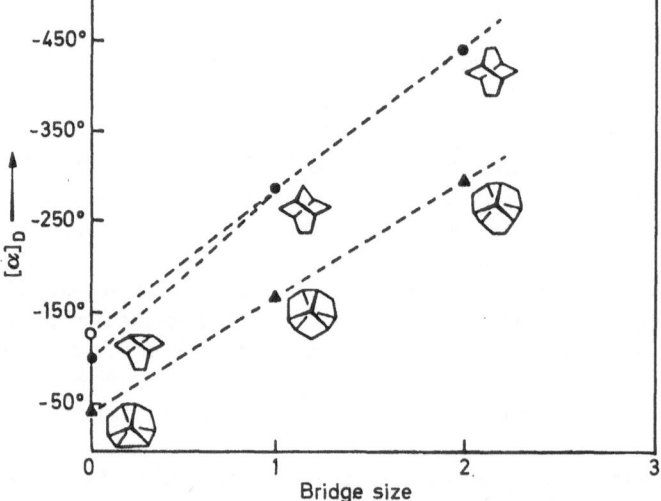

Fig. 25. Correlation of the strain induced by the bridge with the optical rotation [51]

In the series C_2-bismethanotwistane *49*, D_3-twisthomocubane *48* and C_2-bishomocubane *47* the rotation value also decreases linearly with the decrease of the number of the bridging atoms. The rotation values of the cubanes are more than $100°$ lower than those of the twistanes.

Table 6. Chiroptical properties of *51–54*

(+)-tricyclo[4.4.0.^{3,8}]decan-4-one (*51*)	$[\alpha]_D^{25}$ 295.2 (ethanol)	CD (ethanol): $\Delta\varepsilon_{285.5} = +0.88$
(—)-(6R)-tetracyclo[4.4.0.0^{2,9}.0^{3,8}]-decan-5-one (*52*)	$[\alpha]_D^{25}$ 49.4 (methanol)	CD (ethanol): $\Delta\varepsilon_{279} = -8.2$
(+)-(3S)-tricyclo[4.3.1.0^{3,7}]decan-4-one (*53*)	$[\alpha]_D^{25}$ 97.8 (methanol)	CD (ethanol): $\Delta\varepsilon_{300} = +2.7$
(—)-tricyclo[4.4.0.0^{3,8}]decan-4,5-dione (*54*)	$[\alpha]_D^{20}$ —326 (methanol)	CD (dioxane): $\Delta\varepsilon_{445} = -1.26$
		$\Delta\varepsilon_{282} = -1.18$
		$\Delta\varepsilon_{217} = +3.17$

Double bonds in the bridges of the twist helices do not seem to appreciably influence the magnitude of the rotation values. An exception is 4,5-diazatwistaene *43*, the rotation of which is $[\alpha]_D^{RT} = 1165$ [51, 52].

The CD spectrum of *43* shows a negative Cotton effect at 370 nm and a weaker positive Cotton effect at 222 nm. The electronic energy transfer at 370 nm corresponds to a n→π-transition, whereas the band at 222 nm stems from a lower-lying n→π-transition as in twistane *3a* (200 nm, $\Delta\varepsilon = 11.6$) [50].

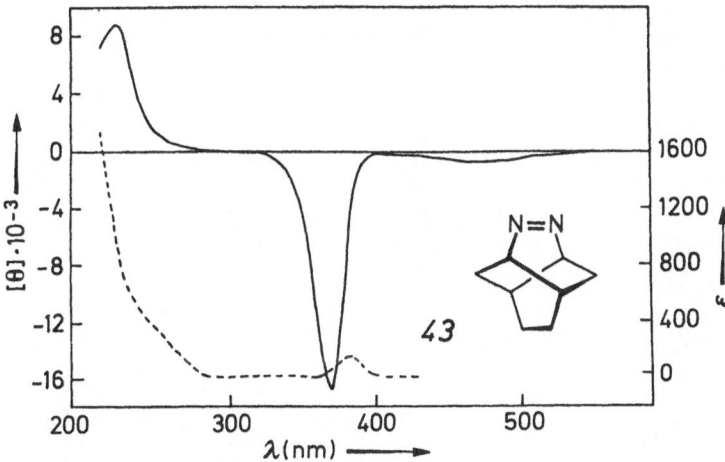

Fig. 26. UV and CD spectra of 4,5-diazatwistaene *43* (in hexane) [51]

2.4 Helical Alkenes

2.4.1 Aliphatically Clamped Alkenes

Introduction of an aliphatic clamp in the (*E*)-position of alkenes leads to twisted double bonds and therewith to helical alkenes like *55–65*.

(*E*)-cyclooctene *55* [63] has hitherto been classified with the regularly planarchiral species [15], but it can also be described under the viewpoint of helicity. Cope succeeded in separating the enantiomers [63] and assigned the absolute configuration by synthesis starting with a chiral educt of known configuration [64]. Scott ascertained the con-

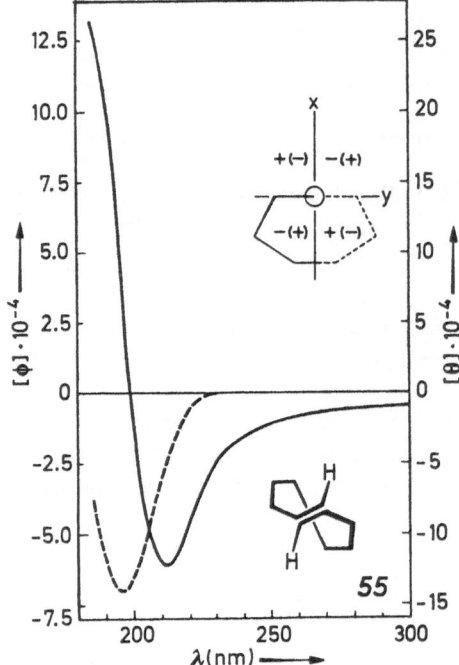

Fig. 27. ORD and CD spectra of (*E*)-(—)-cyclooctene *55* (in cyclohexane). Octant projection along the Z axis; background octant signs in brackets [65]

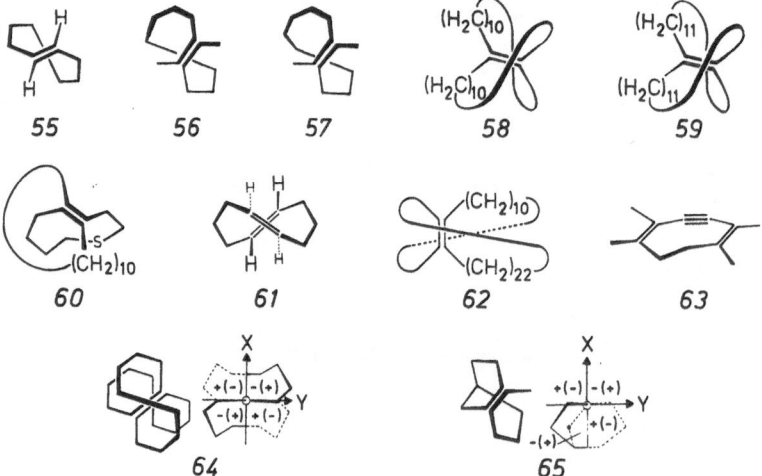

Fig. 28. Examples of aliphatically clamped alkenes *55–65*. Octant projections of (*S*)-(+)-*64*, (*R*)-(—)-*65* according to Scott [65]

figuration by means of an octant projection [65]. The sign and the rotational strengths have been explained by the asymmetric (helical) surrounding of the double bond, which is formed by the loop of the methylene groups. As can be seen from (*E*)-(—)-cyclooctene *55* in Fig. 27, all methylene groups are situated in octants with a negative sign. The ORD and CD spectra of this enantiomer therefore show a negative Cotton effect. The rotation value of (E)-cyclooctene is $[\alpha]_D^{25} = -476$ (in CH_2Cl_2).

Scotts viewpoint [65] can be applied also to the (E)-2-cyclooctene-type molecules 64 and 65, synthesized by Nakazaki [68]. The octant rule predicts the absolute configuration (S) at C atom 5 for the minus rotating enantiomer of bicyclo[3.3.1]non-1(2)-ene (64) ($[\alpha]_D^{30} = -237$, in ethanol, $[\alpha]_D^{30} = -259$, in $CHCl_3$) [69]. For the counterclockwise rotating enantiomer of (R)-(−)-D_2-bicyclo[8.8.0]octadec-1(10)-ene (65), which was prepared only in low optical purity ($[\alpha]_D^{24} = -2.3$, in dioxane) [66], the (R)-configuration is predicted by the octant rule.

(E)-Cyclononene (56) and (E)-cyclodecene (57) are not optically stable. Only at a temperature of −27 °C a rotation value of $[\alpha]_{678}^{-20} = -119$ was measured for (−)-56. The racemization barrier is $\Delta G_{263}^{\neq} = 79.2 \pm 0.8$ kJ/mol. 57 had no measurable optical activity [67].

A rotation value of $[\alpha]_D^{29} = +46.9$ ($CHCl_3$) has been found for [10.10]betweenane 58. The R-configuration of the (+)-modification has been correlated with a negative Cotton effect and compared with the Cotton effect of the 1,2-disubstituted (E)-cycloalkenes [72a]. The syntheses of [11.11]betweenane 59 and of the thia derivative [10.7]betweenane 60 have been carried out, but no chiroptical investigation has been undertaken hitherto. Recently, Nakazaki reported on the synthesis of (R)-(+)-[10.22]betweenane (62) with an optical rotation of $[\alpha] = +27.2$, $[\theta]_{220} = +128$ [72].

1,5-Cyclooctadien-3-ine 63 and (E),(E)-1,5-cyclooctadiene (61) are interesting helical molecules, which proved too reactive (63) [70] and photolabile (61) [71] to be separated into the corresponding enantiomers.

2.4.2 Distorted Alkenes

Helical atropisomers are produced if sterically demanding substituents are fixed at the double bond, which cause a twisting.

Attempts to construct the highly twisted double bond of 66 have not been successful to date [73]. Nevertheless, the less sterically strained tetraneopentylethene 67 has

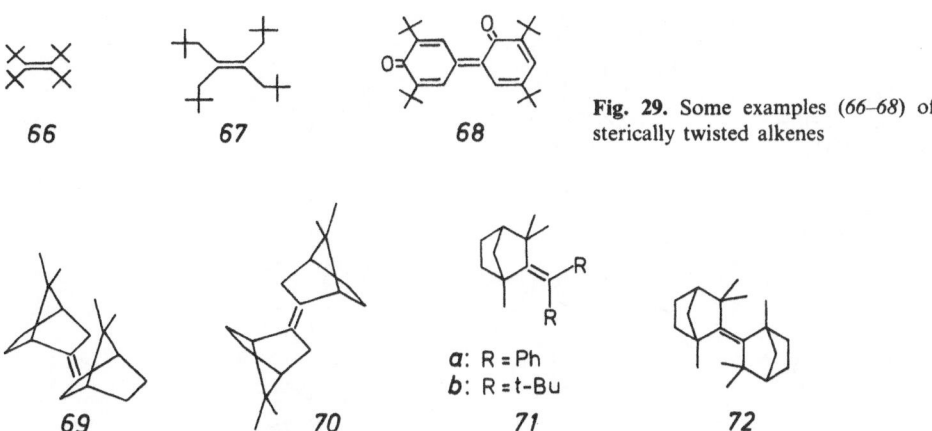

66   **67**   **68**

Fig. 29. Some examples (66–68) of sterically twisted alkenes

69   **70**   a: R = Ph / b: R = t-Bu   **71**   **72**

Fig. 30. Twisted bornanylidenes 69–72 [75−77]

been synthesized [74] — its barrier of racemization is $\Delta G^{\ne}_{145} = 90.8 \pm 2$ kJ/mol. The bis-chinonmethide 68 even has been separated into the enantiomers; the barrier of racemization is $\Delta G^{\ne} = 107$ kJ/mol [78].

The sterically twisted bornanylidenes 69, 70 and norbornylidenes 71, 72 have been separated into the enantiomers and their chiroptical properties have been studied [75–77] (see Table 7).

Table 7. Optical rotations of 69–72 [75–77]

No.	$[\alpha]^T_\lambda$ (solvent)
69	$[-178.2]_{578}$, $[718.1]_{368}$ (n-hexane)
70	$[180.9]_{578}$, $[-723.5]_{365}$ (n-hexane)
71a	$[-360.2]^{25}_D$ (ethanol)
71b	$[230.8]^{23}_D$ (ethanol)
72	$[-240]^{23}_D$ (ethanol)

2.4.3 Helical Aralkenes

β-Methyl-β-arylacrylic acids of the type 73 have a helical character due to their hindered rotation. Adams et al. were successful in separating the enantiomers of 1-α,β-dimethyl-β-(2,4-dimethyl-6-methoxyphenyl)acrylic acid (73) using its brucine salt [79, 81]. The rotational value $[\alpha]^{22}_D$ of a not completely separated sample in n-butanol was determined to be —51. The average half-life of the enantiomers amounts to 73 min at 44 °C, and to 700 min at 22 °C according to Adams [79]. In the framework of studies on molecular asymmetry, Cope et al. investigated (E)-bicyclo[8.2.2]tetra-deca-5,10,12,13-tetraene (74) [82]. The pure enantiomers could not be obtained, but an enantiomeric complex of (S)-(—)-α-methylbenzylamine ($[\alpha]^{25}_D = -35.1$) was isolated by fractional crystallization.

The structural helicity of benzo[9]annulenone (75, X=C=O) in solution has been evaluated by the chiral shift reagent Eu(dcm)$_3$ [83].

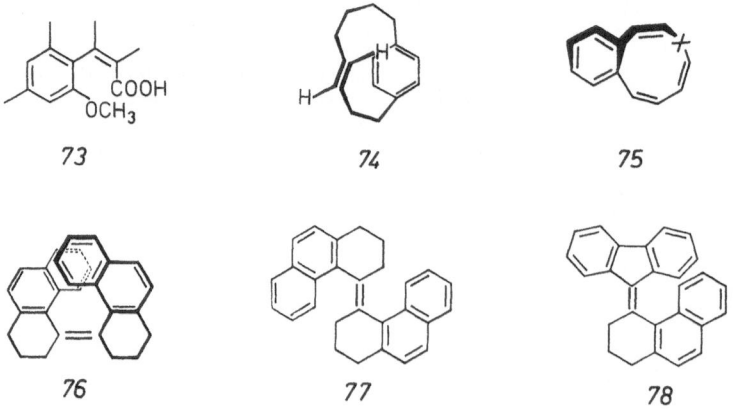

73　　　　74　　　　75

76　　　　77　　　　78

Fig. 31a. Helical Aralkenes 73–78

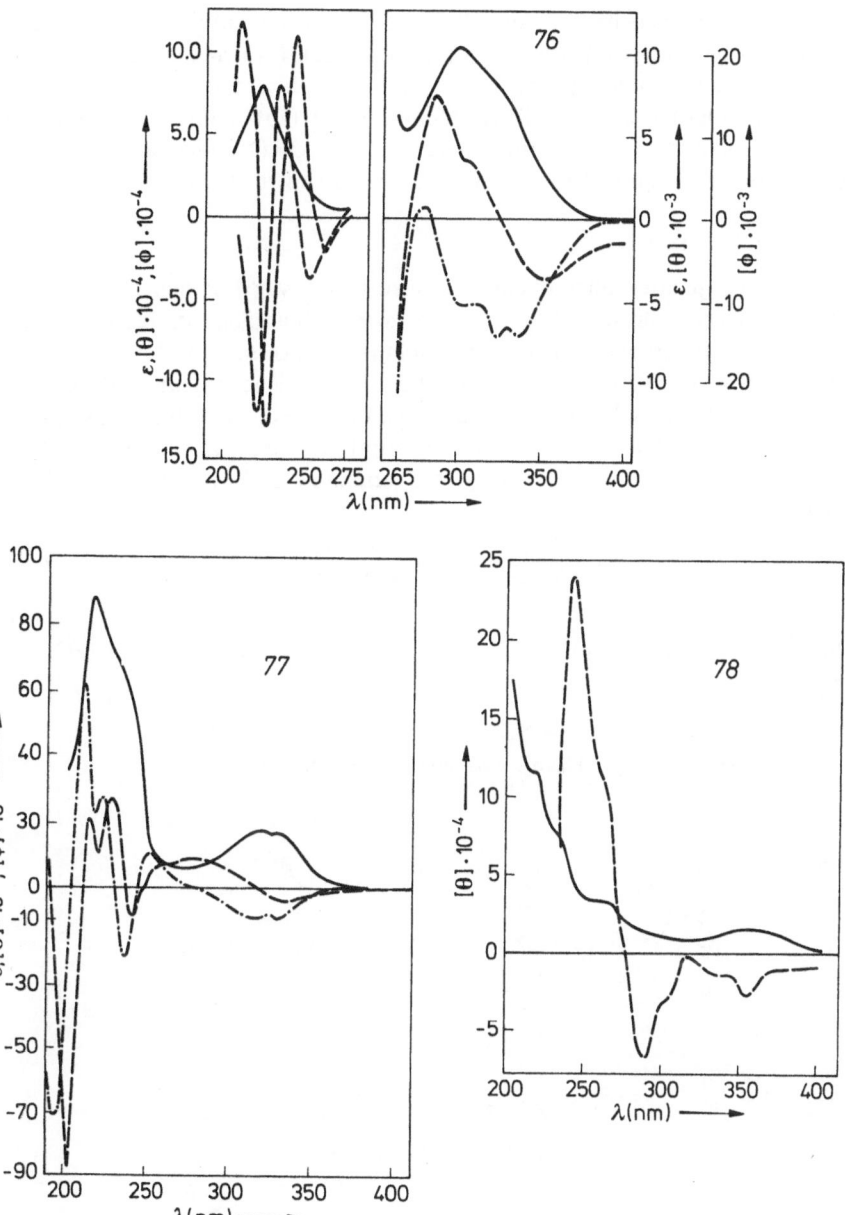

Fig. 31b. I , ORD (----) and CD (-·--·-) spectra of *76–78* [84)]

In the process of synthesizing heptahelicenes, Wynberg et al. [84)] were successful in obtaining (*Z*)- and (*E*)-4,4'-bi-1,1',2,2',3,3'-hexahydrophenanthrylidene (*76*, *77*). The chiroptical properties of these sterically twisted helical aralkenes were studied carefully. The enantiomer separation was carried out using an aluminium oxide column doped with (*S*)-(+)-TAPA. Whereas *76* is structurally related to heptahelicene (v.i.), *77* shows a double helix structure [84]. The rotation value of the enriched enantiomer mixture of (+)-*77* is $[\alpha]^{22}_{578} = +88$ (in n-hexane). (—)-*76*, on the other hand, could be separated completely: $[\alpha]^{22}_{578} = -508$ (in n-hexane).

4-(9-Fluorenyliden)-1,2,3,4-tetrahydrophenanthrene (*78*) exhibits a similar high rotation value of $[\alpha]^{22}_{578} = -770$ (in n-hexane). UV, ORD and CD spectra of the (*Z*)-alkenes *76* and *78* and of the (*E*)-alkene *77* are shown in Fig. 31b [84].

2.4.4 Helical Oligoalkenes

Atropisomeric conjugated oligoalkenes can also be classified as helical compounds. Such structures are represented by *s-cis*- or *s-trans*-butadiene derivatives bearing voluminous substituents which cause anticoplanar conformations.

There is a difference between biphenyl- and 1,3-butadiene systems with respect to symmetry: Atropisomeric biphenyl compounds, because of their dihedral symmetry, need a specific substitution pattern to be chiral. In contrast, nonplanar, helical butadienes belong to the point groups C_2 or C_1 and are chiral without bearing specific substituents.

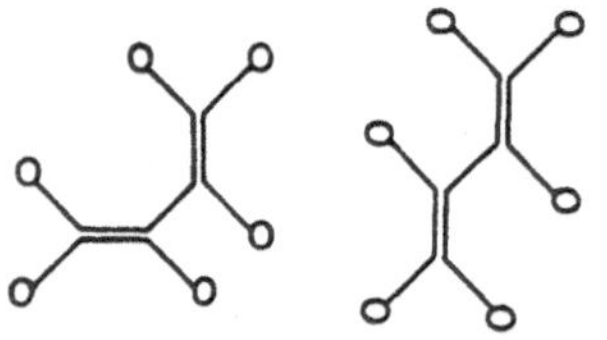

Fig. 32. *s-cis*- and *s-trans*-butadienes

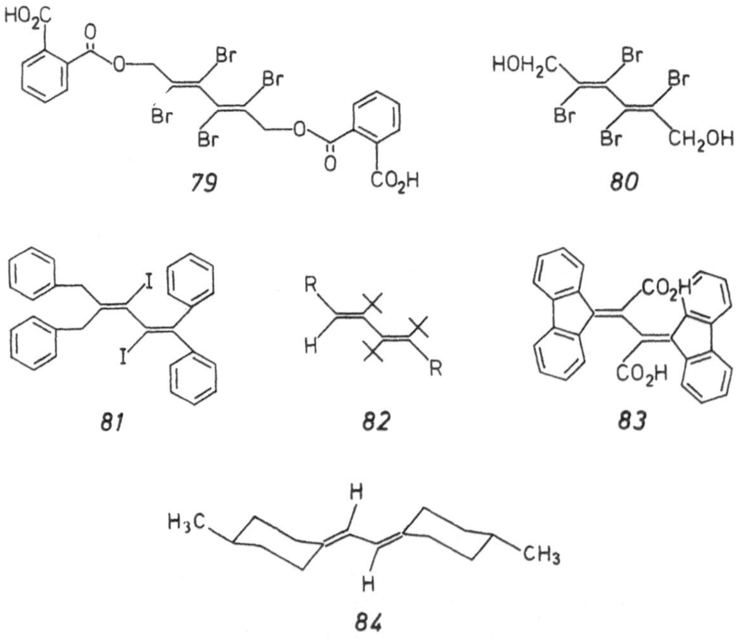

Fig. 33. Helical butadienes *79–84*

Köbrich et al. [85] were successful in separating the enantiomers of *79* via the diastereomeric salts of *79* with D-(+)-α-phenylethylamine by repeated extraction with hot acetone.

The diastereomeric salt obtained showed a specific rotation of $[\alpha]_{546}^{20} = -9.5$ (in methanol). Diluted sulfuric acid released optically active (−)-*79* with $[\alpha]_{546}^{20} = -16.0$ (in acetone). Diluted sodium hydroxide at room temperature led to (*E,E*)-(−)-*80* with a rotation of $[\alpha]_{D}^{20} = -14.3$ (in acetone). The optical active butadienes *79*, *80* are stable as solids, but racemize slowly at room temperature in solution (*79*: $\Delta G_{20}^{\neq} = 96.6 \pm 1.3$ kJ/mol, and *80*: 99.1 ± 1.3 kJ/mol, resp.) [85].

Attempts to synthesize the helical oligoalkene *81*, which is believed to form stable atropisomers, were unsuccessful as yet [85]. *82* has been synthesized, but proved to be rather unstable ($\Delta G^{\neq} = 100.4$ kJ/mol) [78].

As early as 1975, Goldschmitt et al. [86] partially separated the racemate of the fulgenic acid *83*. (+)-Bisdiphenylfulgenic acid was obtained using the brucine salt: $[\alpha]_{D}^{20} = +9.1$ (in CHCl₃). The optically active acid *83* racemizes completely at room temperature in 20 min, at 8 °C in approx. 70 min.

Walborski et al. [88,89] carefully studied the chiroptical properties of (*E*)- and (*Z*)-10,19-dihydrovitamin-D isomers, which have a biaxial *s-trans*-1,3-diene structure. (*aR,aR*)-(+)-bis(4-methylcyclohexyliden)ethane (*84*) shows a rotation value of $[\alpha]_{Hg}^{25} = +37.9$ (in ethanol); $[\theta]_{254} = +14\,500$, $[\theta]_{249} = +22\,400$, $[\theta]_{241} = +20\,200$, $[\theta]_{234} = +7280$.

2.4.5 Helical Allenes

Nakazaki et al. synthesized the doubly bridged allene *85* [90] with an enantiomeric purity of 4%: $[\alpha]_{D}^{18} = +4.3$ (in n-hexane], $[\theta]_{220} = +1.4 \times 10^{2}$, $[\theta]_{231} = -2.0 \times 10^{2}$. Moore et al. in their synthesis of (*R*)-(+)-1,2-cyclononadiene (*86*) [92] reached an enantiomeric purity of 85% with $[\alpha]_{D}^{25} = +138$. Spectroscopic experiments showed

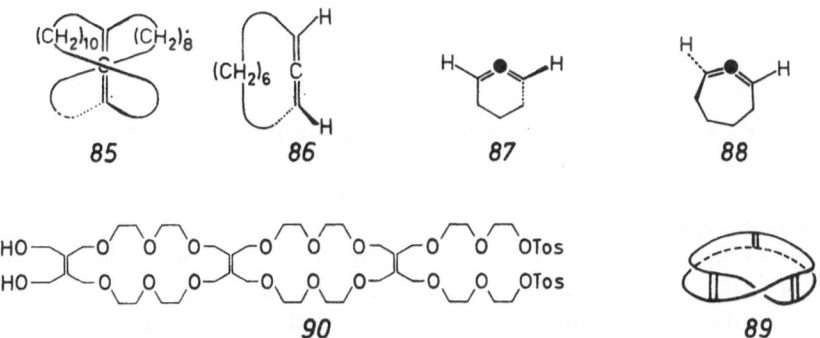

Fig. 34. Helical allenes *85–88* and topological aliphatic helices *89* [87,90]

that 1,2-cyclohexadiene (*87*) is stable in solution at room temperature, but optical activity is lost at 80 °C [93]. It is difficult to prove chirality of 1,2-cycloheptadiene (*88*); higher homologues have not shown chiral properties, to date.

2.4.6 Topological Aliphatic Helices

A new aliphatic helical topology was realized with the synthesis of the first molecular Möbius-band molecule *89*, which was obtained in 57% yield from *90*. The chirality of *89* was proved by the NMR method using (+)-(2,2,2)-trifluoro-9-anthrylethanol as an optically active solvent [87].

3 Aromatic Helices: Helicenes

Helicenes are benzologues of phenanthrene in which a regular cylindrical helix is formed through an all-ortho annellation of the aromatic rings. The helical structure is a consequence of the repulsive steric overlap of the terminal aromatic nuclei.

In 1956, M. S. Newman et al. successfully synthesized the first helicene, hexahelicene (*91*), in a direct way. In the same pioneering paper, they also first described the synthesis of α-2,4,5,7-tetranitro-9-fluorenylidenaminooxy-propionic acid (TAPA, Newman's reagent, *112*), which proved to be a π-acceptor suitable for enantiomer separation of π-donors like hexahelicene.

The IUPAC name of hexahelicene is phenanthro[3,4:d]phenanthrene [95]. In a simplified nomenclature for the helicenes introduced by M. S. Newman, the number of the annellated aromatic rings is set in brackets before the helicene name: hexahelicene = [6]helicene. Today, according to the building blocks composing the helicene, one differentiates carbohelicenes, heterohelicenes, double helicenes, bihelicenyls, metallohelicenes and cyclophanohelicenes. The names "dehydrohelicene" and "circulene" are used for some related compounds (v.i.).

Some of the properties of this attractive class of compounds are summarized shortly as follows: The helicenes as a common feature bear an inherent chiral chromophor and exhibit very high specific optical rotations depending on gross structure, fine geometry, electronic interactions, etc. [95].

3.1 Carbohelicenes

Carbohelicenes, also called all-benzene helicenes, only contain carbon atoms in their skeleton.

These helicene molecules usually have a C_2 axis, which is not necessarily identical with the crystallographic axis. The structure of the helicenes [96] can be described generally in a simplified way by distributing the atoms of the skeleton on three screw lines: one inner helix having n+1 C atoms, one outer helix being composed of 2n C atoms and one middle helix having n+1 C atoms, where n is equal to the number of benzene rings.

Generally, the pitch of the inner helix is smaller than that of the middle helix, which is itself smaller than that of the outer helix. The former is relatively independent of the value of n, the latter two sink with increasing n. Replacing hydrogen atoms of the terminal benzene rings by larger substituents leads to increased steric repulsion. With increasing n, the two outer pitches only change little. The radius of the inner helix also remains approx. constant, whereas the one of the middle and outer helices increases with increasing n. Compared to benzene, the lengths of helicenes are shorter in the inner part of the helix, whereas in the periphery they are elongated.

The strain is not equally distributed inside the helicene molecule: the torsion angle between the inner bonds in the molecule are large (approx. $25 \pm 3°$). This induces a coplanarity, whereas the terminal rings are nearly completely coplanar with bond lengths and angles similar to those of phenanthrenes [95].

[6]Helicene (*91*), the historical and structural mother substance of the helicenes, possesses the space group $P2_12_12_1$. The interplanar angle between the terminal benzene rings is remarkably large: 58.5°. This is due to the repulsion of the facing terminal benzene rings at both ends of the helix [96].

[7]Helicene (*92*) exists in two crystalline modifications: a) point group $P2_1$, monoclinic with an interplanar angle of 30.7° [98], b) point group $P2_1/c$, monoclinic, with an angle of 32.3° [98]. In both modifications one also observes a significant reduction of the angle between the terminal benzene rings.

The tribenzo[f,l-r]heptahelicene (*93*) (Fig. 39, point group Pbca, orthorhombic) [99] differs from heptahelicene in that the bonds at the condensation sites of the outer

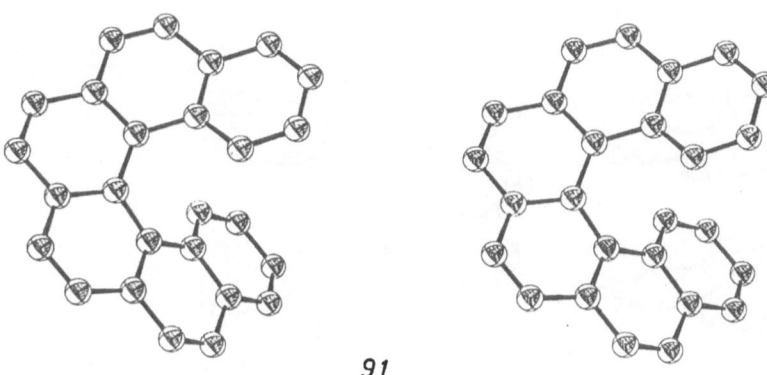

91

Fig. 35. [6]Helicene (*91*) (stereo view)

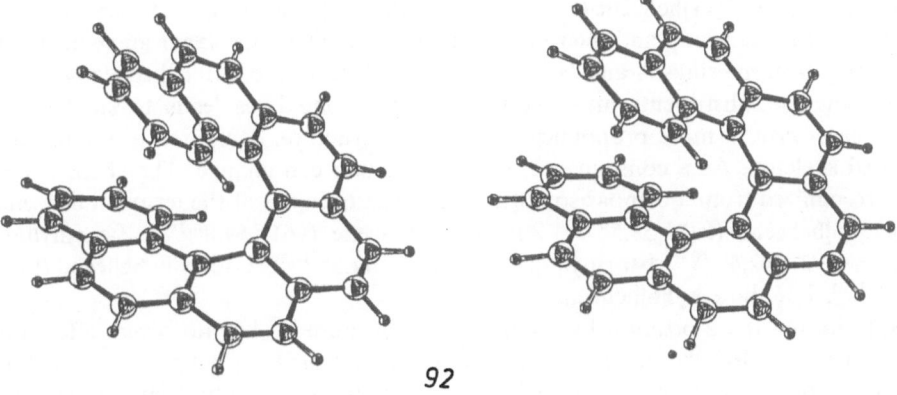

92

Fig. 36. [7]Helicene (*92*) (stereo view)

Table 8. Comparison of pitches of the helicenes *94*, *91* and *100*[95]

Helix	[10]helicene *94*	[6]helicene *91*	1,16-dimethyl[6]helicene *100*
pitch of helix [nm]:			
inner helix	322	326	324
middle helix	362	457	394
outer helix	388	604	402

benzene rings are not shortened, as usually observed; the interplanar angle here is 33.2°.

[10]Helicene (*94*)[100, 101] crystallizes in the point group $P2_1$. It is similar in structure to 1,16-dimethyl[6]helicene (*100*) as is shown by a comparison of their pitches and with that of [6]helicene (*91*) (Table 8).

The pitches of the inner helices correspond to each other, those of the middle and outer helix are quite similar for *94* and *100*. The interplanar angle of the two benzene rings A and H facing each other is only 5°, which means that they are arranged nearly parallel to each other.

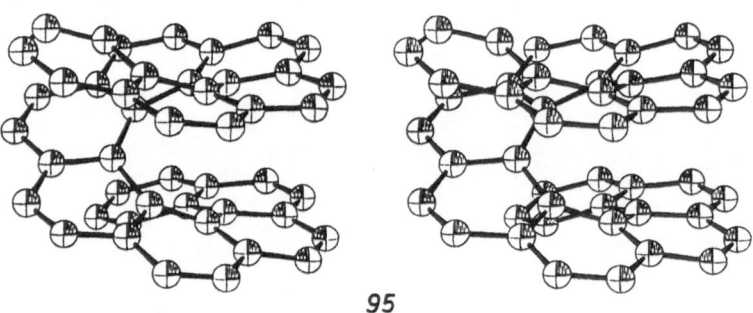

95

Fig. 37. [11]Helicene (*95*) (stereo view)

The structure of [11]helicene (*95*) is similar to that of [10]helicene (*94*). It crystallizes in the same space group and their pitches are similar. The interplanar angle between the facing benzene rings A and H is 4°, also similar to that of [10]helicene (*94*).·

Introducing substituents into hexahelicene (*91*) sometimes leads to smaller, in other cases also to more pronounced changes in structure, compared to the unsubstituted skeleton. As a consequence, the space group can change. The changes are easy recognized from a comparison of the interplanar angles of the terminal benzene rings: [6]helicene (*91*): 58.5°[96], 2-methyl[6]helicene (*96*): 54.8°[104], 1-methyl[6]-helicene (*97*): 42.4°[103], 1-formyl[6]helicene (*98*): 44.5°[103], 1-acetyl[6]helicene (*99*): 42.5°[103], 1,16-dimethyl[6]helicene (*100*): 29.6°[103].

Substituting the 2-position by methyl[102] or bromine[96] has no severe effect on the structure of [6]helicene. The space group remains similar, but the introduction of such substituents leads to a translation of the atoms of the molecule parallel to the C axis; the crystals are not isomorphic[96].

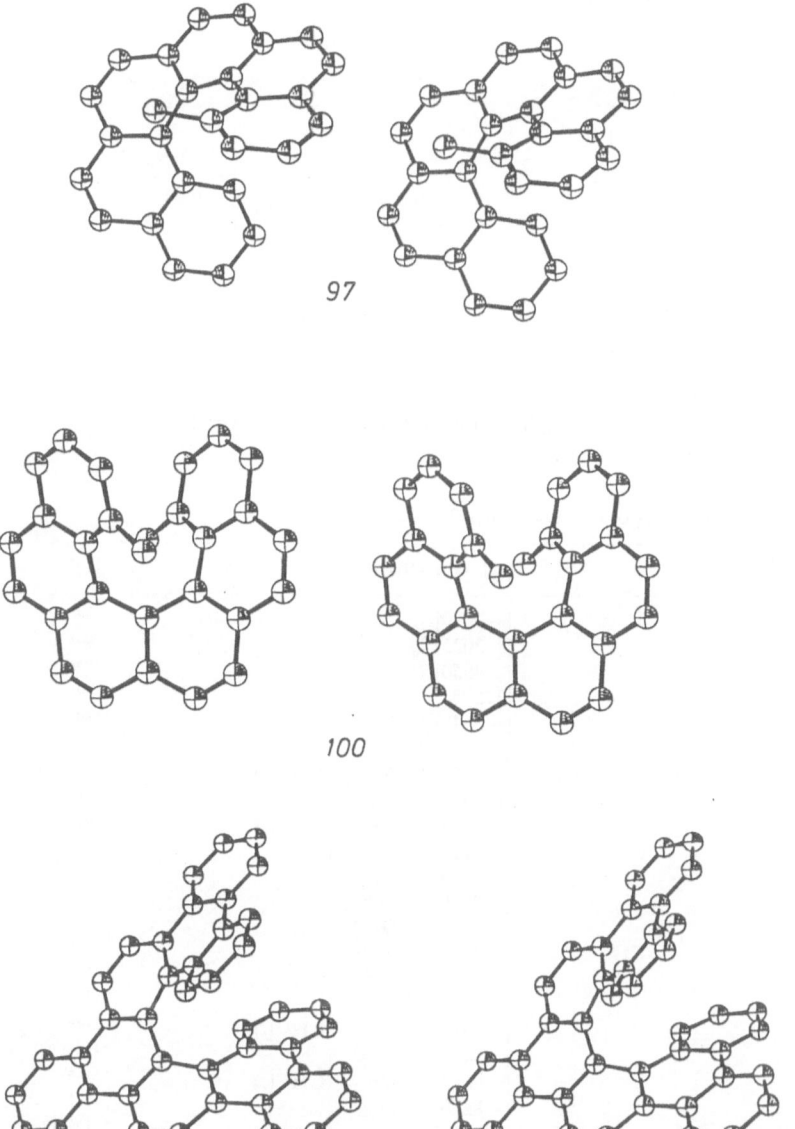

Fig. 38. 1-Methyl- (*97*) and 1,16-dimethyl[6]helicene (*100*) and tribenzo[7]helicene (*93*) (stereo views)

Substitution in 1-position changes the space group and lowers the interplanar angles. The introduction of substituents also leads to a stronger distortion of the inner bonds, compared to the unsubstituted skeleton, as can be concluded from the torsion angles.

Attaching two methyl groups, as in 1,16-dimethyl[6]helicene (*100*) [103], changes the geometry of the helix significantly through steric effects. In hexahelicene, as a main consequence, twisting takes place around the C(21)-C(23) and C(21)-C(19) bonds.

35

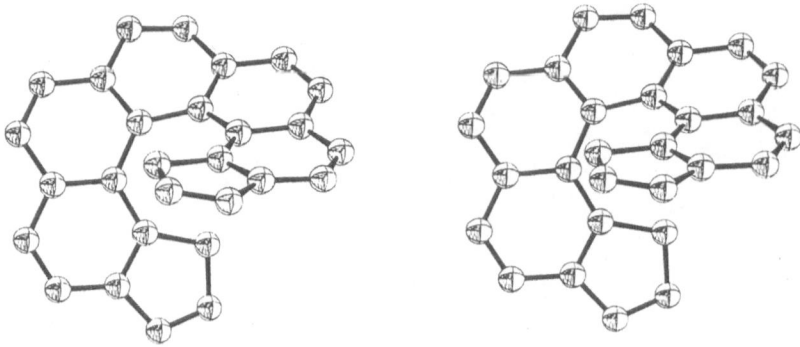

Fig. 39. 1,16-dihydroindeno[5,4-c:4′,5′-g]phenanthrene (*101*) [104] (stereo view)

Table 9. Optical rotation values of carbohelicenes *91–95* and *102–107*

Helicene	$[\alpha]_\lambda^T$ (solvent)	Ref.
[5]helicene (*102*)	$[-1670]_{578}^{26}$	109)
	$[-2025]_{546}^{26}$	109)
	$[-4950]_{436}^{26}$	109)
[6]helicene (*91*)	$[3750]_D^{25}$	110)
	$[-3640]_D^{24}$	120)
	$[-4820]_{546}^{24}$	94)
[7]helicene (*92*)	$[-5900 \pm 200]_{579}^{25}$ (CHCl$_3$)	108, 110)
	$[6200]_D^{25}$	111)
[8]helicene (*103*)	$[-6900 \pm 200]_{579}^{25}$ (CHCl$_3$)	108, 110)
	$[-7170 \pm 100]_{579}^{25}$ (CHCl$_3$)	
	$[-6690 \pm 100]_D^{25}$	
[9]helicene (*104*)	$[-8100 \pm 200]_{579}^{25}$ (CHCl$_3$)	108, 110)
	$[-8150 \pm 100]_{579}^{25}$ (CHCl$_3$)	114b)
	$[-7500 \pm 100]_D^{25}$	114b)
[10]helicene (*94*)	$[-8940 \pm 100]_{579}^{25}$ (CHCl$_3$)	114b)
	$[-8300 \pm 100]_D^{25}$	114b)
[11]helicene (*95*)	$[-9310 \pm 100]_{579}^{25}$ (CHCl$_3$)	114b)
	$[-8460 \pm 100]_D^{25}$	114b)
[13]helicene (*105*)	$[-9620 \pm 100]_{579}^{25}$ (CHCl$_3$)	114b)
	$[-8840 \pm 100]_D^{25}$	114b)
(−)-7-methyl[6]helicene (*106*)	$[-3157]_{589}$	110, 112a)
	$[-3399]_{578}$	112a)
	$[-4185]_{546}$	112a)
	$[-12332]_{436}$	112a)
2-bromo[6]helicene (*107*)	$[-3560]_D^{25}$ (CHCl$_3$)	
	$[-4690]_{546}$	

In the 1,16-dimethyl derivative *100*, the twist is smeared over all C-C bonds of the inner helix core. The interplanar angle, which is much smaller than in the unsubstituted helicene, hints to a parallel arrangement of the facing terminal benzene rings. As a consequence, this also leads to a shorter distance between C(2) and C(15): 403 pm compared to 448 pm in hexahelicene.

In 1,16-indeno[5,4-c:4',5'-g]phenanthrene (*101*) [104] the terminal aromatic rings are spread away from each other, with an interplanar angle of 69.1°. The torsion angle between C(12)-C(11)-C(11^i)-C(12^i) (40.8°) is wide compared to [7]helicene (*92*) (25°). The strain, in contrast to tribenzo[7]helicene (*93*) and 1,16-dimethyl[6]helicene (*100*), seems mainly to be localized in the inner part of the helix.

The helicenes are unique with respect to their very high rotation values, which are a consequence of the inherent chiral chromophor enclosing the whole molecular skeleton. This also attracts interest with respect to the use of helices as chiral inductors for asymmetric syntheses.

ORD, CD and UV spectra of [5]helicene (*102*) [104], [6]helicene (*91*) [106], and [7]- to [9]helicene (*92, 103, 104*) [108] have been measured. The ORD curves of these helicenes have a similar shape; with increasing degree of annellation the Cotton effects are shifted to longer wavelengths. The CD and UV curves show the same tendency.

UV and CD spectra of [5]- to [7]helicene are discussed in the literature [104, 106, 108, 109], including a general theoretical treatment of the correlation of electron transitions. The specification of the enantiomers to the P- and M-series was achieved by making use of their chiroptical properties.

The *"most fascinating and surprising discovery"* in the field of helicenes, according to Martin [110], is their thermal racemization. A pure conformative process is supposed today to be operative [105]. Its barrier is remarkably low, because the necessary molecular deformations are distributed on many bonds of the helicene skeleton. For the transition state, nonplanarchiral conformations are supposed, in which the two terminal benzene rings both are bent to one side [107]. Table 10 gives kinetic data for the racemizations of some carbohelicenes [110].

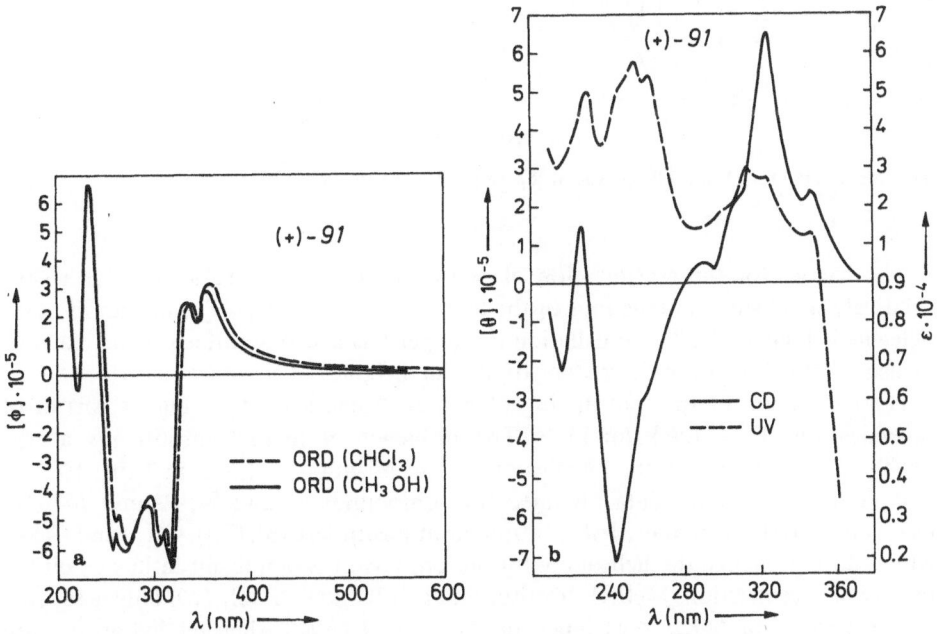

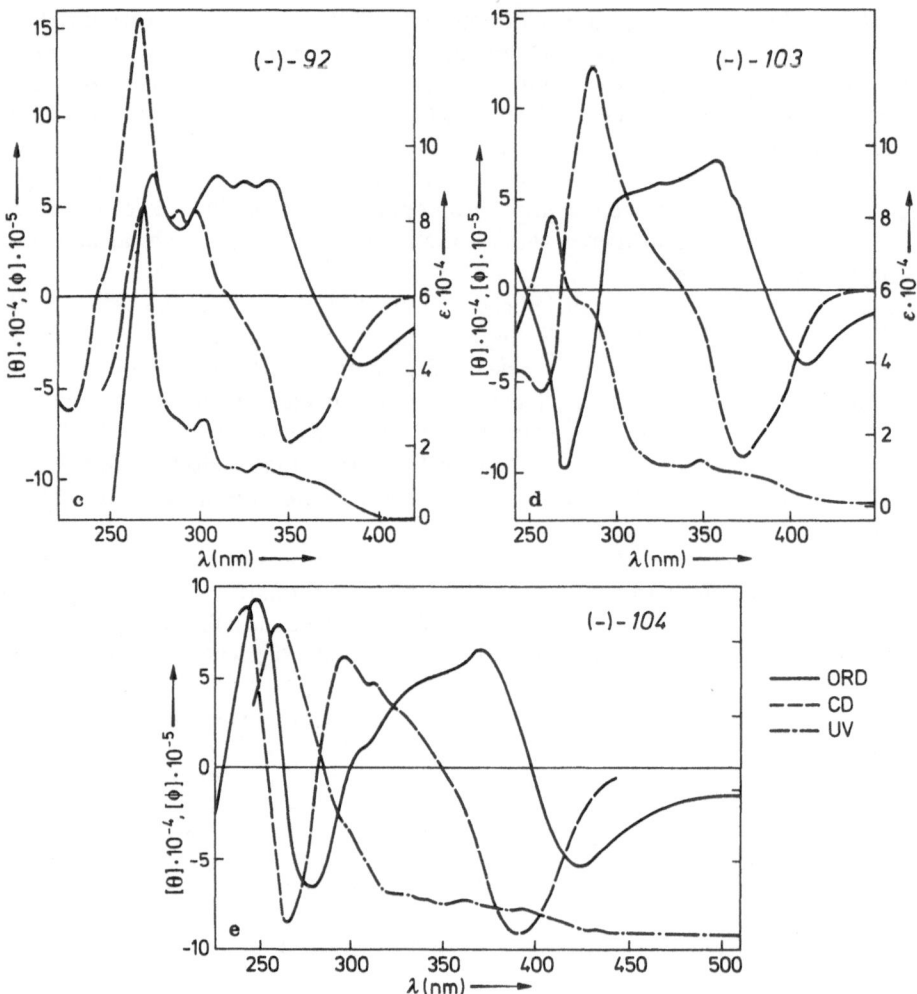

Fig. 40a–e. ORD, CD and UV curves of [6]- to [9]helicene (*91, 92, 103, 104*)

The barrier for the racemization of pentahelicene is low (Table 10) due to the relatively small steric interactions of the terminal benzene rings. The barrier of hexahelicene is higher. Further annellation no longer has a strong influence: the barriers measured for [7]- to [9]helicene are of similar height.

Hydrogenation of the central ring of pentahelicene leads to a higher barrier of racemization: $E_a = 126 \text{ kJ/mol}$ [109]. The influence of methyl substituents at the hexahelicene skeleton on the kinetics of the racemization is shown in Table 10.

A comparison of the data obtained for compounds *91* and *97, 92* and *108–111* (see Table 10) shows that methyl substituents at positions C(3), C(4), C(13) and C(14), which do not change the helical conformation, have no significant influence on the barrier of racemization. Methyl substitution at C(2) and C(5) (cf. *110*) only leads to a minor increase of ΔG_{rac}. A stronger influence is observed when methyl groups are

Table 10. Comparison of racemization barriers of methyl-substituted helicenes with helicenes

Helicene	$\Delta G^{\neq}_{rac}$ (T)		$\Delta H^{\neq}_{rac}$		$\Delta S^{\neq}_{rac}$	$t_{1/2}$	K_{rac} [T (K)]	E_a		T
	[kJ/mol]	[kcal/mol]	[kJ/mol]	[kcal/mol]	[eu]	[min]	[s^{-1}]	[kJ/mol]	[kcal/mol]	[K]
pentahelicene (102)	100.8	24.1	95.8	22.9	− 4.1	62.7	—	98.3	23.8	330
hexahelicene (91)	151.5	36.2	146.4	35.0	− 4.2	13.4	5.6×10^5 (467)	148.9	35.6	495
heptahelicene (92)	174.4	41.7	169.5	40.5	− 3.9	13.4	—	172.0	41.1	568
octahelicene (103)	177.4	42.4	171.5	41.0	− 4.6	31	—	174.0	41.6	566.5
nonahelicene (104)	182.0	43.5	174.5	41.7	− 6.1	123	—	176.9	42.3	566.5
1-methylhexahelicene (97)	183.2	43.8	161	38.5	− 9.8	—	2.5×10^5 (542)	—	—	542
1,14-dimethylhexahelicene (108)	183.2	43.8	161	38.5	—	—	2.4×10^5 (542)	—	—	542
1,16-dimethylhexahelicene (100)	186.5	44.0		—	—	—	1.3×10^5 (543)	—	—	543
1,3,14,16-tetramethyl-hexahelicene (109)	186.5	44.0	158	37.7	− 12.9	—	1.3×10^5 (543)	—	—	543
2,15-dimethylhexahelicene (110)	165.2	39.5	—		—	—	2.6×10^5 (513)	—	—	513
4,13-dimethylhexahelicene (111)	153.9	36.8	—		—	—	6.6×10^5 (469)	—	—	469

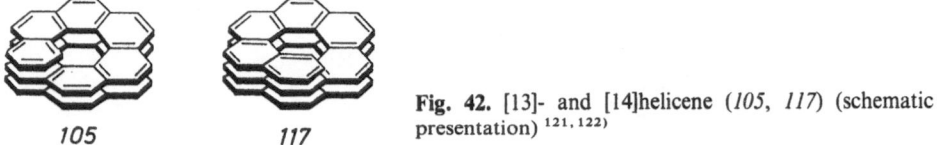

Fig. 41. Reagents for the resolution of enantiomers of helicenes (*112–116*) [117–120]

Fig. 42. [13]- and [14]helicene (*105, 117*) (schematic presentation) [121,122]

105   *117*

introduced at C(1) (cf. *97, 108, 100, 109*). This influence is not essentially increased by a second methyl group at C(16) (*108, 100*). A methyl group in 1-position has more severe consequences than two additional benzene rings (cf. heptahelicene, octahelicene). This seems to be due to a larger entropy loss in the course of racemization of the methyl derivative, because the rotation of the methyl groups is hindered [107].

Resolutions of the racemates of helicenes have been carried out in different ways: Helicenes, which often spontaneously crystallize as conglomerates, have been successfully separated into the enantiomers by Pasteurs mechanical method. Enantiomer separation using the diastereomeric salts prepared from 7-triphenylphosphoniomethyl[6]helicene and D-(—)-dibenzoyltartaric acid [112] was successful, as was the separation of [6]helicene on polymers [114].

Photocyclization using circular polarized light [113] yielded dihydro[5]helicene in a small enantiomeric excess (ee = 3%). Attempts were also made to enantioselectively synthesize helicenes using chiral solvents [115] as well as cholesteric liquid crystals [116]. Excellent enantiomeric excesses (up to 98%) were obtained through temporary introduction of optically active residues like mandelic acid, lactic acid derivatives and (—)-menthyl esters [115].

Enantiomeric pure hexahelicene was used as an optical inducting moiety for the synthesis of [8]- to [11]- and [13]helicenes [116].

Separations of [n]helicene racemates have also been attempted using π-acids, such as 2-(2,4,5,7)-tetranitrofluorenylidene-9-aminooxypropionic acid (TAPA, *112*, see above) [117], its butyric acid analogue TABA (*113*) [117] and binaphthyl-2,2-diyl-hydrogenphosphate (BPA, *115*) [118]. Other employed methods were inclusion chromatography on triacetyl cellulose [119] or helical polymers like (+)-poly(triphenyl-methyl-methacrylate [(+)-PTrMA, *116*] [120].

To date, the highest homologue of carbohelicenes are [13]helicene (*105*) and [14]-helicene (*117*). Both have been separated using columns doped with (+)-TAPA [121, 122].

3.2 Heterohelicenes

Heterohelicenes, apart from the presence of heteroatoms, differ from carbohelicenes through their geometry. Heteroaromatic moieties change the bond angles at the annellation positions. In the case of five-membered rings, a higher annellation degree is necessary to obtain an overlap of the terminal aromatic rings. For all-thiophene helices, which have not yet been synthesized, this means that eight condensed thiophene rings are necessary [95].

The information given in Fig. 43 can be used to design new heterohelicene skeletons, which differ from carbohelicenes in their optical properties.

In benzo[d]naphtho[1,2-d']benzo[1,2-b:4,3-b']dithiophene (*118*) [124], two thiophene units are incorporated.

In contrast to [6]helicene (*91*), the benzene and thiophene rings in heterohelicenes do not deviate strongly from planarity. The bond lengths and angles also are quite similar to non-annellated benzene and thiophene rings. The angles between the planes of the aromatic rings vary from −18.4° to +19.9°. The inner atoms of the

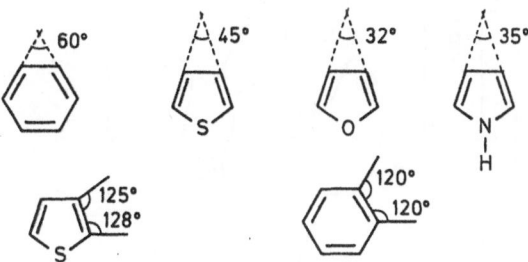

Fig. 43. Comparison of differing angles of benzo-, thiopheno-, furano- and pyrrolo-building units [123]

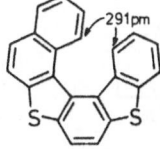

118　　**Fig. 44.** Benzo[d]naphtho[1,2-d']benzo[1,2-b:4,3-b']dithiophene (*118*) [124]

heterohelicenes form an irregular screw. The shortest C-C distance is 291 pm. The absolute configuration [125] has been evaluated by the dipole velocity method and is in accord with the experimental CD spectra. Independently, it has been determined by the X-ray method [125] that (+)-heterohelicene *118* has the right-handed (P) screw structure.

The absolute configuration of other heterohelicenes has been correlated by comparison of their ORD and CD spectra: (+)-heterohelicenes generally have P screw sense.

Nitrogen and oxygen heteroatoms can also be introduced into helicenes. 5,10-Dihydrocarbazolo[3,4-c]carbazole (*119*) [126] was the first helicene synthesized (in 1927). Heterohelicenes containing several nitrogen atoms as well as hydrogenated heterohelicenes have been synthesized by Teuber and Vogel [127]. These authors also studied the consequences of the presence of sp³ centers in helicenes [127].

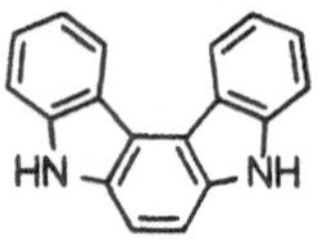

119

Fig. 45. 5,10-Dihydrocarbazolo[3,4-c]carbazole (*119*), the first helicene synthesized (in 1927) [126]

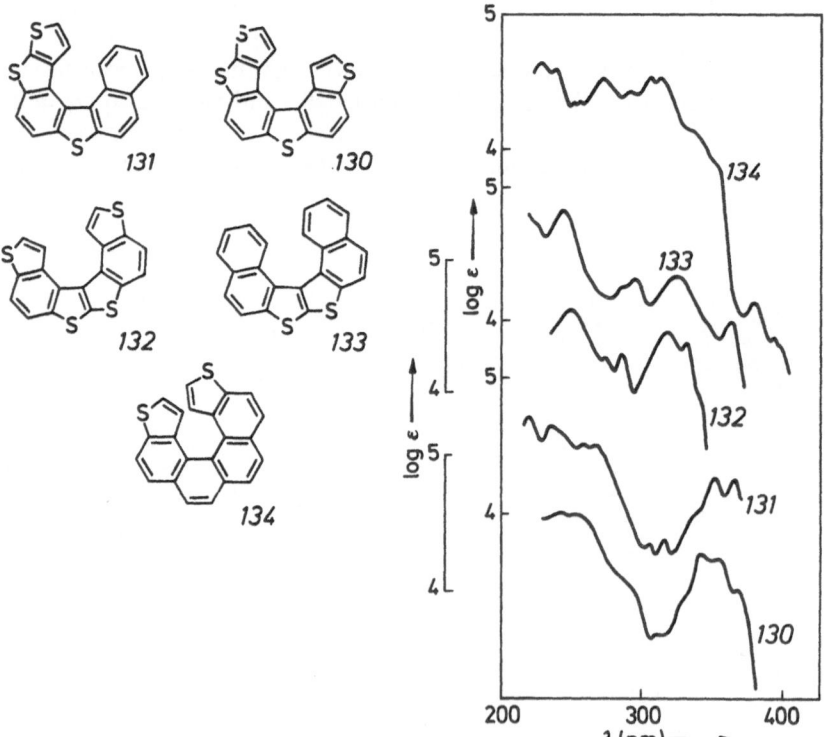

Fig. 46. UV spectra of the heterohelicenes *130–134* (in cyclohexane) [134]

Table 11. Optical rotation values of heterohelicenes *118*, *120–129* and *135*

Heterohelicene		$[\alpha]_\lambda^T$ (solvent)	Ref.
[6]heterohelicenes:			
	118	$[2050]_{541}^{20}$ [a]	128)
	135	$[+3640]_{436}^{20}$ [b]	129)
	120	$[-2287]_{589}^{RT}$ $[+2359]_{589}^{RT}$	130)
[7]heterohelicenes:			
	121	$[-2177]_{589}^{RT}$ $[+2165]_{589}^{RT}$	130)
	122	$[-1784]$ $[+1698]$	130)
	123	$[-2509]$ $[+2371]$	130)
	124	$[+7200]_{436}^{25}$ (trichloromethane) $[2000]_{546}^{20}$ [c]	123) 128)

Table 11. (continued)

Heterohelicene		$[\alpha]_\lambda^T$ (solvent)	Ref.
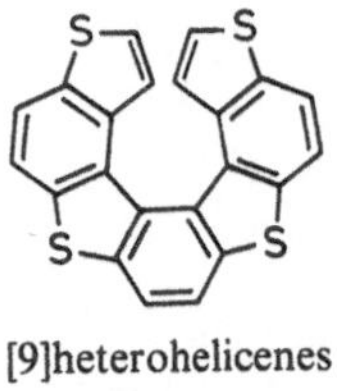	*125*	$[+2990]_{500}^{23}$ $[-2980]_{500}^{23}$	[131]
[9]heterohelicenes:			
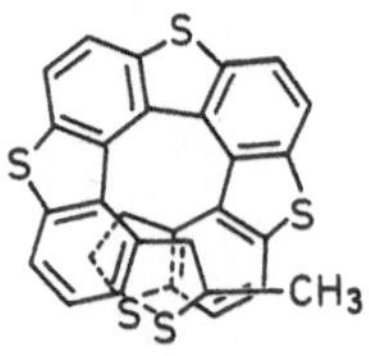	*126*	$[-2726]_{589}^{RT}$ $[+2772]_{589}^{RT}$	[130]
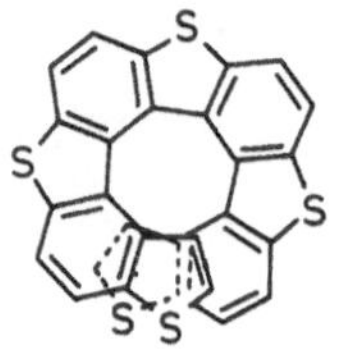	*127*	$[+3760]_{500}^{23}$ $[-3700]_{500}^{23}$	[131]
[11]heterohelicenes:			
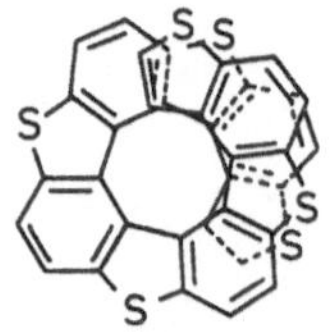	*128*	$[4440]_{500}^{23}$ $[-4550]_{500}^{23}$	[131]
[13]heterohelicenes:			
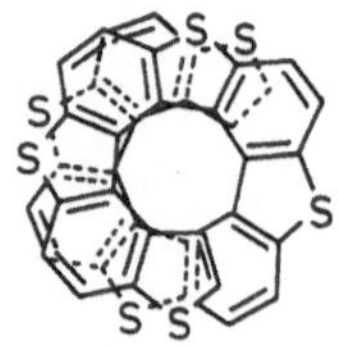	*129*	$[8170]_{500}^{23}$ $[-8290]_{500}^{23}$	[132]

[a] Optical purity is not clearly determined. [b] Highest optical rotation value. [c] Optical purity is not determined.

Like the carbohelicenes the heterohelicenes also show high rotation values; some of them are compared in Table 11.

▶

Fig. 47. ORD and CD spectra of some heterohelicenes [128, 133]

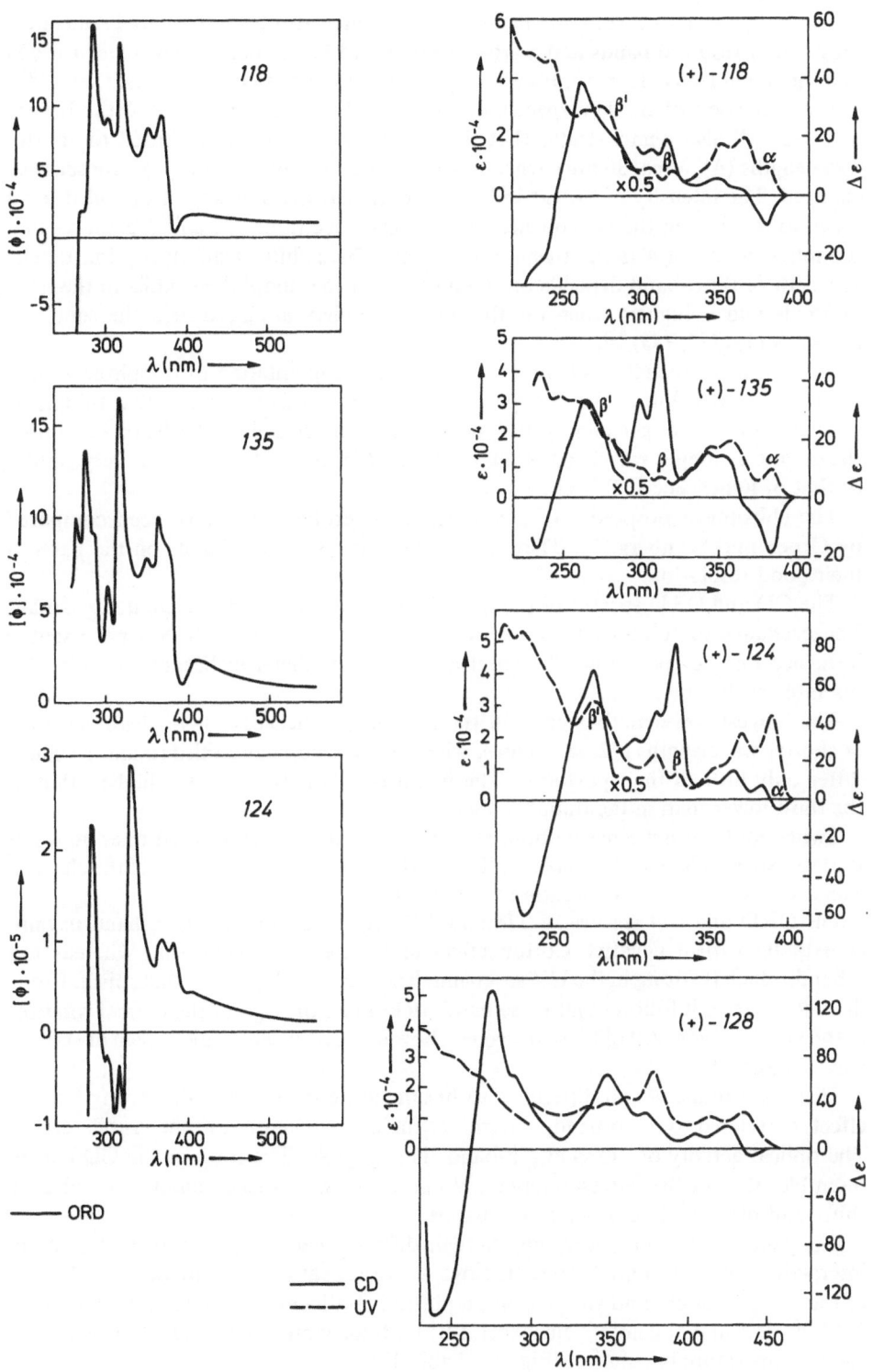

ORD

CD

UV

The UV spectra of carbo- and heterohelicenes have also been compared: the latter show better resolved bands and a more pronounced fine structure. The reason seems to be not completely clear [133]. Figure 46 gives the UV spectra of five thiahelicenes [134].

A comparison of the UV spectra of the heterohelicenes *130–134* and the benzohelicenes [134] also demonstrates that the longest wave α-band is shifted to shorter wavelengths ($\Delta\lambda$ = ca. 30 nm), whereas the second and third band (p, β) are scarcely changed. The intensity of the α-band is much higher, the intensity of the third band much lower than in the benzohelicenes. The UV spectra of *132* and *133* shows that the introduction of a thiophthene unit effects a blue shift of all absorption bands. The shift of the α-band depends on the position of the thiophthene units in the helix molecule and is highest when the thiophthene moiety is placed near the center of the molecule (*132, 133*) [95].

In contrast to the effect of thiophene units, the implantation of thiophthene does not lead to a red shift in the UV spectra. This may be due to the fact that thiophene possesses even stronger conjugative properties than benzene. The helix *134* shows the α-band at longer wavelengths than *118* and *125*, the UV spectrum is very similar to that of [6]helicene (*91*) itself.

The chiroptical properties of a number of heterohelicenes have been compared by Groen and Wynberg [128]. They were able to correlate the bands on the basis of their good resolution.

The CD and ORD spectra of the [6]heterohelicenes *118* and *135* are quite similar. The exchange of terminal benzene rings by thiophene obviously has no essential influence on the spectrum. The spectra of the heptaheterohelicenes *121–125* also resemble each other.

The longest wavelength α-band in the thiophenohelicenes systematically appears at shorter wavelengths ($\Delta\lambda$ ca. 30 nm), whereas the second and third (p and β) bands differ only little in their positions. The intensity of the first band is higher, that of the third lower than in the thiophene series.

Switching from helicenes without terminal overlap to overcrowded ones does not lead to a strong change of the spectra. For a HMO theoretical treatment of the electron excitation we refer to the literature ([133], p. 2970).

The ORD and CD spectra of *118* and *135* are quite similar. The α-band usually goes along with a negative Cotton effect and a negative dichroism, whereas the p-band, which is strong in the UV spectrum, leads to a weak positive dichroism. From the ORD curves it follows that these bands determine the sign of the optical rotation in the visible region. All of these helicenes show a strong negative dichroism at shorter wavelengths [95].

There is a fundamental difference to hexahelicene in so far as the strong Cotton effect, correlated to the p-band, determines the optical rotation in the visible region. The optical activity of the α- and β-bands is negligible. The form of the CD curves resembles that of the corresponding UV curves if one neglects minor red and blue shifts and differing intensities of the bands.

The [6]heterohelicenes, according to their different geometry comprising less steric interactions of the terminal aromatic rings, racemize easier. The kinetic data of some of these helicenes are known: benzo[d]naphtho[1,2-d']benzo[1,2-b:4,3-b']dithiophene (*118*) [129, 135] and thieno[3,2-e]benzo[1,2-b:4,3-b']benzothiophene (*135*) racemize at room temperature in solution (Fig. 48, Table 12) [95].

Table 12. Kinetic data of heterohelicenes *118* and *135*

No.	E_a [kJ/mol] (kcal/mol)	ΔH^* [kJ/mol] (kcal/mol)	ΔS^* [eu]	ΔG^* [kJ/mol] (kcal/mol)	$t_{1/2}$ [25)] [min]
118	108.7 (26.0)	106.2 (25.4)	5.5	99.1 (23.7)	241
135	94.5 (22.6)	92.0 (22.0)	0	92.0 (22.0)	13.0

The racemization of *135* is quicker than that of *118* (approx. 18 times); a difference is found in ΔG^* of 7.11 kJ/mol (1.2 kcal/mol).

Wynberg et al. [136] successully separated the enantiomers of several heterohelicenes (*120–124*) by means of HPLC on TAPA-containing columns. The separation of *121* and *125* on (M)-(−)-BPA-doped columns (in n-hexane) only gave moderate results.

The diaza[7]helicene *136* (Fig. 51) has been separated completely into the enantiomers using M-(−)-BPA [137]. The successful separations of heterohelicenes are comparable to those of the carbohelicenes. The highest arenologues of the heterohelicenes, up to now, are the [13]- and [15]thiophenohelicenes *129* and *137* [132].

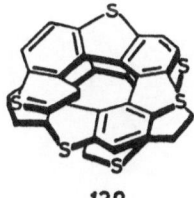

136

Fig. 48. Diaza[7]heptahelicene (*136*) [137]

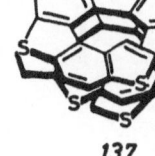

129   *137*

Fig. 49. [13]- and [15]heterohelicenes *129* and *137*. containing seven and eight alternating thiophene units, resp. [132]

3.3 Double Helicenes

The double helicenes can be devided into three groups according to their annellation pattern.

Diphenanthro[3,4-c:3′,4′-1]chrysen (*138*) [141] belongs to type A. Two optically active and one meso structures theoretically are possible. The lower-melting racemic mixture and the higher-melting meso compound were obtained experimentally. The latter is thermodynamically more stable because both helix windings lie on facing molecular parts. The racemic mixture has partly been separated into the enantiomers, which do not racemize upon heating, but rearrange to the meso compound.

Hexaheliceno[3,4-c]hexahelicene (*139*) [140] is an example of type B of the double helicenes. Theoretically one meso and two optically active molecules are expected here. In the synthesis only one singlet isomer, most probably the racemic compound, is formed exclusively. The racemic mixture should be energetically favoured, because the terminal aromatic rings are situated on different sides of the molecule, with respect to the central naphthaline unit [142].

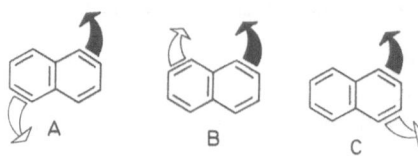

Fig. 50. Types A to C of double helicenes

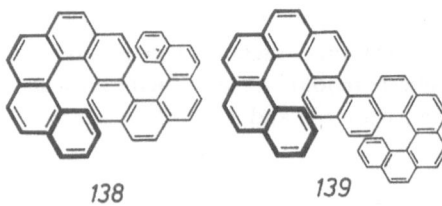

138   *139*

Fig. 51. Diphenanthro[3,4-c:3′,4′-l]chrysen (*138*) and hexaheliceno[3,4-c]hexahelicene (*139*)

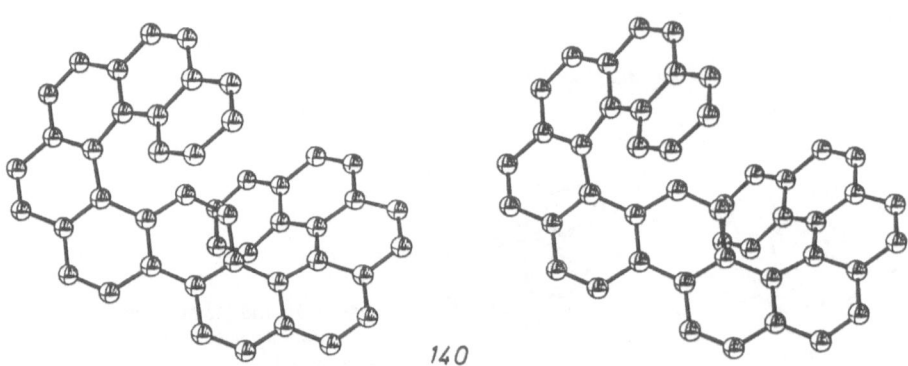

140

Fig. 52. Diphenanthro[4,3-c:3′,4′-o]picen (*140*) (stereo view) [138]

Diphenanthro[4,3-a:3′,4′-o]picen (*140*) [142] can be regarded as the lower benzologue of hexaheliceno[3,4-c]hexahelicene (*139*) [140]. The thermodynamically more stable d,l-compound obtained in the synthesis has been ascertained by X-ray analysis [138].

140 crystallizes monoclinically in the space group $P2_1/c$. Both halves of the molecule are arranged similar to hexahelicene, but the deviation from a perfect symmetry of both molecular halves is appreciable. This may be due to environmental differences in the crystal packing (Fig. 53b).

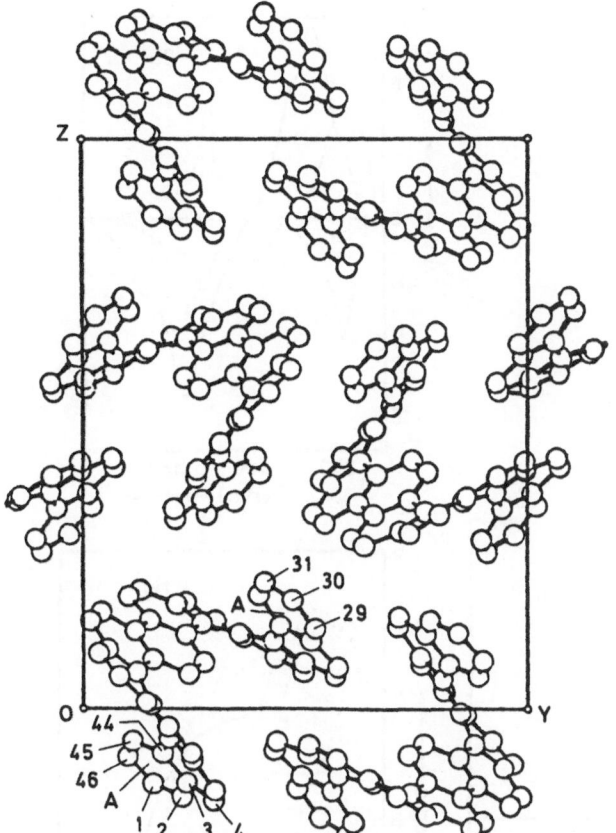

Fig. 53. Crystal packing of diphenanthro[4,3-c:3′,4′-o]-picen (*140*)

The racemate of *140* has been partially separated on a TAPA (*112*)-impregnated silicagel column [142]. The rate of racemization (k = 1.9×10^{-2} min^{-1}) and the half-life ($t_{1/2}$ = 38 min at 210 °C) are of the same order as those of hexahelicene ($t_{1/2}$ = 48 min at 205 °C). This means that the conformational flipping of the one helical part of the molecule is not strongly hindered by the other. An intermediate of this conformational process therefore should be the symmetrical meso form.

The highest specific rotation value is $[\alpha]_{380}$ = 32300 ($[\alpha]_{545}$ = 9900), but with undetermined optical purity. The ORD curve (in CH_2Cl_2) shows maxima at 545, 480 and 380 nm and minima at 500, 450 and 330 nm.

3.4 Helicenes Containing Cyclophane Units

Originally, helicenes bearing a paracyclophane unit, which forces the helicene to extend only in one direction, have been synthesized for the determination and proof of absolute configurations [143, 144].

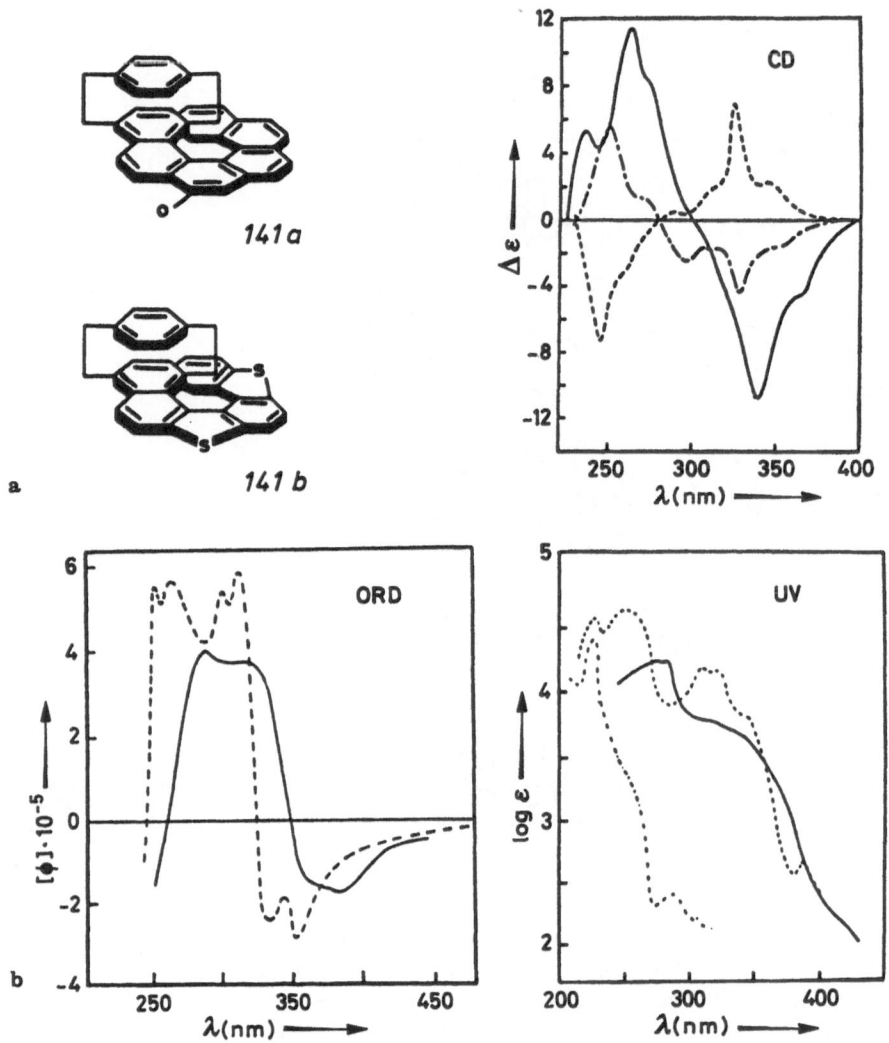

Fig. 54. a Paracyclophano-helicene (*141a*) and paracyclophano-heterohelicene (*141b*), **b** spectroscopic and chiroptical data of *141a* [143, 144]

The following chiroptical data have been evaluated for *141a*: $[\alpha]_{578}^{25} = -2716$ [143], $[\alpha]_{577} = -662$ [144], $[\theta]_{578}^{25} = -12464 \pm 200$. For comparison, the data of (—)-hexahelicene are given: $[\alpha]_{578} = -3750$, $[\theta]_{578} = -12300$. A comparison of the ORD curves of (—)-hexahelicene and (—)-*141a* demonstrates that both these enantiomers have the same sign of screw winding. The same conclusion has been drawn from CD spectra. The UV and ORD curves of *141a* resemble those of the corresponding heterohelicenes and admit a correlation of their configurations.

3.5 Helical Metallocenes

The first helicene synthesized, which was bridged by a ferrocene unit, is based on the pentahelicene skeleton [145]. At its terminal phenylene rings, cyclopentadienyl units were condensed. The di-lithium salt of this compound with iron(III)chloride yielded the helical ferrocene *142* with an uninterrupted conjugation between the two cyclopentadienyl rings (Fig. 55). The cyclopentadienyl moieties are clamped by the iron atom in an angle deviating 19.4° from parallelity.

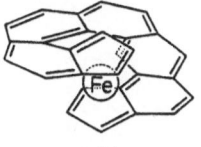

142   **Fig. 55.** The helical pentahelicenophane *142*

3.6 Bihelicenyls

2,2'-Bis-hexahelicyl (*143*) [146a], composed of two hexahelicyl moieties, which are connected through a single bond, occurs in a meso and two racemic configurations: the meso compound prefers a planar conformation at the central C-C single bond. For the d,l isomer such a spatial arrangement is sterically unfavourable; in this case, the helicyl units are twisted around the central single bond. The d,l isomer rearranges at its melting point to give the more stable meso compound.

The syntheses of the "propellicene" *145* [146b], which can be considered as a bis-2,13-pentahelicenyl, and of the dibenzo[def,pqr]tetraphenylene (*146*) [146c] have been carried out, but the configurations or chiroptical properties have not been described, as yet.

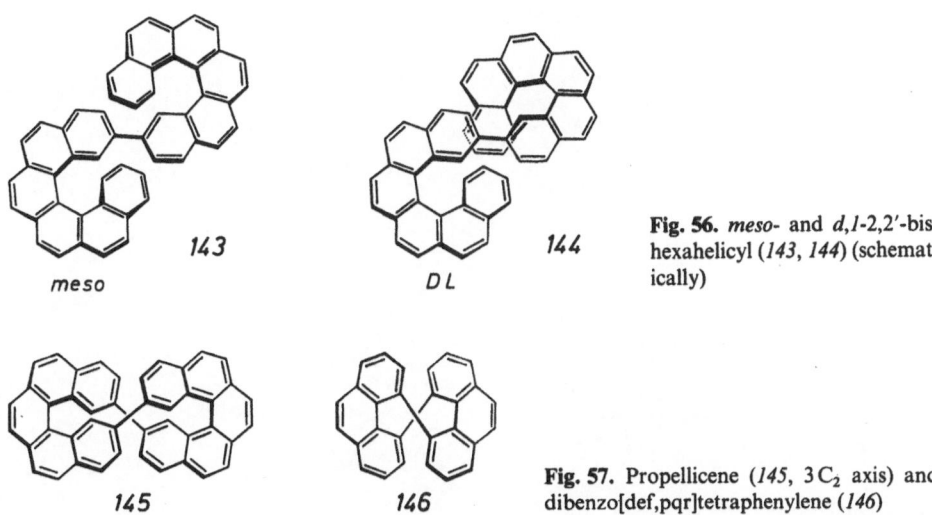

143
meso

144
DL

Fig. 56. *meso-* and *d,l*-2,2'-bis-hexahelicyl (*143*, *144*) (schematically)

145

146

Fig. 57. Propellicene (*145*, 3 C$_2$ axis) and dibenzo[def,pqr]tetraphenylene (*146*)

3.7 Circulenes, Dehydrohelicenes

Circulenes [147] are obtained when the two terminal aromatic rings of a helicene are annellated to each other. A [m]circulene is formed from m-condensed aromatic rings arranged in a closed macroring. Representatives are the bowl-shaped [5]circulene (*147*) ("corannulene") [148] and the planar [6]circulene *148*, the latter better known by the classical name "coronene" [149]. According to the number of aromatic rings composing the circulenes, either planar, boat- or saddle-shaped aromatic skeletons are obtained, the symmetry or helicity of which may be strongly disturbed compared to the helicenes.

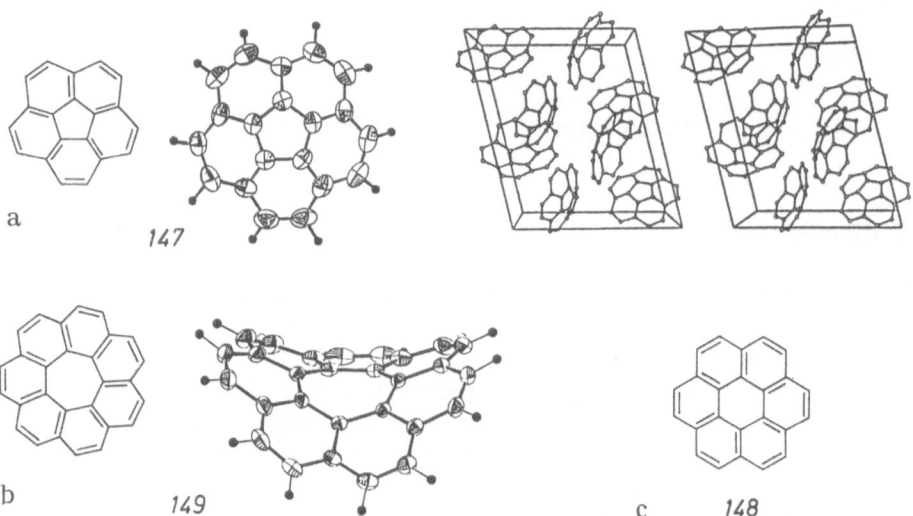

a   *147*

b   *149*   c   *148*

Fig. 58. Circulenes: a) corannulene (*147*, bowl-shaped); b) [7]circulene (*149*, saddle-shaped); c) coronene (*148*, planar)

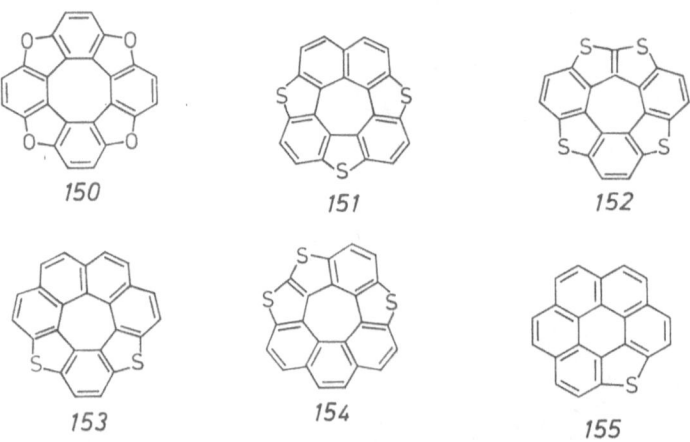

150

151

152

153

154

155

Fig. 59. Heterocirculenes (*150–155*)

The C-C bond distances of the inner core of the coronene molecule (*148*) are 143 pm, the C-C bond distances of the outer perimeter 138 and 141 pm. These distances, which are similar to those observed in benzene itself can be used as standards for the purpose of comparison (v.i.).

[7]Circulene (*149*) [150], described recently, shows a nonplanar saddle-shaped structure in the crystal, as revealed by X-ray analysis. Wynberg et al. synthesized several heterocirculenes, which mainly show a nonplanar corrugated structure (*150–155*).

Attempts were unsuccessful to separate the nonplanar heterocirculenes into its enantiomers using diastereomeric charge-transfer complexation, e.g. (+)-TAPA as a π-acceptor, even at low temperatures (−30 °C, in CS_2) [151]. The partial aromaticity of the inner ring seems to facilitate a planar aromatic transition-state, which, as a consequence, lowers the barrier of racemization.

Dehydrohelicenes are compounds in which both terminal aromatic rings of a (hetero-)helicene are connected through a σ-bond [151].

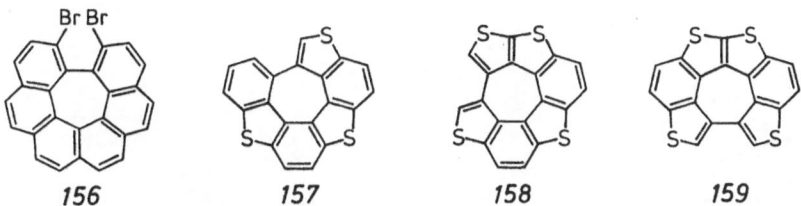

Fig. 60. Dehydro-heterohelicenes *150–159*

As a consequence of the higher conjugation of *156–159*, in comparison to the corresponding helicenes, bathochromic shifts of the α-band of Δλ ca. 30 nm are observed in their UV spectra [152]. Heterocirculenes show red shifts of the α-band of Δλ ca. 10 and 62 nm, compared to the related dehydroheterocirculenes, which have also been explained in terms of a stronger conjugation in the heterocirculenes. The dehydrocarbohelicene *156* has been separated into the enantiomers using HPLC and (+)-TAPA: (−)-*156* showed a rotation value of $[\alpha]_D^{22} = 505$ (in methanol) [150].

4 Araliphatic Helices: Helicenophanes and Arenicenes

Apart from the helicenes and the other above-mentioned helical compounds, it is of interest to gain fundamental knowledge of inherent dissymmetric and asymmetric chromophores. Special attention is devoted to the interaction of spatially overlapping π-electron clowds in the use of helical structure elements for asymmetric synthesis. The use of helical building units in complex ligands and receptor models for the study of chiral recognition processes is also to be considered. Clamped helicenes — helicenophanes, as well as bridged oligoarenes (arenicenes) — can contribute to the understanding of helicity, to the correlation of helicity with chromophores and to the relation of structure and helicity. The low-molecular-weight arenicenes also have been developed as they promise ways to new helical skeletons which can be synthesized and handled easier than the helicenes.

4.1 Helicenophanes

A bridging or clamping of the open ends of helicene molecules, as shown by the above-mentioned [5]- to [14]helicenes, is not necessary to obtain enantiomer stability. Martin et al. [110, 153)] rather followed the aim, with the synthesis of the bridged [7]helicenes *160*, *161* and *163* (Fig. 61), to study the consequences of bridging helicene structures. Conformational changes of the skeleton and changes of the chromophor in the solid state and in solution were expected, and it seemed interesting to compare the bridged compounds with the unbridged "open" helicenes.

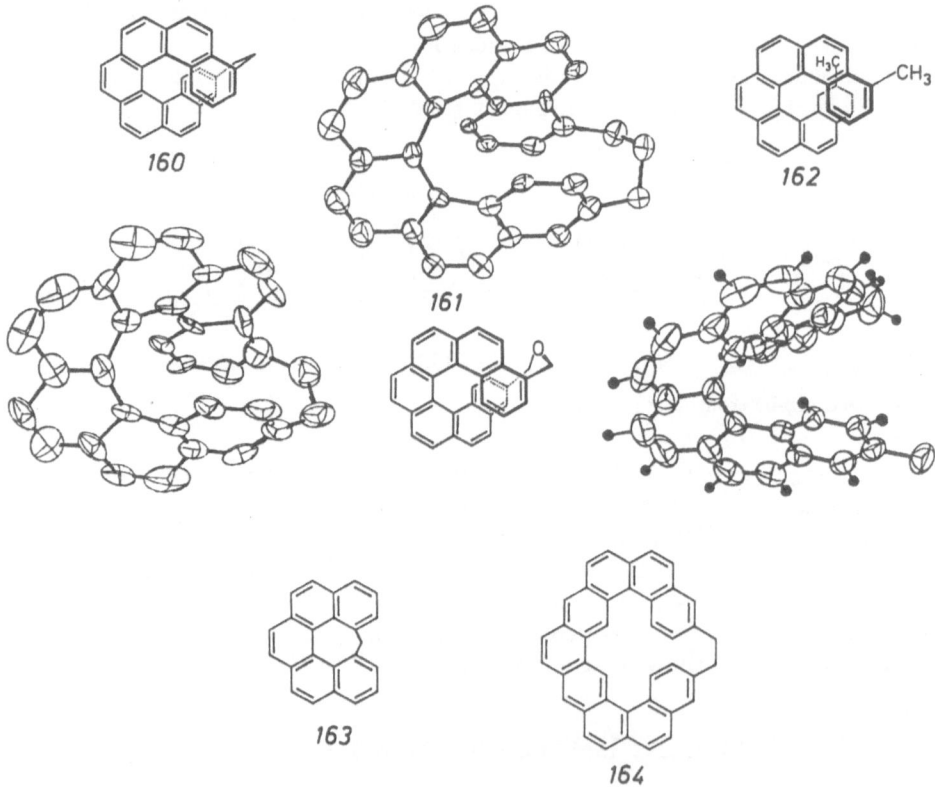

Fig. 61. Helicenophanes *160–164* (above) and X-ray structures of *160–162* (below)

Staab et al. [153b)] synthesized 3,18-ethanodinaphtho[1,2-a;2′,1′-o]pentaphene (*164*); the helicity of *164* has not yet been investigated.

In fact, the electron spectra of the bridged helicenes *160* and *161* show bathochromic effects compared to those of the open 3,15-dimethyl[7]helicene *162* ($\Delta\lambda = 8$–50 nm, in the region of 250–400 nm). The dependency of the pitch from the radius and the number of the condensed aromatic rings is shown in Fig. 62.

The helicenophane *161* forms the shortest pitch combined with a large "radius"

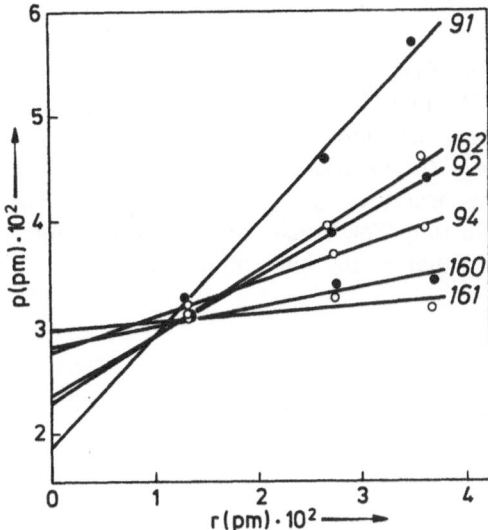

Fig. 62. Correlation between pitch and radii of the helicenes *91, 92, 94* and the bridged helicenes (helicenophanes) *160–162* [153]

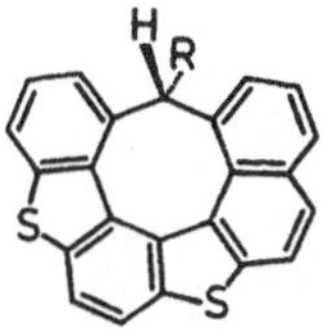

a: R=H
b: R=CH₃

165

Fig. 63. The methano-bridged heterohelicenes *165a, b* [154]

of the helix. Generally, clamping of the helicenes as well as the ortho annellation of additional benzene rings leads to a shortening of the pitch and to a widening of the radius of the helicene. The methano-bridged heterohelicenes *165a,b* (Fig. 63) synthesized by Wynberg et al. [154] analogously show a hypsochromic shift of the α-band in the absorption spectrum compared to the corresponding bands of open-chained alkyl-substituted heterohelicenes.

The chiral crown ethers *166* and *167* respectively containing penta- and hexahelicene anchor groups have been synthesized by Nakazaki et al. [155]. It is of stereochemical interest that the crown ring in *166* has the opposite helicity with respect to the helicene moiety [156], whereas in *167* the screw sense of the crown and the helicene parts of the molecule have the same screw sign. As a consequence of this peculiarity, M-(–)-*166* and M-(–)-*167* show an opposite chiral recognition in the transport of methyl-(±)-phenylglycine (*I*) and (±)-1-phenylethylamine (*II*) through liquid membranes. *166* thereby effects a higher enantiomeric selectivity than *167*: P-(+)-*166* leads to an enantiomeric purity of 77% of (*R*)-*I*, M-(–)-*166* to an enantiomeric purity of 75% for (*S*)-*I*. The enantiomer separations of (±)-*166* and (±)-*167* were carried out using (+)-poly(triphenylmethyl-methacrylate) [*116*, (+)-PTrMA]. Elution with methanol yielded optically pure M-(–)-*166*, P-(+)-*166*, M-(–)-*167* and P-(+)-*167* with rotation values of $[\alpha]_D^{25}$ = –754, +748, –1269 and +1260 (in methanol), respectively. The absolute configurations were correlated through comparison of

55

the circular dichrograms with those of M-(—)-pentahelicene (*102*) [157] and M-(—)-hexahelicene (*91*) [158].

Since Wudl in 1972 [163] published the first synthetic chiral macrocycles of the crown type [164], the groups of Cram [159], Lehn [160], Stoddart [161] and Kellogg [162] studied chiral host compounds. Complex formation and chiral differentiation are investigated for a row of guest molecules [164]. The axial-chiral binaphthyl unit can be considered here as a pro-pentahelicene structure, whereby a condensed aromatic ring of a helicene is replaced by a crown ring. The helicity of the coordinating part (the crown ring) of *168* corresponds to that of the binaphthol unit.

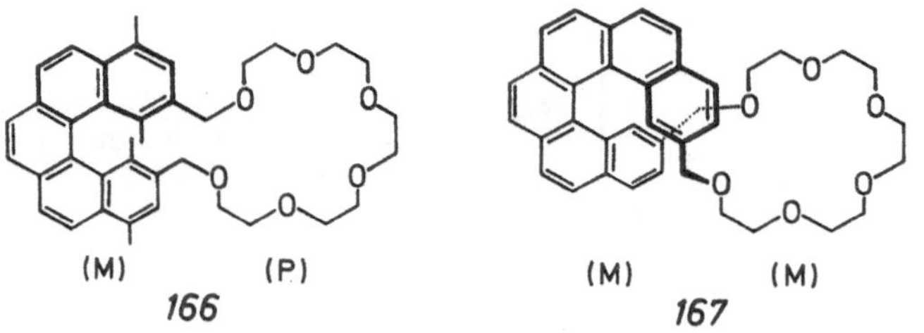

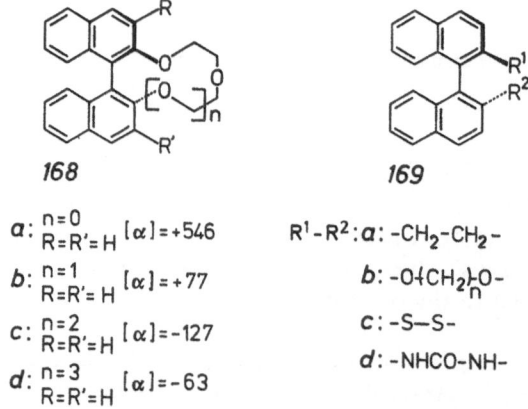

Fig. 64. Penta- and hexahelicene crown ethers *166*, *167*

168   *169*

a: $\begin{matrix} n=0 \\ R=R'=H \end{matrix}$ $[\alpha] = +546$

b: $\begin{matrix} n=1 \\ R=R'=H \end{matrix}$ $[\alpha] = +77$

c: $\begin{matrix} n=2 \\ R=R'=H \end{matrix}$ $[\alpha] = -127$

d: $\begin{matrix} n=3 \\ R=R'=H \end{matrix}$ $[\alpha] = -63$

e: $\begin{matrix} n=4 \\ R=R'=H \end{matrix}$ $[\alpha] = -77$

R^1-R^2: a: $-CH_2-CH_2-$

b: $-O(CH_2)_nO-$

c: $-S-S-$

d: $-NHCO-NH-$

Fig. 65. The binaphthyl derivatives *168a–e* and *169a–d*

The addition of optically active bridged biaryl molecules such as *169a–d* to nematic mesophases like, e.g. 4-cyano-4'-n-pentylbiphenyl or a mixture of cyclohexane derivatives (ZLT 1167), leads to an induction of cholesteric mesophases [165a, b].

A heterocyclic pentahelicene structure hides behind the cyclic binaphthyl phosphoric acid derivatives *170a, b*, which have been applied as reagents for the separation of enantiomers: $[\alpha]_D^{RT} = +530$ (c = 1.35, in methanol). The enantiomer separation of *170* was achieved using the cinchonine salt [165c].

Yamamoto et al. [166] described an enantioselective synthesis of d-limonene (*171*) [and (+)-β-bisabolene, respectively] using binaphthol as an inductor. The transition state produced by the aluminum compound indicates a pentahelicene-analogous structure which induces one of the two enantiomers favourably: "metal anchimeric assistance", metal neighbouring group effect.

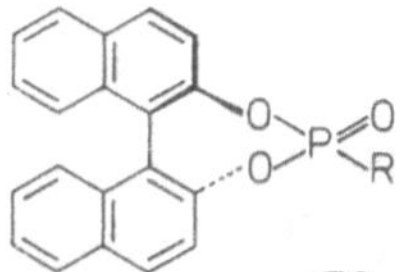

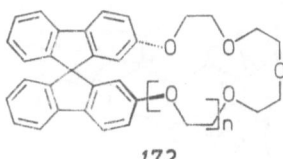

Fig. 66. Cyclic binaphthyl phosphoric acid derivatives (BPA, 170a, b)

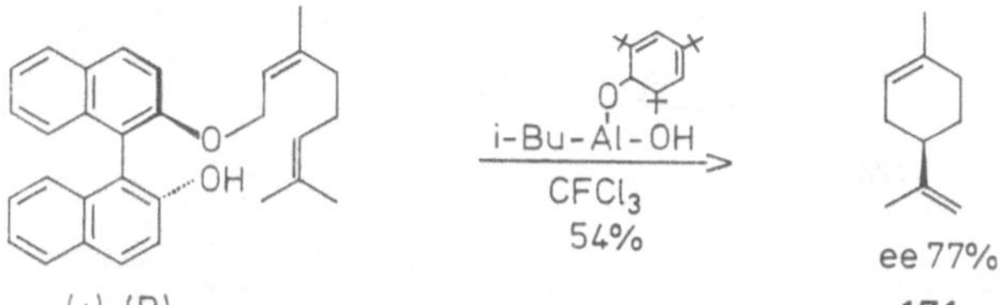

170

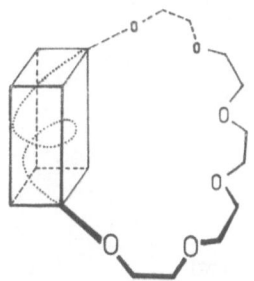

(+)-(R)

i-Bu-Al-OH

CFCl₃

54%

ee 77%

171

Fig. 67. Helical transition state in chiroselective reactions

The stereoselectivity of 9,9'-spirofluorene crown compounds *172* to lipophilic salts of biologically relevant α-amino alcohols [R'CH(OH)CH(NHR)R''] was presented by Prelog et al. [167]: The axial-chiral moiety exhibiting a heptahelicene-analogous structure led to a stabilization of the crown ring and, so far, to an optical induction transmitted by the coordinating crown ether skeleton.

172

Fig. 68. Optically active 9,9'-spirofluorene crown *172*

Fig. 69. General principle: insertion of a rigid helical segment to fix an achiral flexible chain helically

Kurt P. Meurer and Fritz Vögtle

4.2 Araliphatic Propellers

A series of propeller-shaped structures *173–179* were synthesized and investigated spectroscopically [170–172]. Multiple-layered paracyclophanes like *178* reach their screw shape by means of several aliphatic clamps [171b, 172a].

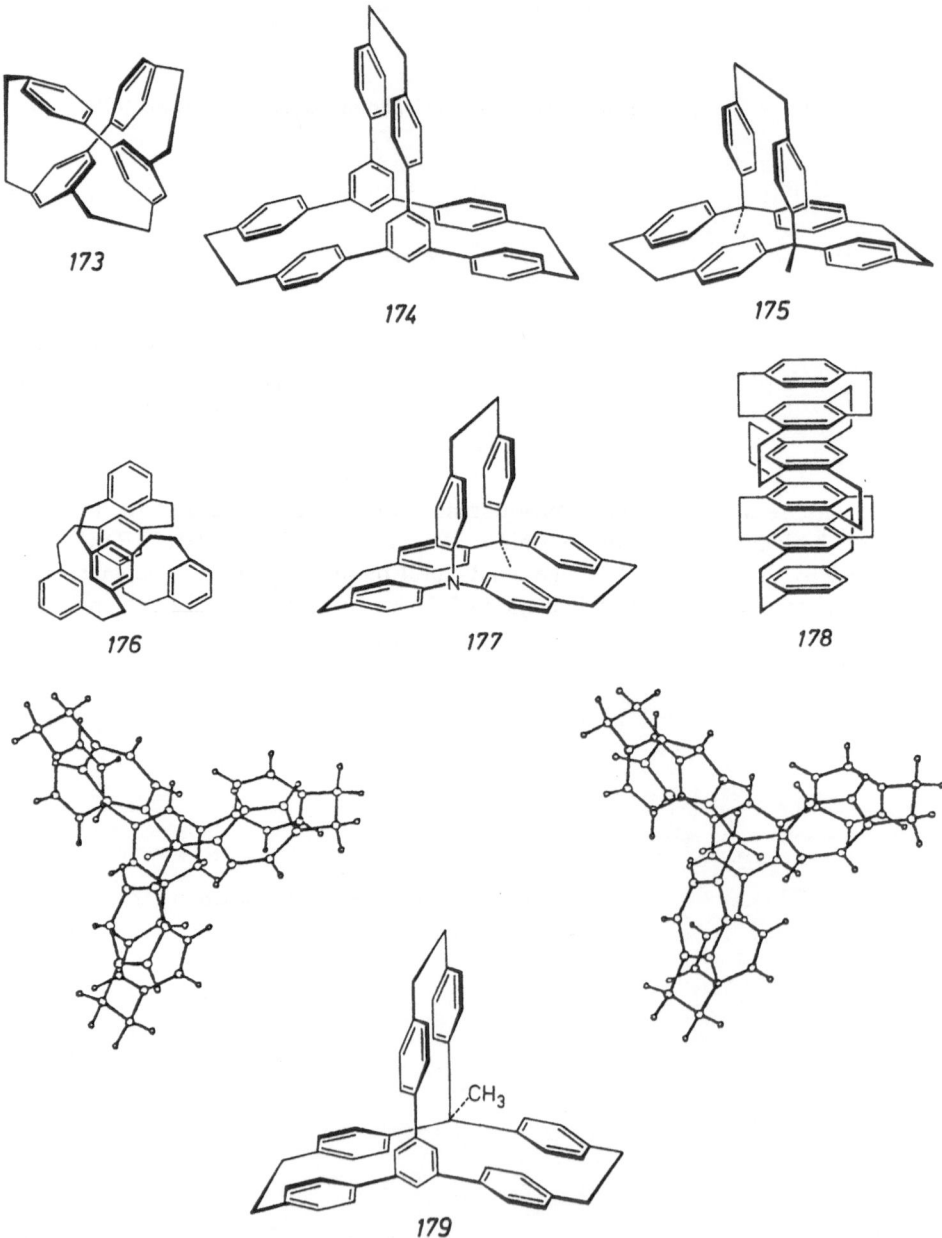

Fig. 70. Araliphatic propellers *173–179* and X-ray analysis of *179* (stereo view)

4.3 Arenicenes: Aliphatically Bridged Oligophenyl Assemblies

By removing structural parts from helicenes, new helices have been developed, which no longer are composed of fully condensed aromatic rings. They gain rigidity by bridges between the terminal aromatic rings. They can be considered as helicenoides, or more precisely as phenylenicenes, if phenylene moieties are the building blocks. When other aromatic rings are present, the term arenicene can be used as a family name to characterize this type of helical molecules.

Here, helicity does not stand for ortho-annelated arene rings; clamping of the ends of aromatic ring assemblies leads to a helical structure of both the aromatic and aliphatic components (Fig. 71).

Fig. 71. Structure of phenylenicenes (arenicenes); one or more of the shaded arene rings may be protruding upwards

4.3.1 Pentaphenylenicenes

The P ⇌ M inversion of the m,m″-clamped quinquephenyls *180a–e* has been investigated by means of dynamic (low temperature) ^{1}H-NMR spectroscopy. The barriers are as follows: *180a*: 43.9 kJ/mol at $T_c = -70.5\,°C$, *180d*: 44.8 kJ/mol at $T_c = -70\,°C$, *180e*: 48.1 kJ/mol at $T_c = -55\,°C$. The helical macrocycles *180a–f* therefore are conformatively flexible and are resolvable into their enantiomers at low temperature [168, 173].

More rigid helices are encountered in the series of the o,p″-clamped pentaphenylenicenes *181a–f* [173], the o,p″″/p,o″″- and o,o″″-clamped pentaphenylenicenes *182–184*.

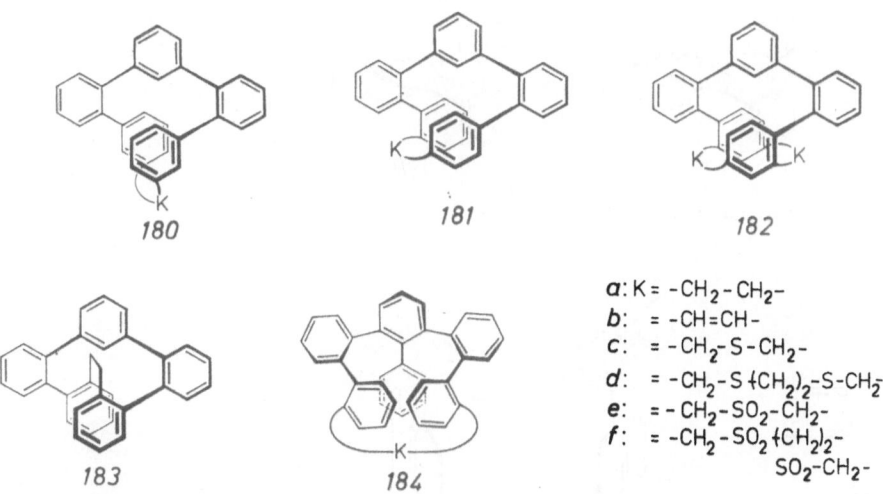

a: K = $-CH_2-CH_2-$
b: = $-CH=CH-$
c: = $-CH_2-S-CH_2-$
d: = $-CH_2-S\,(CH_2)_2-S-CH_2^-$
e: = $-CH_2-SO_2-CH_2-$
f: = $-CH_2-SO_2\,(CH_2)_2-$
        SO_2-CH_2-

Fig. 72. Pentaphenylenicenes

The ^{1}H-NMR spectra of the latter are temperature independent up to 190 °C [168]. A rotation of the p-phenylene rings has been excluded as well as a Plus/Minus interchange of the rigid helices. Enantiomer separations of these hydrocarbons therefore should be possible. They have been attempted using (+)- and (−)-TAPA (112); but the π-complex formation of these noncondensed hydrocarbons seemed not to be sufficient to achieve a simple chromatographic separation or enrichment. Nevertheless, separations using more efficient methods seem promising. On the other hand, these phenylenicenes might be used as unfunctionalized hydrocarbons to test the efficiency of new optically active column materials.

4.3.2 Tetraphenylenicenes

Tetraphenylenicenes [174–176] are composed of four terminally, bridged benzene rings. 190 shows a relatively low energy barrier for the interconversion of helical conformations: $\Delta G_c^{\neq} = 79.3$ kJ/mol at $T_c = 115$ °C. 189 is conformationally more stable: the ring inversion barrier is as high as 92 kJ/mol (^{1}H-NMR, $T_c = 137$ °C). An attempt to separate the enantiomers using a cellulose triacetate column was not yet successful [175].

187 scarcely shows helical stability, but for 186 the barrier was determined to be $\Delta G_c^{\neq} = 83.7$ kJ/mol at $T_c = 140$ °C; 185 indicates higher conformative stability

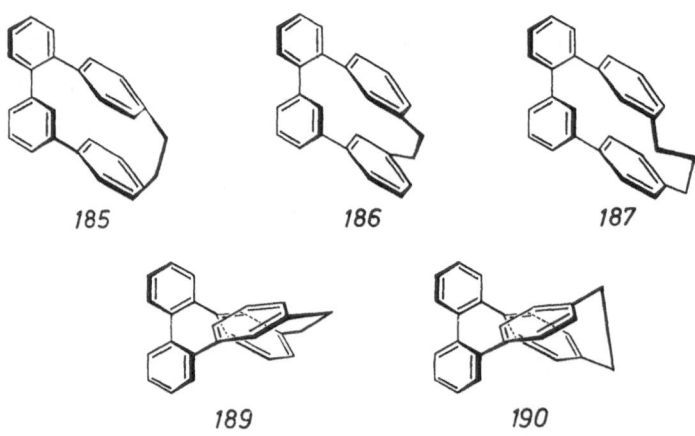

Fig. 73. Tetraphenylenicenes 185–190

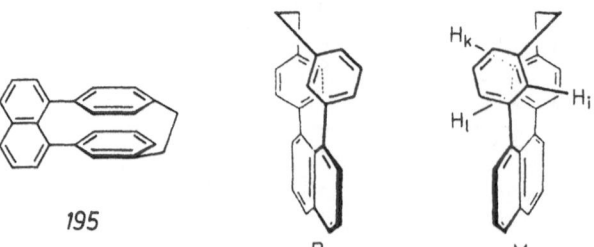

Fig. 74a. M and P configuration of 195

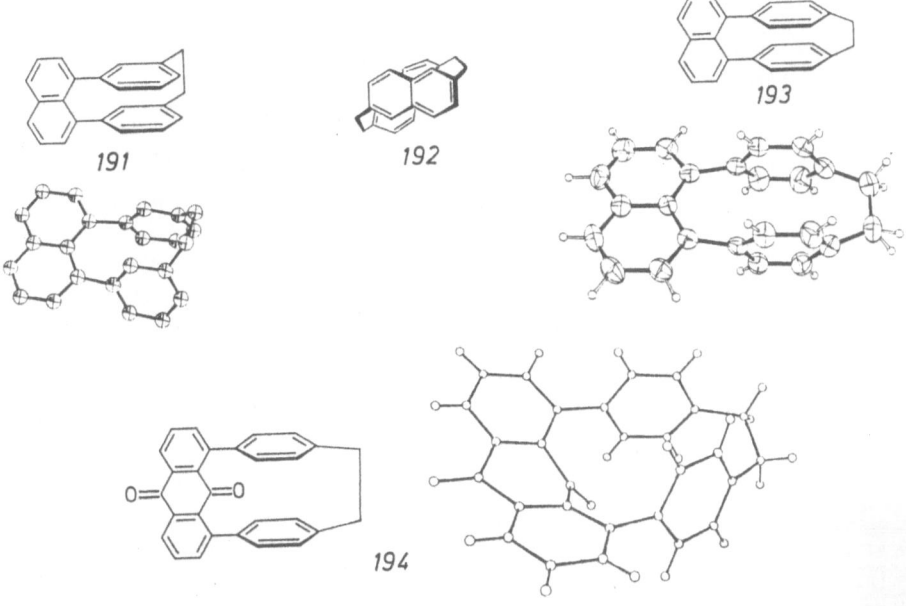

Fig. 74b. Helically deformed [2.2]naphthalinophanes[3.2]meta-, [3.2]para- and [5.2]paracyclophanes (*191–194*), and X-ray structure of *191, 192, 194*

($\Delta G_c^{\neq}$ = 88 kJ/mol). These helices have not yet been separated into the enantiomers [175].

The enantiomers of the [3.2]meta- (*191*), [3.2]metapara- (*195*) and [3.2]paracyclophanes (*193*) were successfully enriched using TAPA-doped column material only in the case of *192* and *195* [176]. The barrier of racemization was measured by polarimetric detection (*195*): E_A = 80 kJ/mol, $\Delta G_{20}^{\neq}$ = 93.6 kJ/mol; $t_{1/2}$ = 11 min at T = 40 °C. The helical structure of *191, 193* and *194* was proved for the crystalline state by X-ray analysis [176].

Tetraanthranilides like *196* [184] also exhibit helical conformations. The [n.n]-vespirenes (n = 6–8)   *a–c* [177], which also can be considered as araliphatic helices

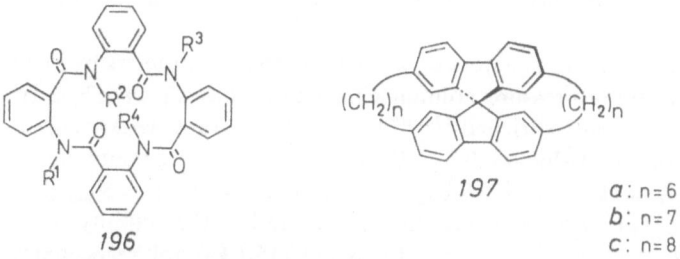

a: n = 6
b: n = 7
c: n = 8

Fig. 75a. Tetraanthranilides *196* and [n.n]vespirenes *197a–c*

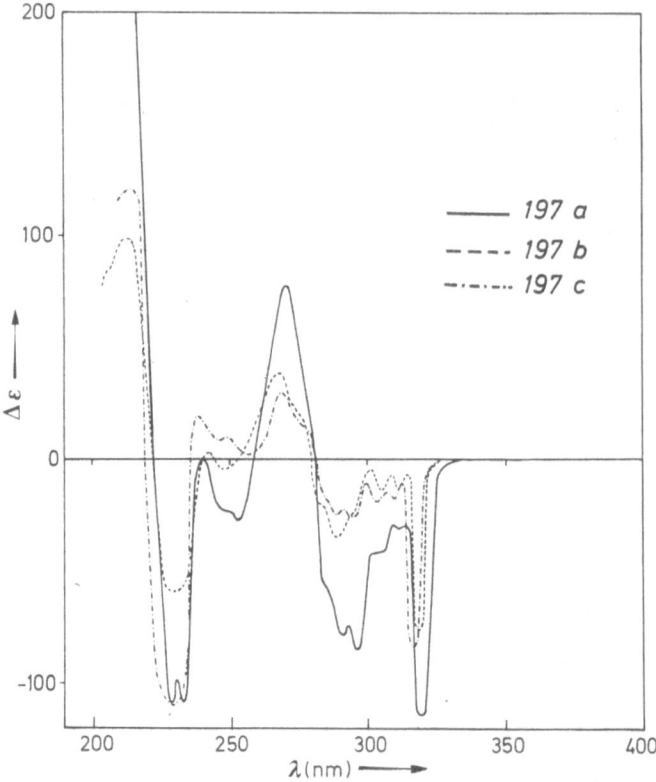

Fig. 75b. CD spectra of the vespirenes *197a–c*

(propellers) show correspondingly high rotation values: $[\alpha]_D = -600$, $[\alpha]_{320} = -380000^{48)}$.

4.3.3 Triphenylenicenes

Clamped phenylenicenes containing three benzene rings on account of their large ring strain and of their good synthetic availability belong to the best investigated family of araliphatic helices. Detailed studies have been made regarding their geometry (NMR, X-ray), enantiomer stability, separation of enantiomers and chiroptical properties.

The helical compounds *198a* and *199* were separated into the enantiomers by inclusion chromatography (medium-pressure column chromatography on cellulose triacetate). *200* was enantiomerically enriched [178, 179]. $(-)$-*200* was only weakly enriched: $[\alpha]_D^{RT} = 20$ [180]. The rotation volues of *198–200* are high in relation to their helicity: $[\alpha]_{max}^{20} = [3210]_{436}$ (*198a*, in ethanol), $[2096]_{436}$ (*199*, in ethanol) and $[-65]_{546}$ (*200*, in acetone), respectively. The barriers of racemization underline the stability of the optical antipodes of *198a–200*: $\Delta G_{INT} = 125.5$, 121.6 and 115.1 kJ/mol, respectively. The circular dichrograms reveal high $\Delta\varepsilon$ values for the pure enantiomers $(+)$-*198a/199*

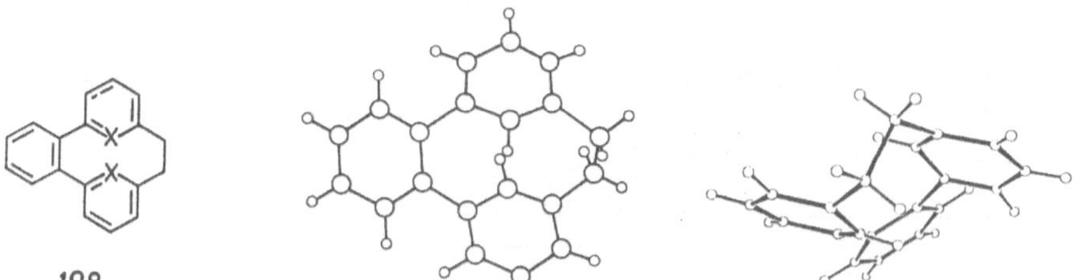

198

a: X = CH
b: X = N

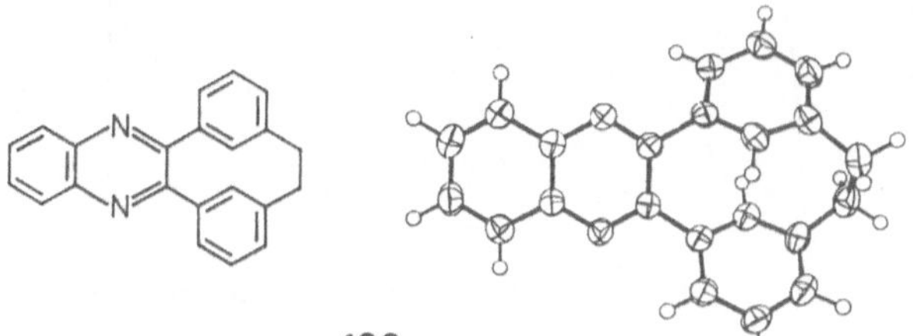

199

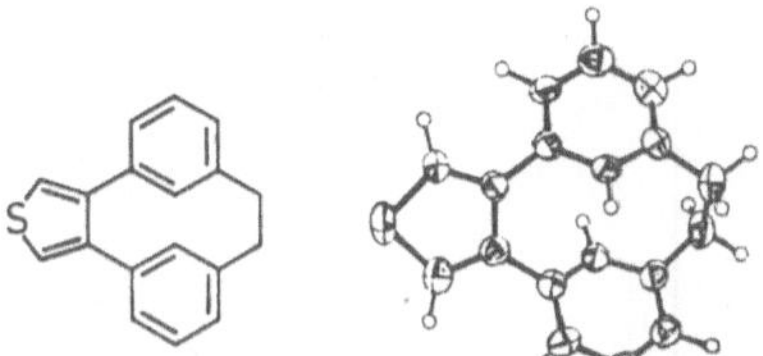

200 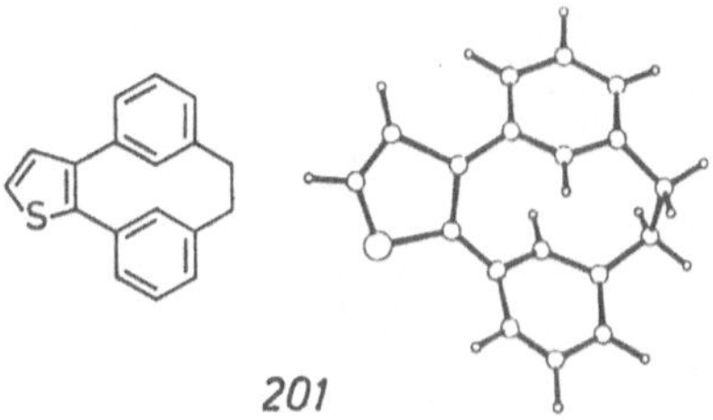   201

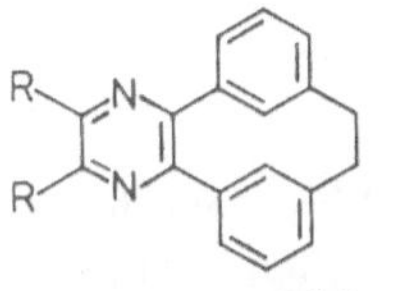

202   203   204   205
a: R = Phenyl
b: R = H

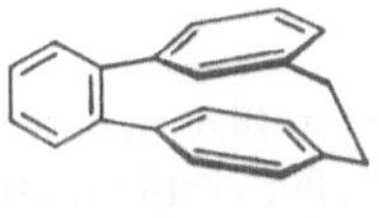

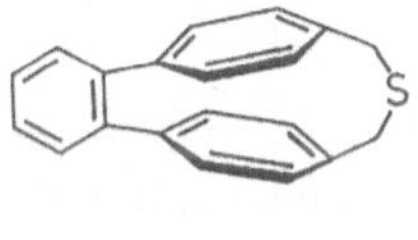

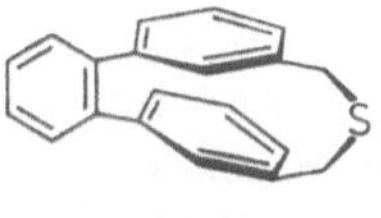

Fig. 76. Structures of triacenicenes *198–201* and *202–205*

and (−)-*198a/199*: Δε = +80 to +90 and −80 to −90, respectively. Three CD maxima are observed for these helices. Thus, it is possible to correlate the absolute conformation in this group of compounds with some certainty. On the other hand, the absolute configuration of (+)-*198*, (+)-*199* and (+)-*200* has been proved by the Bijvoet method [179a]: P-(+)-*198a–200*, M-(−)-*198a–200* [179b]. It seems possible to propose an absolute configuration of some further three-core helices of this family on the basis of their CD spectra. *203* was found to be much more flexible ($\Delta G_c^{\neq}$ = 76 kJ/mol) than *204* and *205* ($\Delta G_c^{\neq}$ = 37 kJ/mol for *205*) [181].

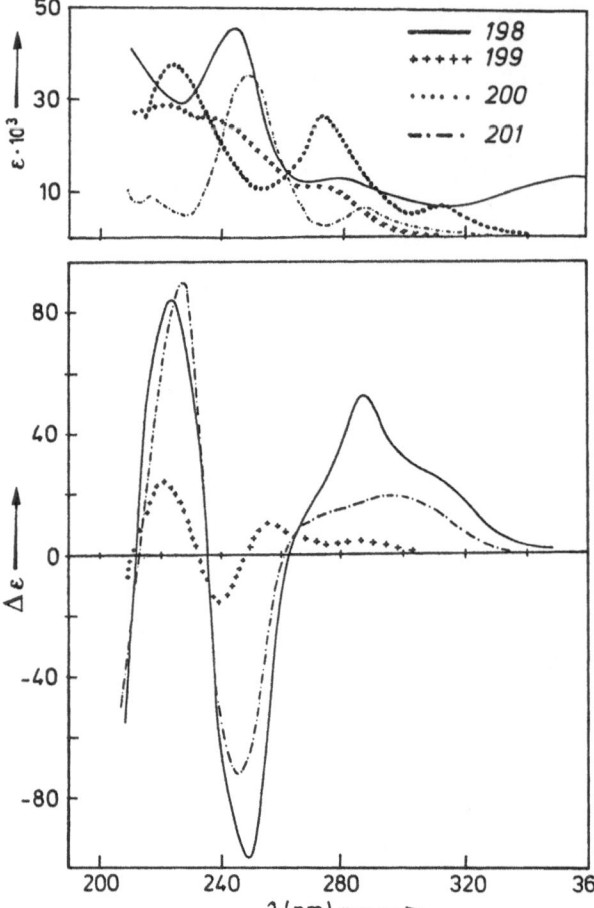

Fig. 77. UV and CD spectra of *198–201*

The tri-o-thymotids *206–209* spontaneously crystallize as solvent adducts and form enantiomorphic crystals, which contain the propeller conformation of the host molecule [183]. The (+)-CHCl$_3$ adduct of *206*, e.g., shows a rotation value of $[\alpha]_D^{294.6K} = -83$, the (−)-*206* CHCl$_3$ adduct $[\alpha]_D^{294.4K} = -77$. The half-lifes are $t_{1/2} = 34$ min at 274.6 K, 12.4 min at 283.3 K and 2.4 min at 294.6 K, respectively. The activation energy of the racemization of *206* has been estimated to be 93 kJ/mol [184]. The barrier for the ring inversion of *209a* and *209b* amounts to 85–88 kJ/mol [185]. The barrier for the ring inversion of *208* is as high as 112 kJ/mol [186]. A separation into the enantiomers seems to be promising.

Collet et al. [187] started chiroptical investigations of (C$_3$)-cyclotriveratrylene [188]. The absolute configurations of (C$_3$)-(2H$_3$)-cyclotribenzylene were found to be P-(+)-*210* and M-(−)-*210*, respectively. The barrier of the "crown-to-crown" interconversion was measured as 110.9 kJ/mol. The circular dichrograms of *210a–e* were correlated using the exciton optical activity of isotope-substituted representatives (Fig. 79, Table 13).

206
207
208
209 a
209 b
Me
S
O
N

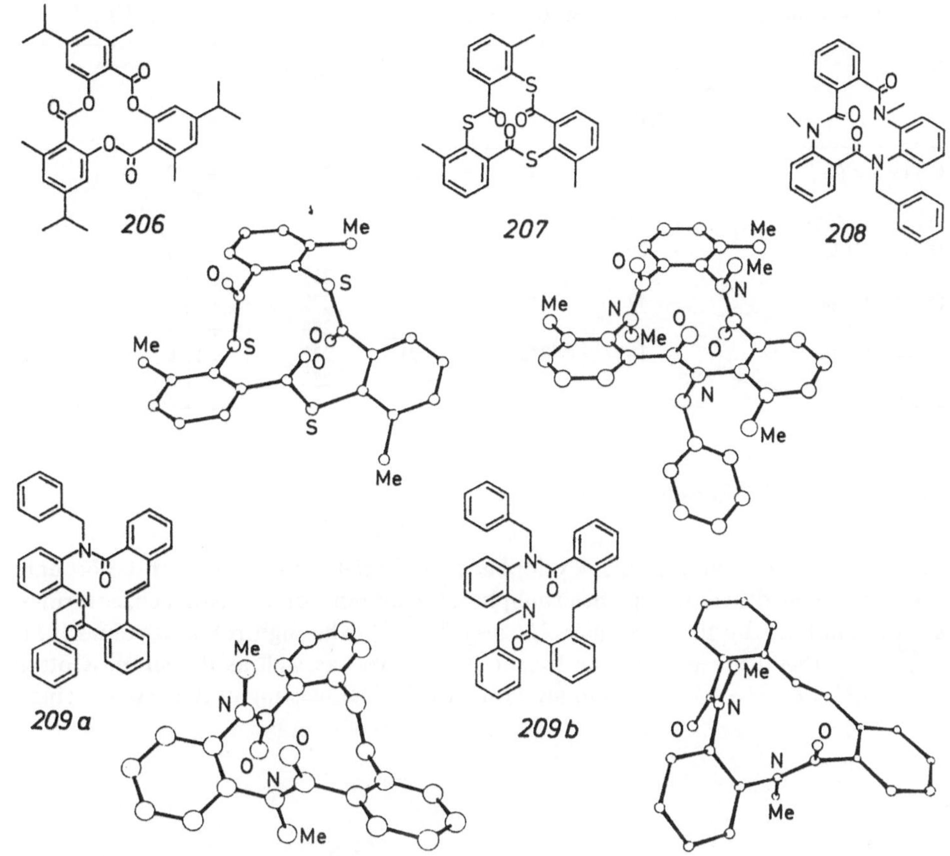

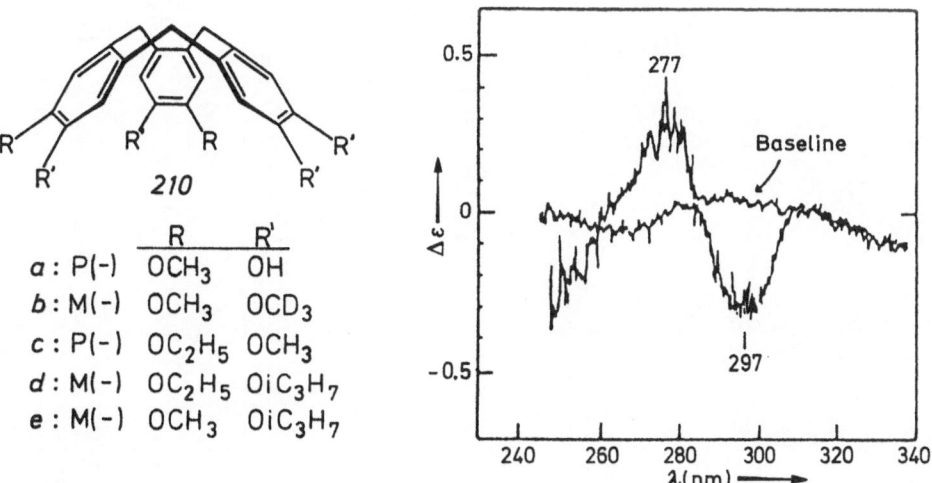

Fig. 78. Tri-o-thymotids *206–209* and X-ray structures [182–186]

R
R'
210
R R'
a : P(-) OCH3 OH
b : M(-) OCH3 OCD3
c : P(-) OC2H5 OCH3
d : M(-) OC2H5 OiC3H7
e : M(-) OCH3 OiC3H7
0.5
0
-0.5
Δε
277
297
Baseline
240 260 280 300 320 340
λ(nm)

Fig. 79. CD spectra of (C_3)-cyclotriveratrylene derivatives *210a–e* [187]

Table 13. Circular Dichroism Spectra of (C_3)-Cyclotriveratrylene Derivatives *210a–e* [187]

No.	$[\alpha]_D$			B_{1u}	
	(CHCl^3)	λ	Δε	λ	Δε
(P)-(—)-*210a*	−253	208	−47	236	+ 3.0
				249	−11
(M)-(—)-*210b*	— · 3.2			250	—
(P)-(—)-*210c*	— 20			233	− 5.5
				252	+ 4.1
(M)-(—)-*210d*	— 47	213	−33	230	+ 4.5
				248	+ 1.7
(M)-(—)-*210e*	— 14	213	−21	232	+12.5
				253	− 5.0

4.4 Helical[2.2]Metacyclophanes

The helicity of double-core metacyclophanes and related molecules can be brought about through different, e.g. heteroaliphatic, clamping of the two benzene rings, as is exemplified by the molecules *211* and *212* [189]. The high rotation values [*211*: $[\alpha]_{365}^{RT} = -1500$ (dioxane)] of the (pure) enantiomers as well as the strong Cotton effects (*211*: Δε = −60.5, −80.6) underline the helical twisting of the benzene rings.

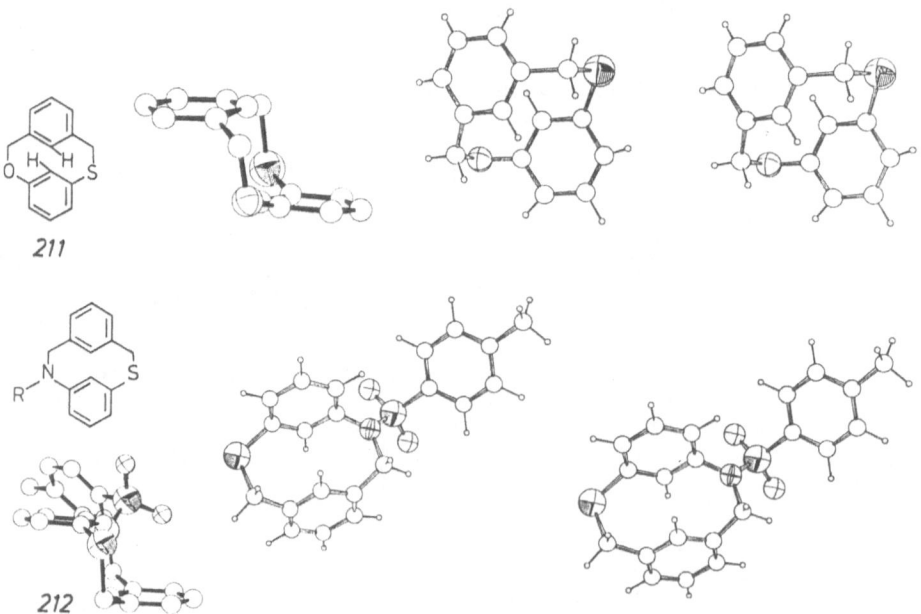

Fig. 80. Structures of the helical [2.2]metacyclophanes bearing heterocyclic clamps (*211, 212*) and X-ray structures [189]

The enantiomers of *212* were enriched through inclusion chromatography on cellulose triacetate. They show the rotation values: (*212*) $[\alpha]^{RT}_{365} = -57$ (CD: $\Delta\varepsilon = -7.9$ and $+4.1$).

A comparison with the CD spectra of the triphenylenicenes *198–200* leads to the conclusion that the absolute configuration of (+)-*211* and (+)-*212* is P, and that of (−)-*211* and (−)-*212* is M. The stability of the enantiomers is advantageous with regard to eventual applications: $\Delta G_{INT} = 123.3$ kJ/mol at T = 100 °C (*211*, ΔG_{INT} = 130.7 kJ/mol at T = 102 °C (*212*).

The helical structure of *213* has been proven by X-ray analysis [190], but a separation of the enantiomers has not been achieved hitherto, as the low chemical stability leads to benzpyrene through oxidation or disproportionation. The phane *214*, which

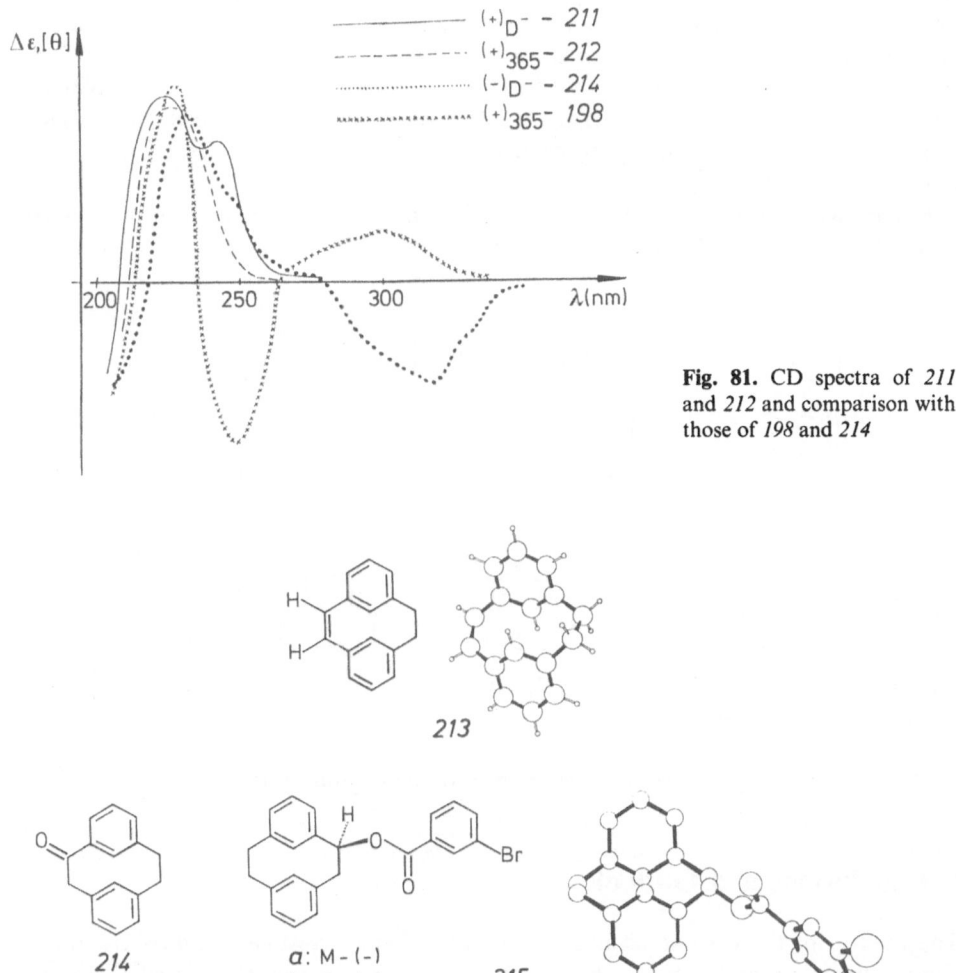

Fig. 81. CD spectra of *211* and *212* and comparison with those of *198* and *214*

213

214

a: M – (−)
b: P – (+)

215

Fig. 82. Helical carba[2.2]metacyclophanes *213–215*

is somewhat more stable, should have a weakly helical structure on account of its carbonyl function in one of the two bridges, but it can be interpreted approximately as planarchiral [191]. A separation of the enantiomers has been achieved here through fractionized crystallization of the diastereomers (*215*). The stability of the ring inversion is comparable to that of the carba[2.2]metacyclophane *214*: ΔG_{150} = -104.8 kJ/mol. The CD spectrum of *214* shows Cotton effects at $\lambda = 320$ nm, $[\theta]_{max} = -36.7$, $\lambda = 231$ nm, $[\theta]_{max} = +63.1$ [192]. The rotation values are high ($[\alpha]_D^{25} = 439.3$, c = 1.19, CHCl$_3$), but they are in the range of other planarchiral compounds [195, 196]. The absolute configuration was found by comparison of the isomer *215* to *214*, and has been proved using Bijvoet's method: P corresponds to (+), M to (−).

Studies of the double-core, aliphatically clamped [3.2]metacyclophanes by dynamic NMR spectroscopy were conducted as early as 1968 [193]: For *216*, $\Delta G_c^{\neq}$ was found to be 54.8 kJ/mol at $T_c = -8$ °C, 67.8 kJ/mol at $T_c = -45.5$ °C. These [3.2]meta-cyclophanes exhibit higher conformational flexibilities than their [2.2]metacyclophane analogues. This statement also holds for the *anti*-[4.1]metacyclophanes *217* and *218*, the barriers of which are $\Delta G_c^{\neq} = 82.1$ kJ/mol at $T_c = 115-120$ °C [194]. The planar-chiral model compounds *219–222*, on the other hand, are conformatively more rigid: $\Delta G_{INT} = 107.2$ and 131 kJ/mol, respectively (*220*, *219*) $[\alpha]_{365}^{RT} = 6$ in CHCl$_3$ for *219*, $[\alpha]_{365}^{RT} = 808$ for *220*) [195].

For future investigations of structures lying in between planarchiral and helical molecules, the type of double-core molecules seems to be well suitable [195, 196].

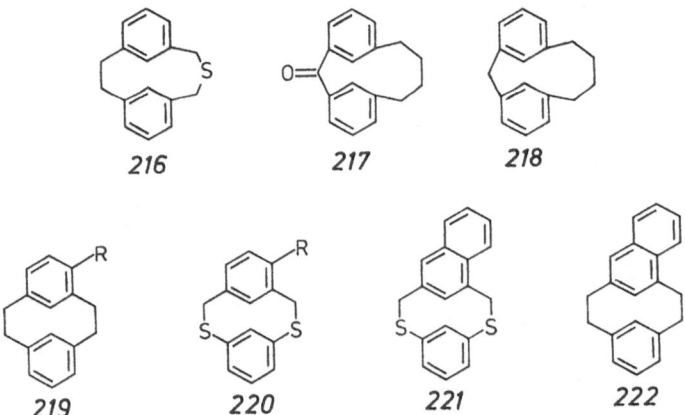

Fig. 83. Some low-molecular-weight helical and planarchiral molecules (*216–222*)

5 Conclusions and Outlook

The interest of the organic chemist to synthesize and investigate higher-structured chiral substances and reagents has grown. This interest was stimulated in recent years, on the one hand, through the fast progress of structural knowledge in bio-chemistry and macromolecular chemistry, and on the other hand, through advances in synthesis, e.g. organometallic and stereoselective methods. In nature, even complicated centrochiral, axialchiral and helical compounds are synthesized. Apart from the high-molecular-weight bio-helices mentioned in the introduction, the appearance of axially chiral molecules like skyrin[1], a pigment from several Penicillium species, proves that nature, apart from chirality centers and helices, also induces and uses axial chirality [197]. Pharmacologically efficient substances, e.g. ( )-colchicin[2] show P configuration.

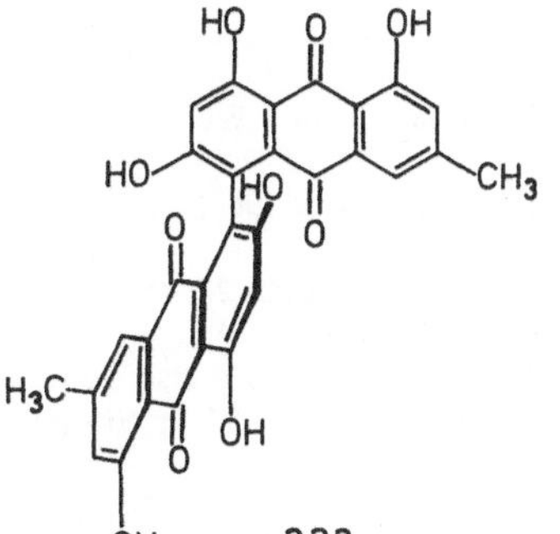

Fig. 84. Axialchirality of skyrin (*223*) [197] and colchicin (*224*)

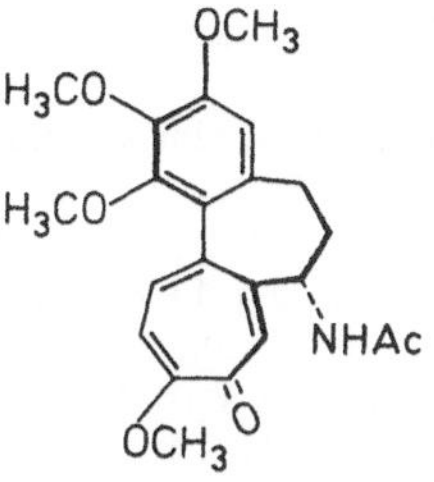

The (bio-)chemical synthesis of suitable axial-chiral, planarchiral and helical skeletons in enantiomeric pure form will be a challenging experimental field also in the future. Heterocyclic helical structures seem to be reasonably easy to synthesize today, as shown especially in Section 3.4, and they may possibly have some advantages in special applications as "chiron" components [198] compared to centro-chiral substances taken from the chiral pool: Helical molecules due to their strongly different shape in the three directions of space usually cause high rotation values, strong CD absorptions, intensive bathochromic shifts and charge-transfer interactions. Helical molecules influence the formation of helical conformations in biological systems, liquid crystalline media [165b] and in the initiation of isotactic polymerization (chiral polymeric cavities or niches). Their application as asymmetric inductors, especially as asymmetric catalysts or as part of macrocyclic host/guest complexes, seems attractive. The efficiency of asymmetric inductions can possibly be strengthened through coordination sites in the molecular helix framework. Appropriate substituents could facilitate the helix to be conformationally changed or stabilized. The possibility to intercalate guest molecules in between helix windings seems of interest for inclusion chemistry as well as for mimicking biochemical processes. We are confident that helicity, which in organic chemistry for some time has been considered somewhat esoteric and has been standing in the background compared to centrochiral molecules, in future will face more attention.

[1] *223*: $[1400]_D^{RT}$ (methanol; $\Delta\varepsilon_{485} = 18.65$, $\Delta\varepsilon_{260.5} = 39.6$)
[2] *224*: $[-443]_D^{24}$ (H_2O).

6 Acknowledgement

We thank Dr. U. Lehmann and Dr. M. Wittek-Hufnagel for contributions to Section 3. We are grateful to Dipl.-Chem. A. Aigner for critical advice and Dipl.-Chem. J. Franke and J. Ahrendt for the design of stereo- and other drawings. We also thank the Deutsche Forschungsgemeinschaft and the Fonds der Chemischen Industrie for financial support of our own studies.

7 References

1. a) Cf. Price, C.: Die räumliche Struktur organischer Moleküle, Verlag Chemie, Weinheim 1971, p. 87;
   b) Cf. Rawn, J. D.: Biochemistry, Harper and Row Publ. Inc., New York 1983, pp. 95, 103, 143. 151–157, 307, 340–345
2. a) Freudenberg, K., Schaaf, E., Dumpert, G., Plock, T.: Naturwissenschaften 27, 850 (1959);
   b) Hui, Y., Russell, J. C., Whitten, D. G.: J. Am. Chem. Soc. 105, 1374 (1983)
3. Watson, J. D., Crick, F. H. C.: Nature 171, 737 (1953); ibid. 171, 964 (1953); Watson, J. D.: The Double Helix, Atheneum, New York 1968. Stryer, L.: Biochemistry, 2nd ed., W. M. Freeman and Company, San Francisco, 1981, p. 560, 585–587
4. Franklin, R. E., Gosling, R. G.: Nature 172, 156 (1953)
5. a) Lehninger, A. L.: Biochemistry, World Publishers, Inc., New York, 1970;
   b) Saenger, W.: Nachr. Chem. Tech. Lab. 30, 10 (1982)
6. Jungermann, K., Möhler, H.: Biochemie, Springer Verlag, Berlin 1980
7. Mahler, H. R., Cordes, E. H.: Biological Chemistry, Harper and Row, New York 1966
8. Bailey, J. A.: The Nature of Collagen, Compr. Biochem. 26B, 297 (1968);
   Traub W., Fiez, K. A.: Advan. Protein Chem. 25, 243 (1971)
9. Ross, R., Bornstein, P.: Sci. Amer. 224 (6), 44 (1971)
10. Rastetter, W. H., Erickson, T. J., Venutti, M. C.: J. Org. Chem. 46, 3579 (1981)
11. Okamoto, Y., Honda, S., Okamoto, I., Yuki, H., Murata, S., Noyori, R., Takaya, H.: J. Am. Chem. Soc. 103, 6971 (1981);
    Okamoto, Y., Shohi, H., Yuki, H.: J. Polym. Sci. Polym. Lett. Ed., 19, 451 (1981)
12. Elias, H. G.: Makromoleküle, Hüthig und Wepf Verlag, Basel 1971
13. a) Saenger, W.: Umschau 1974, 635; Schlenk, W.: Liebigs Ann. Chem. 1973, 1145; 1156; 1179; 1195
    b) Sheldrick, W. S.: J. Chem. Soc. Perkin Trans. II, 1976, 1457;
    c) Lehner, H., Braslwasky, S. E., Schaffner, K.: Angew. Chem. 90, 1012 (1978); Angew. Chem., Int. Ed. Engl. 17, 948 (1978); Braslwasky, S. E., Holzwarth, A., Lehner, H., Schaffner, K.: Helv. Chim. Acta 61, 2219 (1978), and further publications
14. Neumüller, O. A.: Römpps Chemie-Lexikon, Franck'sche Verlagsbuchhandlung, Stuttgart 1972
15. Cahn, R. S., Ingold, C., Prelog, V.: Angew. Chem. 78, 413 (1966); Angew. Chem., Int. Ed. Engl. 5, 385 (1966)
16. Brewster, J. H.: Top. Curr. Chem. 47, 29 (1974)
17. Adachi, K., Mamura, K., Nakazaki, M.: Tetrahedron Lett. 1968, 5467
18. Tichy, M., Sicher, J.: ibid. 1969, 4609
19. Brewster, J. H.: Top. Stereochem. 2, 1 (1967)
20. Brewster, J. H.: Tetrahedron Lett. 1972, 4355
21. Brewster, J. H.: J. Am. Chem. Soc. 81, 5475, 5483, 5493 (1959)
22. Pino, P.: Advan. Polymer Sci. 4, 393 (1965)
23. Pino, P., Ciardelli, F., Zandomeneghi, M.: Ann. Rev. Phys. Chem. 21, 561 (1970)
24. Prelog, V., Helmchen, G.: Angew. Chem. 94, 614 (1982); Angew. Chem., Int. Ed. Engl. 21, 567 (1982)
25. Mislow, K., Siegel, J.: private communication. Mislow, K., Siegel, J.: J. Am. Chem. Soc. 106, 3319 (1984). Dagani, R.: Chem. Engin. 1984, June 11, 21

26. a) Mislow, K., Gust, D., Finocchiaro, D., Boettcher, R. J.: Top. Curr. Chem. *47*, 1 (1974);
    b) Farina, M., Morandi, C.: Tetrahedron *36*, 1819 (1974);
    c) Mislow, K.: Acc. Chem. Res. *9*, 26 (1976);
    d) Biali, S. E., Rappoport, Z.: J. Am. Chem. Soc. *106*, 477 (1984); .
    e) Krow, G.: Top. Stereochem. *5*, 31 (1970);
    f) Pindur, U., Akgün, E.: Pharm. in uns. Zeit. *12*, 135 (1983)
27. Mislow, K., Glass, M. A. W.: J. Am. Chem. Soc. *83*, 2780 (1961); Mislow, K., Hopps, A. B., Simon, E., Wahl, Jr., G. H.: ibid. *86*, 1710 (1964)
28. Bastiansen, O.: Acta Chem. Scand. *3*, 408 (1949)
29. Ermer, O., Gerdil, R., Dunitz, J. D.: Helv. Chim. Acta *54*, 2476 (1971); Ginsburg, D.: Acc. Chem. Res. *2*, 121 (1969)
30. Dilthey, W., Hertig, G.: Ber. Dtsch. Chem. Ges. *67*, 2004 (1934); Bart, J. C. J.: Acta Crystallogr. *24B*, 1277 (1968)
31. Barbour, R. H., Freer, A. A., Mac Nicol, D. D.: J. C. S. Chem. Commun. *1983*, 362
32. Sundarahugam, M., Jensen, L. H.: J. Am. Chem. Soc. *88*, 198 (1966)
33. Kessler, H., Moosmayer, A., Rieker, A.: Tetrahedron *25*, 287 (1969)
34. Rieker, A., Kessler, H.: Tetrahedron Lett. *1969*, 1227
35. Sabacky, M. J., Johnson, S. M., Martin, J. C., Paul, J. C.: J. Am. Chem. Soc. *91*, 7542 (1969)
36. Schuster, I. I., Colter, A. K., Kurland, R. J.: ibid. *90*, 4679 (1968); Rakshys, Jr., J. W., McKinley, S. V., Freedman, H. H.: ibid. *92*, 3518 (1970)
37. a) Wittig, G., Braun, A., Christen, A. J.: Angew. Chem. *79*, 721 (1967); Angew. Chem., Int. Ed. Engl. *6*, 700 (1967); Wittig, G., Braun, A., Christen, A. J.: Liebigs Ann. Chem. *751*, 17 (1971);
    b) Heß, H., Musso, H.: ibid. *1979*, 431;
    c) Hess, H., Burger, G., Musso, H.: Angew. Chem. *90*, 645 (1978); Angew. Chem., Ind. Ed. Engl. *17*, 612 (1978)
38. Boettcher, R. J., Gust, D., Mislow, K.: J. Am. Chem. Soc. *95*, 7157 (1973)
39. Hayes, K. S., Nagumo, M., Blount, J. F., Mislow, K.: ibid. *102*, 2773 (1980)
40. Okamoto, Y., Yoshima, E., Hatada, U., Mislow, K.: unpublished results; Mislow, K.: private communication
41. Tanaka, K., Toda, F.: Tetrahedron Lett. *1980*, 2713
42. a) Vögtle, F.: Liebigs Ann. Chem. *728*, 17 (1969);
    b) Leach, D. N., Reiss, J. A.: J. Org. Chem. *43*, 2484 (1978); Leach, D. N., Reiss, J. A.: Tetrahedron Lett. *1979*, 4501
43. Mannschreck, A., Talvitie, A., Fischer, W., Snatzke, G.: Monatsh. Chem. *114*, 101 (1983)
44. Christensen, A., Geise, H. J., Van der Veken, B. J.: Bull. Soc. Chim. Belg. *84*, 1173 (1975)
45. Carter, R. E., Sandström, J.: J. Phys. Chem. *76*, 642 (1972)
46. Isaaksson, R., Liljefors, T.: J. Chem. Soc. Perkin II, *1981*, 1344
47. Gangl, D., Magnus, D., Dass, L., Arnold, E. V., Clardy, J.: J. Am. Chem. Soc. *102*, 2134 (1980)
48. Dale, J.: Stereochemie und Konformationsanalyse, Verlag Chemie, 1. Aufl., Weinheim 1978, p. 124
49. Adachi, K., Naemura, K., Nakazaki, M.: Tetrahedron Lett. *1968*, 5467
50. Tichy, M.: ibid. *1972*, 2001
51. Jenkins, J. A., Doehner, R. E., Paquette, L. A.: J. Am. Chem. Soc. *102*, 2131 (1980)
52. Askani, R., Schwertfeger, W.: Chem. Ber. *110*, 3046 (1977)
53. Nakazaki, M., Narmura, K., Nakahara, S.: J. Org. Chem. *43*, 4745 (1978)
54. Tichy, M., Sicher, J.: Tetrahedron Lett. *1969*, 4609
55. Capraro, H. G., Ganter, C.: Helv. Chim. Acta *63*, 1347 (1980)
56. Ackermann, P., Tobler, H., Ganter, C.: ibid. *55*, 2731 (1972)
57. a) Tichy, M., Hamsikova, E., Blaha, V.: Collect. Czech. Chem. Commun. *41*, 1935 (1976);
    b) Askani, R., Eichenauer, H., Köhler, J.: Chem. Ber. *115*, 748 (1982);
    c) Archelas, A., Furstoß, R., Waegell, B., Le Petit, J., Deveze, L.: Tetrahedron *40*, 355 (1974)
58. Nakazaki, M., Naemura, K., Kondo, Y.: J. Org. Chem. *41*, 1229 (1976)
59. Nakazaki, M., Naemura, K., Sugano, Y., Katacka, Y.: ibid. *45*, 3232 (1980), and literature cited therein
60. Nakazaki, M., Naemura, K., Kadowacki, H.: ibid. *41*, 3725 (1976)
61. Naemura, K., Nakazaki, M.: Bull. Chem. Soc. Jpn. *46*, 888 (1973)

62. Tichy, M.: Collect. Czech. Chem. Commun. *39*, 2673 (1974)
63. Cope, A. C., Ganellia, C. R., Johnson, Jr., H. W., Van Anker, T. V., Winkler, H. J. S.: J. Am. Chem. Soc. *85*, 3276 (1963)
64. Cope, A. C., Mehta, A. S.: ibid. *86*, 1268 (1964)
65. Scott, A. J., Wrixon, A. D.: Tetrahedron *26*, 3695 (1970)
66. Nakazaki, M., Yamamoto, K., Maeda, M.: J. Org. Chem. *45*, 3229 (1980)
67. Cope, A. C.. Banholzer, K., Keller, H., Pawson, B. A., Whang, J. J., Winkler, H. J. S.: J. Am. Chem. Soc. *87*, 3644 (1965)
68. Nakazaki, M., Naemura, K.. Nakahara, S.: J. C. S. Chem. Commun. *1979*, 82
69. Nakazaki, M., Naemura, K., Nakahara, S.: J. Org. Chem. *44*, 2438 (1979)
70. Meier, A., Echter, T., Peterson, H.: Angew. Chem. *90*, 997 (1978); Angew. Chem., Int. Ed. Engl. 17; 942 (1978)
71. Cope, A. C., Whitesides, G. M., Goe, G. L.: J. Am. Chem. Soc. *89*, 7136 (1967)
72. a) Marshall, J. A., Lewellyn, M.: ibid. *99*, 3508 (1977); Marshall, J. A., Bierenbaum, R. E., Chung, K. H.: Tetrahedron Lett. *1979*, 2081; Marshall, J. A., Black, T. H., Shone, R. L.: ibid. *1979*, 4737; Marshall, J. A., Chung, K. H.: J. Org. Chem. *44*, 1566 (1979); Marshall, J. A., Black, T. H.: J. Am. Chem. Soc. *102*, 7581 (1980); Marshall, J. A.: Acc. Chem. Res. *13*, 213 (1980)
    b) Marshall, J. A., Flynn, K. E.: J. Am. Chem. Soc. *105*, 3360 (1983); ibid. *106*, 773 (1984);
    c) Nickon, A., Zurer, P. S. J.: Tetrahedron Lett. *1980*, 3527
73. Cullen, E. R., Guziec, Jr., F. S., Hollander, M. I., Murphy, C. J.: Tetrahedron Lett. *1981*, 4563
74. Olah, G. A., Surya Prakash, G. K.: J. Org. Chem. *42*, 580 (1977)
75. Back, T. G., Barton, D. H. R., Britten-Kelly, M. R., Guziec, Jr., F. S.: J. Chem. Soc. Perkin I, *1976*, 2079
76. Back, T. G., Barton, D. H. R., Britten-Kelly, M. R., Guziec, Jr., F. S.: ibid. *1974*, 1794
77. Wynberg, H., Lammertsma, K., Hulshof, L. A.: Tetrahedron Lett. *1975*, 3749
78. Döpke, W.: Dynamische Aspekte der Stereochemie organischer Verbindungen, Akademie-Verlag, Berlin 1979, p. 123–128
79. Adams, R., Mecorney, J. W.: J. Am. Chem. Soc. *67*, 798 (1945)
80. Adams, R., Ludington, R. S.: ibid. *67*, 794 (1945)
81. Adams, R., Gross, W. J.: ibid. *64*, 1786 (1942)
82. Cope, A. C., Pawson, B. A.: ibid. *90*, 636 (1968)
83. Anastassiou, A. G., Hasan, M.: Tetrahedron Lett. 1983, 4279
84. Wynberg, H., Feringa, B.: J. Am. Chem. Soc. *99*, 602 (1977); Feringa, B., Wynberg, H.: Rec. Trav. Chim. Pays-Bas *97*, 249 (1978)
85. Rösner, M., Köbrich, G.: Angew. Chem. *86*, 775 (1974); Angew. Chem., Int. Ed. Engl. *14*, 708 (1974)
86. Goldschmidt, St., Riedle, R., Reichardt, A.: Liebigs Ann. Chem. *604*, 121 (1957)
87. Walba, D. W., Richards, R. M., Haltiwanger, R. C.: J. Am. Chem. Soc. *104*, 3219 (1982)
88. Duraisamy, M., Walborski, H. M.: ibid. *105*, 3265 (1983)
89. Duraisamy, M., Walborski, H. M.: ibid. *105*, 3270 (1983)
90. Nakazaki, M., Yamamoto, K., Maeda, M.: Chem. Lett. *1981*, 1035.
    Nakazaki, M., Yamamoto, K., Maeda, M., Sato, O., Tsutsui, T.: J. Org. Chem. *47*, 1435 (1982).
    Nakazaki, M., Yamamoto, K., Naemura, K.: Top. Curr. Chem., in press
91. Review: Rossi, R., Diversi, R.: Synthesis *1973*, 25
92. Moore, W. R., Anderson, H. W., Clark, S. D., Ozretich, T. M.: J. Am. Chem. Soc. *93*, 4932 (1971)
93. Balzi, M., Jones, W. M.: ibid. *102*, 7607 (1980); Wentrup, C., Gross, G., Maquestian, A., Flammang, R.: Angew. Chem. *95*, 551 (1983); Angew. Chem., Int. Ed. Engl. *22*, 542 (1983); Wittig, G., Fritze, P.: Liebigs Ann. Chem. *711*, 82 (1968)
94. Newman, M. S., Lednicer, D.: J. Am. Chem. Soc. *78*, 4765 (1956)
95. a) IUPAC-Nomenclature A-21, S-22, B-3, B-4. Lehmann, U., Ph. D. thesis, Univ. Bonn 1982;
    b) Wittek, M., Ph. D. thesis, Univ. Bonn 1982
96. Navaza, I., Tsoucaris, G., Le Bas, G., Navaza, A., de Rango, C.: Bull. Soc. Chim. Belg. *88*, 863 (1979)

97. de Rango, C., Tsoucaris, G.: Cryst. Struct. Commun. *2*, 189 (1973)

98. Beurskens, P. T., Beurskens, G., van den Hark, Th. E. M.: ibid. *5*, 241 (1976); van den Hark, Th. E. M., Beurskens, P. T.: ibid. *5*, 247 (1976)

99. van den Hark, Th. E. M., Noordik, J. H., Beurskens, P. T., ibid. *3*, 443 (1974)

100. Le Bas, G., Navaza, A., Mauguen, Y., de Rango, C.: ibid ⸮ ⸮⸮⸮ ⸮⸮⸮⸮⸮

101. Le Bas, G., Navaza, A., Knossow, M., de Rango, C.: ibid. *5*, 713 (1976)

102. Doesburg, H. M.: ibid. *9*, 137 (1980)

103. van den Hark, Th. E. M., Noordik, J. H.: ibid. *2*, 643 (1973)

104. Dewan, J. C.: Acta Crystallogr. *B37*, 1421 (1981)

105. Martin, R. H., Marchant, M. J.: Tetrahedron *30*, 347 (1974)

106. Newman, M. S., Darlak, R. S., Tsai, L.: J. Am. Chem. Soc. *89*, 6191 (1967)

107. Borkent, J. H., Laarhoven, W. H.: Tetrahedron *34*, 2565 (1978)

108. Martin, R. H., Marchant, M. J.: ibid. *30*, 343 (1974)

109. Goedicke, Ch., Stegemeyer, H.: Tetrahedron Lett. *1970*, 937

110. Martin, R. H.: Angew. Chem. *86*, 727 (1974); Angew. Chem., Int. Ed. Engl. *13*, 649 (1974)

111. Brown, A., Kemp, C. M., Marson, S. F.: J. Chem. Soc. (A) *1971*, 645

112. a) Newman, M. S., Chen, C. H.: J. Org. Chem. *37*, 1312 (1972);
    b) Lindner, A. J., Kitschke, B.: Bull. Soc. Chim. Belg. *88*, 831 (1979)

113. Scholz, M., Mühlstadt, M., Dietz, F.: Tetrahedron. Lett. *1967*, 665;
    Goedicke, C., Stegemeyer, H.: Chem. Phys. Lett. *17*, 492 (1972);
    Bromberg, A., Muszkat, K. A., Fischer, E.: Isr. J. Chem. *10*, 765 (1972);
    Fischer, E., Fischer, G., Knittel, T.: J. C. S. Chem. Commun. *1972*, 84

114. a) Martin, R. H., Vanest, J. M., Gorsane, M., Libert, V., Pecher, J.: Chimia *29*, 343 (1975);
    b) Martin, R. H., Libert, V.: J. Chem. Res. (S) *1980*, 130; J. Chem. Res. (M) *1980*, 1940.

115. Cuppen, Th. J. H. M., Laarhoven, W. H.: J. Chem. Soc., Perkin Trans. II, *1978*, 315

116. Peter, R., Jenny, W.: Helv. Chim. Acta *49*, 2123 (1966); cf. ref. [112a]; see also ref. [112b]

117. Mikes, F., Boshardt, G., Gil-Av, E.: J. C. S. Chem. Commun. *1976*, 99

118. Mikes, F., Boshardt, G.: ibid. *1978*, 173
    op den Brouw, P. M., Laarhoven, W. H.: Rec. Trav. Chim. Pays-Bas *97*, 265 (1978)

119. Häkli, H., Mintas, M., Mannschreck, A.: Chem. Ber. *112*, 2028 (1979);
    Hesse, G., Hagel, R.: Liebigs Ann. Chem. *1976*, 996

120. Okamoto, Y., Yuki, H.: Resolution by Optically Active Poly(triphenylmethyl-methacrylate) in: Asymmetric Reactions and Processes in Chemistry (E. L. Eliel, S. Otsuka, Eds.), ACS Symp. Series 185, 1982;
    Okamoto, Y., Okamoto, I., Yuki, H.: Chem. Lett. *1981*, 835;
    Yuki, H., Okamoto, Y., Okamoto, I.: J. Am. Chem. Soc. *102*, 6356 (1980)

121. Mikes, F., Boshardt, G., Gil-Av, E.: J. Chromatogr. *112*, 205 (1970)

122. Martin, R. H., Morren, G., Schuster, J. J.: Tetrahedron Lett. *1969*, 3683

123. Wynberg, H.: Acc. Chem. Res. *4*, 65 (1971)

124. Stulen, G., Visser, G. J.: J. C. S. Chem. Commun. *1969*, 965;
    Yamada, K., Yamada, T., Kawazura, H.: Chem. Lett. *1978*, 933;
    Konno, M., Saito, Y., Yamada, K., Kawazura, H.: Acta Crystallogr. *B36*, 1680 (1980)

125. a) Groen, M. B., Stulen, G., Visser, G. J., Wynberg, H.: J. Am. Chem. Soc. *92*, 7218 (1970);
    b) Wynberg, H., Groen, M. B.: ibid. *90*, 5339 (1968)

126. Fuchs, W., Niszel, J.: Ber. Dtsch. Ges. *60*, 209 (1927)

127. Teuber, H.-J., Vogel, L.: Chem. Ber. *103*, 3319 (1970)

128. See ref. [125]

129. Wynberg, H., Groen, M. B.: J. C. S. Chem. Commun. *1969*, 964;
    Wynberg, H., Cabell, M.: J. Org. Chem. *38*, 2814 (1973)

130. Numan, H., Helder, R., Wynberg, H.: Rec. Trav. Chim. Pays-Bas *95*, 211 (1976)

131. Nakagawa, H., Ogashiwa, S., Tanaka, H., Yamada, K., Kawazura, H.: Bull. Chem. Soc. Jpn. *54*, 1903 (1981)

132. Yamada, K., Ogashiwa, S., Tanaka, H., Nakagawa, H., Kawazura, H.: Chem. Lett. *1981*, 343

133. Groen, M. B., Wynberg, H.: J. Am. Chem. Soc. *93*, 2968 (1971)

134. Dopper, J. H., Oudman, D., Wynberg, H.: ibid. *95*, 3692 (1973)

135. Review: Kawazura, H., Yamada, K.: Yuki Gosei Kagaku Kyokai Shi. *34*, 111 (1976) [Chem. Abstr. *85*, 21158c (1976)]
136. Numan, H., Helder, R., Wynberg, H.: J. Royal. Neth. Chem. Soc. *95*, 211 (1976)
137. Rau, H., Schuster, O.: Angew. Chem. *88*, 90 (1976); Angew. Chem., Int. Ed. Engl. *15*, 114 (1976)
138. Marsh, W., Dunitz, J. D.: Bull. Soc. Chim. Belg. *88*, 847 (1979)
139. Martin, R. H., Eyndels, Ch., Defay, N.: Tetrahedron *30*, 3339 (1974)
140. Laarhoven, W. H., de Jong, M. H.: Rec. Trav. Chim. Pays-Bas *92*, 651 (1973)
141. Laarhoven, W. H., Cuppen, Th. H. J. M.: ibid. *92*, 553 (1973)
142. Laarhoven, W. H., Cuppen, Th. H. J. M., Nivard, R. J. F.: Tetrahedron *30*, 3343 (1974)
143. Tribout, J., Martin, R. H., Dogle, M., Wynberg, H.: Tetrahedron Lett. *1972*, 2839
144. Nakazaki, M., Yamamoto, K., Maeda, M.: Chem. Lett. *1980*, 1553
145. Katz, T. J., Pesti, J.: J. Am. Chem. Soc. *104*, 346 (1982)
146. a) Laarhoven, W. H., Veldhuis, R. G. M.: Tetrahedron Lett. *28*, 1823 (1972)
    b) Thulin, B., Wennerström, O.: Acta Chem. Scand. *B30*, 688 (1976); Kosower, E. M., Dodiuk, H., Thulin, B., Wennerström, O.: ibid. *B31*, 526 (1977);
    c) Leach, D. N., Reiss, H. A.: J. Org. Chem. *43*, 2484 (1978); Thulin, B., Wennerström, O.: Tetrahedron Lett. *1977*, 929; Kauffmann, T., Lexy, H.: Angew. Chem. *90*, 804 (1978); Angew. Chem., Int. Ed. Engl. *17*, 755 (1978)
147. Hellwinkel, D.: Chem.-Ztg. *94*, 175 (1980); see also: Clar, E.: Polycyclic Hydrocarbons, Academic Press, New York, N.Y. 1964
148. Barth, W. E., Lawton, R. G.: J. Am. Chem. Soc. *88*, 380 (1966);
    Barth, W. E., Lawton, R. G.: ibid. *93*, 1730 (1971)
149. Scholl, R., Meyer, K.: Ber. Dtsch. Chem. Ges. *65*, 902 (1932);
    Clar, E., Zander, M.: J. Chem. Soc. *1957*, 4616;
    Davy, J. R., Reiss, J. A.: J. C. S. Chem. Commun. *1973*, 806;
    Robertson, J. M., White, J. G.: J. Chem. Soc. *1945*, 607
150. Yamamoto, K., Harada, T., Nakazaki, M., Naka, T., Kai, Y., Harada, S., Kasai, N.: J. Am. Chem. Soc. *105*, 7171 (1983)
151. Dopper, J. H., Wynberg, H.: J. Org. Chem. *40*, 1957 (1975)
152. Dopper, J. H., Oudman, D., Wynberg, H.: ibid. *40*, 3398 (1975)
153. a) Joly, M., Defay, N., Martin, R. H., Declerq, J. P., Germain, G., Soubrier-Payen, B., van Meersche, M.: Helv. Chim. Acta *60*, 537 (1977);
    b) Staab, H. A., Diederich, F., Caplar, V.: Liebigs Ann. Chem. *1983*, 2262
154. Numan, A., Wynberg, H.: Tetrahedron Lett. *1975*, 1097
155. Nakazaki, M., Yamamoto, K., Ikeda, T., Kitsuki, T., Okamoto, Y.: J. C. S. Chem. Commun. *1983*, 787
156. The crown ether of (M)-(−)-*166* has P-helicity, that of (M)-(−)-*167* M-helicity
157. Goedicke, C. H., Stegemeyer, H.: Tetrahedron Lett. *1970*, 937
158. Lightner, D. A., Helfelfinger, D. T., Power, J. W., Frank, G. W., Trueblood, K. N.: J. Am. Chem. Soc. *92*, 7218 (1970);
    Brickel, W. S., Brown, A., Kemp, C. M., Mason, S. F.: J. Chem. Soc. (A) *1971*, 756;
    Hug, W., Wagniere, G.: Tetrahedron *28*, 1241 (1972)
159. Kyba, E. P., Siegel, M. G., Sousa, L. R., Sogah, G. D. Y., Cram, D. J.: J. Am. Chem. Soc. *95*, 2691 (1973), and further publications [164]
160. Lehn, J. M., Sirlin, C.: J. C. S. Chem. Commun. *1978*, 949, and further publications
161. Curtis, W. D., Laidler, D. A., Stoddart, J. F., Jones, G. H.: J. Chem. Soc., Perkin Trans. I, *1977*, 1756, and further publications [164]
162. de Vries, J. G., Kellogg, R. M.: J. Am. Chem. Soc. *101*, 2759 (1979)
163. Wudl, F., Gaeta, F.: J. C. S. Chem. Commun. *1972*, 107
164. Review: Jolley, S. T., Bradshaw, J. S., Izatt, R. M.: J. Heterocycl. Chem. *19*, 3 (1982)
165. a) Gottarelli, G., Hilbert, M., Samori, B., Solladié, G., Spada, G. P., Zimmermann, R. G.: J. Am. Chem. Soc. *105*, 7318 (1983);
    b) Solladié, G., Zimmermann, R. G.: Angew. Chem. *96*, 335 (1984); Angew. Chem., Int. Ed. Engl. *23*, 348 (1984);
    c) Jacques, J., Fouquey, C., Viterbo, R.: Tetrahedron Lett. *1971*, 4617
166. Sakane, S., Fujiwara, J., Maruoka, K., Yamamoto, H.: J. Am. Chem. Soc. *105*, 6154 (1983)

167. Prelog, V., Mutak, S.: Helv. Chim. Acta 66, 2274 (1983);
     Prelog, V., Mutak, S., Kovacevic, K.: ibid. 66, 2279 (1983)
168. Vögtle, F., Hammerschmidt, E.: Angew. Chem. 90, 293 (1978); Angew. Chem., Int. Ed. Engl.
     17, 268 (1978);
     Hammerschmidt, E., Ph. D. Thesis, Univ. Bonn 1980
169. Vögtle, F., Steinhagen, G.: Chem. Ber. 111, 205 (1978)
170. a) Hohner, G., Vögtle, F.: ibid. 110, 3052 (1977);
     b) Karbach, S., Vögtle, F.: ibid. 115, 427 (1982); Karbach, S.: Ph. D. Thesis, Univ. Bonn 1982
171. a) Winkel, J., Vögtle, F.: Tetrahedron Lett. 1979, 1561;
     b) Review: Keehn, P. M., Rosenfeld, S. M.: Cyclophanes, Vol. I and II, Academic Press, New
     York, N.Y. 1983;
     c) Mislow, K.: Angew. Chem. 70, 683 (1958). Akimoto, H., Shiori, T., Iitaka, Y., Yamada, S.:
     Tetrahedron Lett. 1968, 97
172. a) Nakazaki, M., Yamamoto, K., Tanaka, S.: Chem. Commun. 1972, 433; Nakazaki, M.,
     Yamamoto, K., Tanaka, S., Kametani, H.: J. Org. Chem. 42, 287 (1977);
     b) Nakazaki, K., Toga, T.: ibid. 45, 2553 (1980); ibid. 46, 1611 (1981)
173. Hammerschmidt, E., Vögtle, F.: Chem. Ber. 112, 1785 (1979); ibid. 113, 3550 (1980)
174. Vögtle, F., Atzmüller, M., Wehner, W., Grütze, J.: Angew. Chem. 89, 338 (1977); Angew. Chem.,
     Int. Ed. Engl. 16, 325 (1977)
175. Wittek, M., Vögtle, F.: Chem. Ber. 115, 2533 (1982);
     Wittek, M.: Ph. D. Thesis, Univ. Bonn 1982
176. Bieber, W., Vögtle, F.: Chem. Ber. 112, 1919 (1979);
     Irngartinger, H., Goldmann, A.: ibid. 116, 536 (1983);
     Irngartinger, H., Hekeler, J.: ibid. 116, 544 (1983);
     Haenel, M., Staab, H. A.: Tetrahedron Lett. 1970, 3585;
     Haenel, M., Staab, H. A.: Chem. Ber. 106, 2203 (1973); ibid. 111, 1789 (1978)
177. Ref. [48], p. 41; Haas, G., Hulbert, P. B., Klyne, W., Prelog, V., Snatzke, G.: Helv. Chim. Acta
     54, 491 (1971);
     Haas, G., Prelog, V.: ibid. 52, 1202 (1969);
     Sagiv, J., Yogev, A., Mazur, Y.: J. Am. Chem. Soc. 99, 6861 (1977)
178. a) Hammerschmidt, E., Vögtle, F.: Chem. Ber. 113, 1125 (1980); Wittek, M., Vögtle, F.,
     Stühler, G., Mannschreck, A., Lang, B. M., Irngartinger, H.: ibid. 116, 207 (1983);
     b) Newkome, R. W., Roper, J. M., Robinson, J. M.: J. Org. Chem. 45, 4380 (1980)
179. a) Vögtle, F., Palmer, M., Fritz, E., Lehmann, U., Meurer, K., Mannschreck, A., Kastner, F.,
     Irngartinger, H., Huber-Patz, U., Puff, H., Friederichs, F.: ibid. 116, 3112 (1983);
     b) Vögtle, F., Aigner, A., Puff, H., Franken, S.: ibid. 118, (1985), in press.
180. Aigner, A.: Ph. D. Thesis, Univ. Bonn 1985
181. Klieser, B.: Ph. D. Thesis, Univ. Bonn 1983
182. Powell, H. M.: Nature 170, 155 (1952);
     Newman, A. C. D., Powell, H. M.: J. Chem. Soc. 1952, 3747;
     Baker, W., Gilbert, B., Ollis, W. D.: Ibid. 1952, 1443
183. Ollis, W. D., Sutherland, I. O.: J. C. S. Chem. Commun. 1966, 402;
     Downing, A. P., Ollis, W. D., Sutherland, I. O.: J. Chem. Soc. B 1970, 24
184. Hoorfar, A., Ollis, W. D., Stoddart, J. F.: J. Chem. Soc., Perkin Trans. I, 1982, 1721
185. Ollis, W. D., Stephanatou, J. S., Stoddart, J. F.: ibid. 1982, 1715, and further publications
186. Edge, S. J., Ollis, W. D., Stephanatou, J. S., Stoddart, J. F.: J. Chem. Soc. Perkin Trans. I,
     1982, 1701
187. Collet, A., Jacques, J.: Tetrahedron Lett. 1978, 1265;
     Collet, A., Gottarelli, G.: J. Am. Chem. Soc. 103, 5912 (1981);
     Collet, A., Gabard, J.: J. Org. Chem. 45, 5400 (1980)
188. Lüttringhaus, A., Peters, K. C.: Angew. Chem. 78, 603 (1966); Angew. Chem., Int. Ed. Engl.
     5, 593 (1966)
189. Meurer, K., Vögtle, F., Mannschreck, A., Stühler, G., Puff, H., Roloff, A.: J. Org. Chem. 49,
     3484 (1984)
190. Blaschke, H., Boekelheide, V.: J. Am. Chem. Soc. 89, 2747 (1967);
     Taylor, D.: Austr. J. Chem. 31, 1235 (1978)
191. Gschwend, H. W.: J. Am. Chem. Soc. 94, 8430 (1972)

192. Mislow, K., Brzechffa, M., Gschwend, H. W., Puckett, R. T.: ibid. *95*, 621 (1973)
193. Sato, T., Wakabayashi, M., Kainosho, M., Hata, K.: Tetrahedron Lett. *1968*, 4185;
Sato, T., Wakabayashi, M., Hata, K., Kainosho, M.: Tetrahedron *27*, 2737 (1971)
194. Atzmüller, M., Vögtle, F.: Chem. Ber. *111*, 2547 (1978); ibid. *112*, 138 (1979)
195. Vögtle, F., Meurer, K., Mannschreck, A., Stühler, G., Puff, H., Roloff, A., Sievers, R.: ibid.
116, 2630 (1983)
196. Glotzmann, C., Langer, E., Lehner, H., Schlögl, K.: Monatsh. Chem. *105*, 907 (1974);
Kainradl, B., Langer, E., Lehner, H., Schlögl, K.: Liebigs Ann. Chem. *766*, 16 (1972)
197. Ogihara, Y., Kobayashi, N., Shibata, S.: Tetrahedron Lett. *1968*, 1881;
Oertel, B.: Ph. D. Thesis Univ. Bonn 1984; and personal communication; Thomson, R. H.:
Naturally Occurring Quinones, Academic Press, New York, N.Y. 1971
198. Henessian, S.: Total Synthesis of Natural Products, The Chiron Approach, Pergamon Press,
New York, N.Y. 1983

Isotopic Separation in Systems with Crown Ethers and Cryptands

Klaus G. Heumann

Institut für Anorganische Chemie der Universität Regensburg, Universitätsstraße 31, 8400 Regensburg, FRG

Table of Contents

1 Introduction

During the last few years, stable isotopes have gained more and more importance as labeling substances in natural sciences as well as in many fields of applied sciences. Furthermore, the selective separation of radionuclides is a frequent problem which has to be solved, e.g. during the recycling process of nuclear fuel materials. Therefore, it is not surprising that attempts have been made to develop chemical systems which are capable of enriching stable isotopes or separating individual radioactive isotopes selectively. For the enrichment of stable isotopes, one of the main aspects is that chemical systems should be much cheaper than physical separation methods such as mass separators, diffusion cells etc., which are normally used today.

Because crown ethers and cryptands are good complex-forming compounds for most of the metal ions [1-6], isotopic separations are obviously best investigated by means of chemical reactions with cyclic polyethers. This is due to the fact that other complex-forming substances have shown isotopic separations for metals. For example, calcium isotopic enrichment has been obtained in chromatographic experiments with a strongly acidic cation exchanger and α-hydroxyisobutyrate [7], as well as with citrate or EDTA [8] as complex-forming agent. Using ion exchangers with complex-forming anchor groups, e.g. Dowex Al which contains iminodiacetic acid groups, calcium isotopic enrichment was also observed [9]. Additionally, for lithium and calcium, it was shown that liquid/liquid extraction chromatography with complex agents produces isotopic separations [10, 11].

Isotopic separations in systems where crown ethers or cryptands are used have been carried out with the alkali and alkaline earth metals in particular. This can be explained by the special property of the cyclic polyethers to form stable complexes with the corresponding ions. Among these metal ions, the enrichment of stable calcium isotopes is of particular interest because labeling experiments with stable calcium isotopes in medicine are of pressing importance at the present time [12]. In 1976, Jepson and DeWitt first described a separation between ^{44}Ca and ^{40}Ca in a liquid/liquid extraction system using a cyclic polyether [13]. In May 1977, worldwide attention was drawn to the work of a French research group at the International Nuclear Conference in Salzburg where this group reported about a chemical enrichment of ^{235}U with crown ethers [14, 15]. Although no details of the French procedure for the ^{235}U-enrichment are yet known, different patent applications (for example see ref. [16]) show the great importance of cyclic polyethers for possible isotopic separations.

2 Fundamentals of Isotopic Effects and Isotopic Separations in Chemical Systems

2.1 General Aspects

The differences in the properties of the isotopes of an element are called isotopic effect. The origin of an isotopic effect is always a physical phenomenon which can be attributed to the isotopic masses in chemical isotopic fractionations [17]. For example,

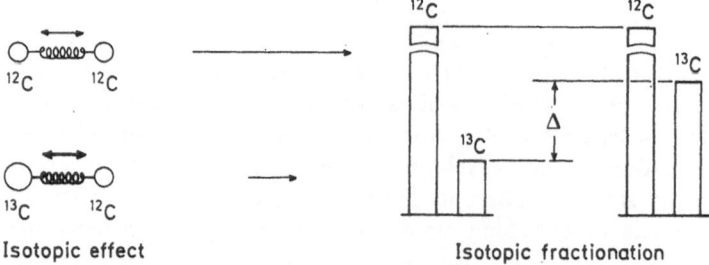

Isotopic effect   Isotopic fractionation

Fig. 1. Schematic representation of the relationship between an isotopic effect (a physical phenomenon) and the occurrence of isotopic fractionation (an observable quantity)

the vibrational energy of a diatomic molecule depends on the masses of the two atoms, and therefore, the vibrational energy is different for isotopic molecules. Usually, the consequence of an isotopic effect is called an isotopic fractionation. That means, an isotopic effect causes an isotopic fractionation which is elucidated in Fig. 1.

In principle, isotopic separations in chemistry are classified into kinetic fractionations and equilibrium fractionations [17, 18]. Whereas the first type of fractionation is due to different reaction rates of the isotopic molecules, the equilibrium fractionation can be attributed to the difference in the energetic state of the isotopic molecules. Hence, it follows that isotopic separations in systems with complex-forming substances should be observed either if complexing or decomplexing reactions of metal isotopes show different reaction rates, or if the energetic state of one metal isotope in the complex compound is not equal to that of another isotope.

Usually, the ion exchange rate between a solvated metal ion and the ion fixed in the complex compound is very fast. This is the reason why most of the cation complexes with cyclic polyethers have high exchange rates which are in the order of 10^3 s^{-1} [19, 20]. Although the exchange rate for metal ions with a bicyclic cryptand is significantly lower than with monocyclic crown ethers [21-24], the kinetic isotopic fractionation during complexing or decomplexing should not be applicable for an isotopic separation. Therefore, the isotopic fractionation within a chemical equilibrium is described in the following chapters.

To make use of an isotopic fractionation for an enrichment of isotopes, a different isotopic distribution has to be carried out in two phases which can be easily separated. On the one hand, polyethers are soluble in various solvents [21]. On the other hand, these compounds can also be fixed as anchor groups at organic or inorganic matrices [25-28]. Therefore, as a heterogeneous exchange reaction for metal ions, liquid/liquid as well as liquid/solid systems can be used in connexion with polyethers.

2.2 Theoretical Description of Equilibrium Isotopic Fractionations

The isotopic exchange reaction of a light isotope (index 1) and a heavy isotope (index 2) between two molecules A and B can be described as

$$nA_1 + mB_2 \rightleftharpoons nA_2 + mB_1 .$$ (1)

The essential calculations on isotopic fractionations have their origin in the considerations of Bigeleisen and Mayer [29] as well as in those of Urey [30]. In the meanwhile, the literature on isotopic fractionations has been summarized in several reviews (Ref. [31-33]) and in a number of monographs (Ref. [17, 34-37]).

From the statistical thermodynamics for the equilibrium constant K of the isotopic exchange reaction (1),

$$K = \left(\frac{Q(A_2)}{Q(A_1)}\right)^n \Big/ \left(\frac{Q(B_2)}{Q(B_1)}\right)^m$$

(2)

follows where Q is the partition functions of the molecules

$$Q = \sum_i g_i \times e^{-E_i/k \times T}.$$

(3)

k   Boltzmann's constant
T   absolute temperature
E_i energy states of the molecule
g_i statistical factor for the energy state E_i.

The ratio Q_2/Q_1 for one molecule is given in Eq. (4):

$$\frac{Q_2}{Q_1} = \frac{S_1}{S_2} \times \left(\frac{M_2}{M_1}\right)^{3/2} \times \sum e^{-E_2/k \times T} \Big/ \sum e^{-E_1/k \times T}$$

(4)

S   symmetry number (number of non-distinguishable forms which can be obtained by rotation of the molecule around all axes of symmetry; for example H_2: S = 2, HD: S = 1)
M   molecular masses

For non-linear polyatomic molecules, Eq. (4) can be replaced by Eq. (5) if the temperature is so high that $k \times T$ is large compared to the separations of the rotational energy levels, as it is in most cases, and if the vibrational energy levels are sufficiently close to harmonic:

$$\frac{Q_2}{Q_1} = \frac{S_1}{S_2} \times \left(\frac{A_2 \times B_2 \times C_2}{A_1 \times B_1 \times C_1}\right)^{1/2} \times \left(\frac{M_2}{M_1}\right)^{3/2}$$

$$\times \prod_{i=1}^{3n-6} \frac{e^{-u_{2i}/2} \times (1 - e^{-u_{1i}})}{e^{-u_{1i}/2} \times (1 - e^{-u_{2i}})}.$$

(5)

u_i is given by:

$$u_i = h \times v_i/k \times T.$$

(6)

A, B, C   principle moments of inertia of the polyatomic molecule
h   Planck's constant
v   frequency

The right side of Eq. (5) contains the product of the u-functions for the $3n - 6$ vibrations of an n-atomic molecule and this molecule's principle moments of inertia. In most cases the principle moments of inertia are unknown. According to Bigeleisen and Mayer, Eq. (5) can be replaced by a reduced partition function ratio according to Teller and Redlich's product rule:

$$\ln\left(\frac{S_2}{S_1}\right) \times f = \frac{1}{24} \times \left(\frac{\hbar}{k \times T}\right)^2 \times \sum_{i=1}^{3n-6} \left(\frac{1}{m_{1i}} - \frac{1}{m_{2i}}\right) \times a_{ii}. \tag{7}$$

$m_{1,2}$ atomic masses of the light and heavy isotope
a_{ii} force constant for the motion of the atom (Cartesian coordinates)
f in Eq. (7) is given as:

$$f = \frac{Q_2}{Q_1} \times \prod_{k=1}^{n} \left(\frac{m_{1k}}{m_{2k}}\right)^{3/2}. \tag{8}$$

Unknown values on the right side of Eq. (7) are only the force constants. These constants are connected with the oscillation frequencies of the molecules. Therefore, the equilibrium isotopic fractionation is principally determinable if the corresponding spectroscopic data is available. Whereas this data for low-molecular isotopic molecules is very often known, it is unknown for high-molecular species. From the reduced partition function it is possible to deduce the following principles about the magnitude and trend of an isotopic fractionation:

1. Isotopic fractionations appear if there is a difference in the vibrational energy of isotopic molecules.
2. Mostly, it is valid that $u_{2i} < u_{1i}$ and, therefore, $\frac{S_2}{S_1} \times f > 1$. From this it follows that the value of an isotopic fractionation usually is > 1.
3. The isotopic fractionations of the different degrees of freedom are additive and cumulative.
4. Isotopic molecules with a comparable structure have more degrees of freedom for vibration if the number of atoms in the molecule is higher. Therefore, the reduced partition function is larger in such molecules where the number of atoms is higher.
5. Because of the following relation: $\ln\left(\frac{S_2}{S_1}\right) \times f \sim (m_2 - m_1)$, the isotopic fractiona-

   tion increases with increasing relative mass difference. From this it follows that an equal mass difference in isotopes causes a larger isotopic fractionation in the case of light elements than in the case of heavy elements.
6. Because of the approximation $\lim_{T \to \infty} \left(\frac{S_2}{S_1}\right) \times f = 1$, the isotopic fractionation

   decreases with increasing temperature, to disappear with very high temperatures.

2.3 Ionic Radius Model

General experience has shown that the magnitude of an isotopic fractionation significantly depends on the type of chemical bond of the investigated species in the reaction. Usually, a larger isotopic fractionation is found in exchange reactions where isotopes are covalently bound than in the case of an ionic bond. As an example, the fractionated precipitation of calcium carbonate from a calcium hydrogen carbonate solution by means of a step-by-step desorption of CO_2 will be discussed:

$$Ca^{2+}_{(aq)} + 2HCO^-_{3(aq)} \rightleftharpoons CaCO_{3(solid)} + H_2O + CO_2. \tag{9}$$

During this fractionated precipitation one does not find any isotopic fractionation for the isotope ratio $^{48}Ca/^{40}Ca$ (ratio with the largest mass difference of stable calcium isotopes) out of the standard deviation of the isotope ratio measurement (0.03 % per mass unit). In contrast, the $^{13}C/^{12}C$-ratio shows a relative enrichment up to 5 % in the different $CaCO_3$ fractions compared with the same ratio in the hydrogen carbonate solution [38]. The difference in the fractionation effect is due to the covalent bond of carbon ($^{13}C/^{12}C$: relative mass difference 8 %) instead of the ionic bond of the calcium ions ($^{48}Ca/^{40}Ca$: relative mass difference 20 %).

In the meanwhile, a great number of calculations of isotopic fractionations for exchange equilibria with low-molecular species has been done, showing good agreement with experimental data [30, 39]. In a few cases it was also possible to calculate isotopic fractionations of high-molecular compounds with very high expenditure by using computers. According to Stern and Wolfsberg, these calculations start from the principle that parts of the molecule which are more than two bonds away from the reaction centre do not contribute significantly to the magnitude of the isotopic fractionation [40]. This type of calculations cannot be applied to electrolyte systems without problems.

During the last few years, the ionic radius model has sufficed as a simple means of explaining isotopic fractionations in ion exchange reactions. One example for such exchange reactions is the cation exchange in systems with crown ethers and cryptands. In many cases the ionic radius model allows the prediction of the trend and the relative magnitude of an isotopic separation without high expenditure by means of conclusions based on analogy. The basis of this model is the assumption of slightly different ionic radii of the isotopes (or more correct: The isotopes in an electrolytic system show chemical properties which correspond to different ionic radii).

Glueckauf [41] as well as Dickel and coworkers [42, 43] were the first who showed that, for lithium isotopes in a system with a strongly acidic cation resin, the isotopic fractionation effect is given by the logarithmic ratio of the activity coefficients of the isotopes. Linear relations between the average activity coefficient and the crystallographic ionic radius are known, e.g. for alkalis and alkaline earth elements [44]. Furthermore, it is known that the solvation of an ion, which depends on the crystallographic ionic radius, strongly influences the activity coefficient [45, 46]. Up to now, the only direct experimental data for a difference of crystallographic ionic radii of isotopes has been presented by Thewlis for lithium. In this investigation, the lattice spacings of ^6LiF and ^7LiF have been determined. Hence, a difference of $r(^6Li^+) - r(^7Li^+)$ of 4×10^{-2} pm follows [47]. Based on Einstein's model of a crystalliz-

Table 1. Calculated ionic radius differences [48]

Isotopes to be compared	Ionic radius difference [pm]
$^6Li/^7Li$	5.0×10^{-2}
$^{22}Na/^{24}Na$	4.8×10^{-3}
$^{24}Mg/^{26}Mg$	3.3×10^{-2}
$^{40}Ca/^{42}Ca$	4.6×10^{-3}
$^{17}F/^{19}F$	1.4×10^{-4}
$^{35}Cl/^{37}Cl$	1.9×10^{-5}

ed solid, which was first applied for the lithium isotopes by Covington and Montgomery, Knyazev has calculated a number of ionic radius differences for metal and non-metal isotopes [48]. A selection of these results is summarized in Table 1. In all cases a smaller ionic radius is calculated for the heavier isotope compared with the lighter isotope, as was also found in the direct experimental data for lithium. The radiographically determined lithium isotope radii agree well with the calculated values.

If the different radii of the isotopic ions are responsible for the isotopic fractionation in an ion exchange equilibrium, there must also be a difference in the solvatation of the isotopes. Plotting the crystallographic (Pauling's) radii of homologeous elements against the hydration number of these ions, one gets a straight line for the alkali metals and alkaline earth elements, as is shown in Fig. 2 [46]. According to the previous considerations, the marked point for the calcium ion in Fig. 2 has to be split into different points for the isotopes as is shown on the right side of this Fig.

An indirect argument for a different hydration of the calcium isotopes according to Fig. 2 was given by Heumann and Lieser in 1972 [49]. An enrichment of $^{48}Ca^{2+}$ and $^{44}Ca^{2+}$ as opposed to $^{40}Ca^{2+}$ was found in the first fractions and a complementary depletion was found in the last fractions during the elution of calcium chloride with

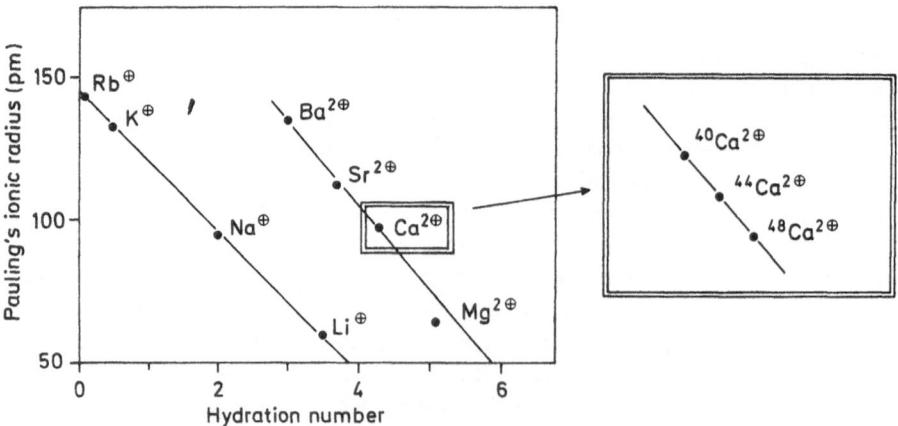

Fig. 2. Relation between hydration number of alkali and alkaline earth ions and Pauling's ionic radius with an assumed isotopic shift

HCl of different concentrations, using a column filled with the strongly acidic cation resin Dowex 50-X12. The results obtained for the isotope ratio $^{44}Ca/^{40}Ca$ dependent on the eluted calcium amount $\Delta m/m$ are given in Fig. 3 for the elution with 3M HCl. The chemical properties of the calcium isotopes can be explained in analogy to the characteristics of the alkaline earth elements at a Dowex 50 exchanger. For the alkaline earth ions, the elution sequence from a Dowex 50-filled column is as follows: Mg^{2+}, Ca^{2+}, Sr^{2+}, Ba^{2+}. The ions with the larger hydrated radius — i.e. with the smaller Pauling's radius — are enriched in the solution phase of the heterogeneous exchange system. An enrichment of the heavy isotopes in the first fractions of the elution band means that these isotopes are preferably in the solution phase and, therefore, should have the larger hydrated ionic radius. In the meanwhile, it has been possible to confirm the results for calcium in other exchange systems [50–52]. Using comparable exchange reactions, Lee and coworkers obtained analogous results for the lithium isotopes [53–56]; Heumann and coworkers for the anionic chloride system with the isotopes ^{35}Cl and ^{37}Cl [57–60].

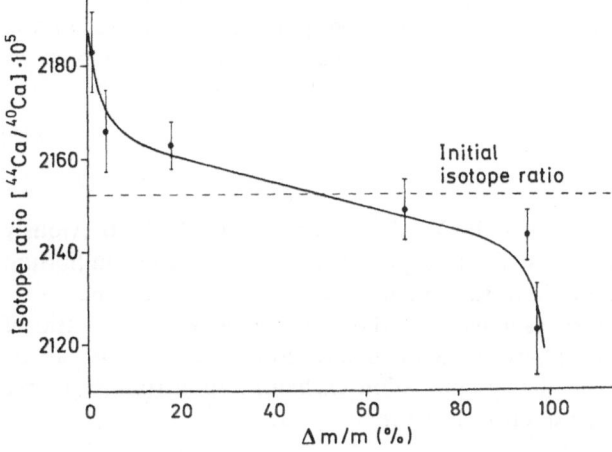

Fig. 3. Isotope ratio $^{44}Ca/^{40}Ca$ dependent on the eluted calcium amount $\Delta m/m$ using exchange chromatography (Dowex 50-X12, 3 M HCl)

The heterogeneous exchange equilibrium of two isotopic metal ions $^{1}Me^{+}$ and $^{2}Me^{+}$ between a phase A (e.g. an ion exchanger) and a phase B (e.g. a solution) is described as:

$$^{2}Me_{A}^{+} + {}^{1}Me_{B}^{+} \rightleftharpoons {}^{1}Me_{A}^{+} + {}^{2}Me_{B}^{+} .\tag{10}$$

The thermodynamic equilibrium constant K_{a} corresponding to Eq. (10) is given as:

$$K_{a} = \frac{a_{A}^{1} \times a_{B}^{2}}{a_{A}^{2} \times a_{B}^{1}} = \frac{c_{A}^{1} \times c_{B}^{2}}{c_{A}^{2} \times c_{B}^{1}} \times \frac{f_{A}^{1} \times f_{B}^{2}}{f_{A}^{2} \times f_{B}^{1}} ,\tag{11}$$

85

a activity, c concentration, f activity coefficient
with

$$\frac{c_A^1 \times c_B^2}{c_A^2 \times c_B^1} = K_c \equiv \alpha \tag{12}$$

and

$$\alpha = 1 + \varepsilon . \tag{13}$$

α isotopic separation factor
ε (elementary) isotopic fractionation effect
If Eqs. (11)–(13) are combined, it follows:

$$\alpha = 1 + \varepsilon = K_a \times \frac{f_A^2 \times f_B^1}{f_A^1 \times f_B^2} . \tag{14}$$

Equation (12) shows that α is determinable by measuring the isotope ratio $^1Me/^2Me$ in phase A and in phase B.

Taking the logarithm of Eq. (14) and making use of the approximation $\ln(1 + \varepsilon) \approx \varepsilon$, which is valid for $\varepsilon \ll 1$, and assumed that $K_a = 1$ then it follows

$$\varepsilon \approx \ln \left(\frac{f_A^2 \times f_B^1}{f_A^1 \times f_B^2} \right) . \tag{15}$$

Equation (15) is the derivation of Glueckauf's [41] and Dickel's [42,43] previously mentioned statement which says that the isotopic fractionation of one equilibrium stage is given in the logarithmic ratio of the activity coefficients of the isotopes.

Experimental conditions can be established without problems where the ratio of activity coefficients f^2/f^1 in one phase is equal to one. For example, this can be obtained in a diluted solution in the system cation exchanger resin (phase A)/free solution (phase B). In that case Eq. (15) can be reduced to:

$$\varepsilon \approx \ln \left(\frac{f_A^2}{f_A^1} \right) . \tag{16}$$

Glueckauf has derived Eq. (17) for the ε-values of alkali metals from equations which contain the correlation between the activity coefficient and the molality of the solution and the ionic radius of the dissolved ion. Equation (17) results in a direct correlation between the isotopic fractionation effect ε and the ionic radius difference of the isotopes:

$$\varepsilon = \frac{0.27 \times m_A}{(r_{Me^+})^2} \times (r^1 - r^2) . \tag{17}$$

m_A exchanger molality
r_{Me^+} ionic radius of Me^+
$r^{1,2}$ ionic radius of isotope 1 and 2

Table 2. Observed and calculated ε-values of lithium in ion exchanger resin systems with different crosslinking (divinyl benzene content) [41,53]

DVB content [%]	Approx. exchanger molality [mol × kg^{-1}]	$\varepsilon \times 10^3$	
		observed	calculated
4	3	1.0	0.9
8	5.5	1.6	1.7
12	7.5	2.7	2.3
16	9	3.7	2.7
24	12.5	3.8	3.8

Using the value for $r(^6Li^+) - r(^7Li^+) = 4 \times 10^{-2}$ pm determined by Thewlis [47], it was possible to calculate ε-values for lithium isotopes with Eq. (17) in systems with Dowex 50 resins. This was done for resins with different contents of divinyl benzene (DVB) which means that the resin's crosslinking is differently. The calculated ε-values show good agreement with the experimental data (Table 2) [41,53]. Additionally, other experiments which have investigated the dependence of the isotopic fractionation on the crosslinking of the ion exchanger, confirm Eq. (17) [53,58]. The measured dependence of the isotopic separation in ion exchange chromatography of cations [50, 51,56,61] as well as of anions [57,59,60,62] on the concentration of the solution used for elution can also be attributed to the correlation given in Eq. (15).

If complex formation takes place in one phase of a heterogeneous system, e.g. when crown ethers or cryptands (ligand L) are present, Eq. (10) has to be modified:

$$^2Me_A^+ + {}^1MeL_B^+ \rightleftharpoons {}^1Me_A^+ + {}^2MeL_B^+ . \tag{18}$$

For every isotope the following partial reactions are valid:

$$^1Me_{sol}^+ + L_{sol} \rightleftharpoons {}^1MeL_{sol}^+ , \tag{19 a}$$

$$^2Me_{sol}^+ + L_{sol} \rightleftharpoons {}^2MeL_{sol}^+ . \tag{19 b}$$

The origin of isotopic fractionations in ion exchange reactions corresponding to Eq. (18) can be attributed to energy differences of the isotopes in phase A, to different complex stability constants for the isotopes (Eq. (19a) and (19b)), or to the sum of both effects. Because the stability constant of crown ether and cryptand complexes with metal ions strongly depends on the cavity diameter of the polyether, the cationic radius of the metal in particular is responsible for the magnitude of the stability constant. If one plots the logarithm of the stability constant K_s of the alkaline earth cryptands {[2.2.1] Me^{2+}} and {[2.2.2] Me^{2+}} in aqueous solutions [63] versus Pauling's ionic radius, Fig. 4 is obtained. With a given ionic radius difference in isotopes, the log K_s-values should be different, and in that way an isotopic fractionation should be observed in such a system. The isotopic shift of log K_s is shown for calcium on the right side of Fig. 4. It is obvious in this diagram that the curve for the cryptand [2.2.1] is flatter than for the cryptand [2.2.2] at the point of calcium. This should bring about differences in the isotopic fractionation, using [2.2.1] and [2.2.2] cryptands.

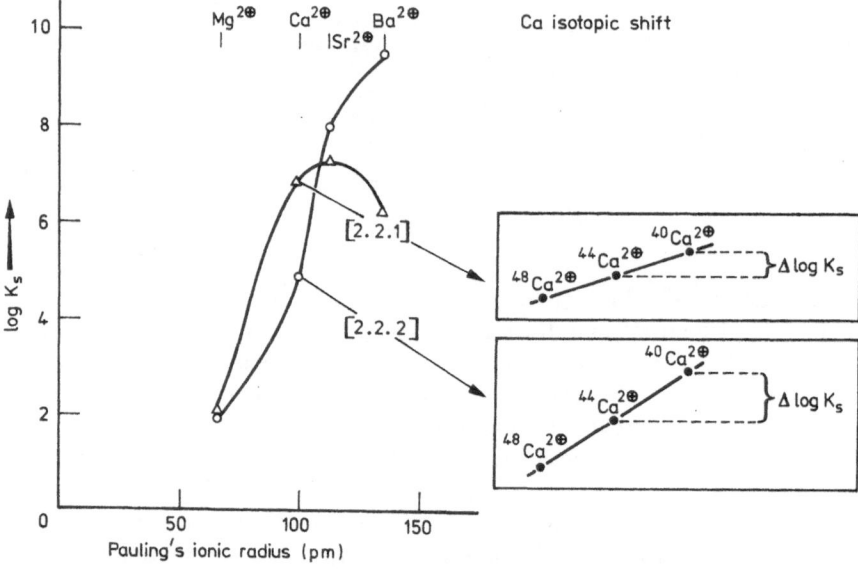

Fig. 4. Stability constants of alkaline earth cryptates dependent on the metal ionic radius with an assumed isotopic shift

2.4 Determination of the Elementary Fractionation Effect in Heterogeneous Exchange Systems

A heterogeneous system where two phases with different isotopic compositions can be separated is the precondition for the application of an isotopic fractionation to enrich isotopes. As mentioned before, the isotopic separation factor α for an exchange reaction corresponding to Eq. (10) can be determined by measuring the isotope ratio in phase A and phase B.

Usually, the isotopic separation within one equilibrium stage is very small, and therefore, this effect is very difficult to measure. To solve this problem one can either transfer the whole effect into one phase, or one can apply a multiplication process (see Chap. 2.5). By means of a mass balance it was shown that the isotopic shift in one phase (e.g. phase A) increases, the smaller the mass ratio becomes for phases A/B. In the other phase (phase B) the isotope ratio approximates the initial ratio which is equal to the ratio of the starting material [49] This procedure has the advantage that, except for the initial isotope ratio, only that of the phase with the smaller sample amount has to be analyzed for the determination of α. One disadvantage is the relatively small amount of enriched isotopes in the phase to be analyzed.

For homogeneous systems the determination of stability constants corresponding to Eq. (19a) and (19b) are of particular interest. The fundamental possibility of determining such constants indirectly by the distribution analysis of complexed and non-complexed species between two phases is discussed in ref. [64-66] (see also Chap. 4.3.1.2).

2.5 Multiplication Processes

2.5.1 Fundamentals

The investigations of isotopic separations of metals in ion exchange equilibria hitherto have shown elementary fractionation effects in the range of $\varepsilon = 10^{-5} - 5 \times 10^{-4}$ per mass unit [7, 8, 50-56, 61, 67-75]. Relatively small fractionation effects are observed for the heavy elements, e.g. for uranium [72-75].

ε-Values of about 10^{-3} and more are found for lithium [41, 43, 53-56, 76-78] but also for other metals in systems with crown ethers or cryptands [13, 64, 79]. Because of the comparatively high isotopic separation factors in ion exchange equilibria with crown ethers and cryptands, special interest in these systems for isotopic enrichments has quite recently been shown.

Because isotopic fractionation effects are in the magnitude of some few per thousands, they can only be analyzed very inaccurately (see Chap. 3) in one equilibrium stage. Therefore, in many experiments a higher total separation effect is established by multiplication processes. From these results the elementary fractionation effect ε can be calculated under certain presuppositions. In principle, a multiplication of the elementary fractionation effect is possible with multi-stage batch experiments (cascade) or with chromatographic systems.

2.5.2 Separation Cascade

The characteristic value for a single-stage separation system is the isotopic separation factor α which is defined by Eq. (12). Because α deviates not very much from one, a number of separation stages must be connected in series for a sufficient total isotopic enrichment. The total isotopic separation factor S, using N equilibrium stages, is given by:

$$S = \alpha^N. \tag{20}$$

In a cascade system, the amount of the enriched compound becomes smaller with an increasing number of separation stages. Therefore, one has to change the separation units continuously from a large scale to a smaller one within an experiment or one has to use a system with uniform separation units where serveral units are arranged side by side in the first stages (separation cascade). A schematic diagram of such a separation cascade for the distribution of the isotopes between phases A and B is shown in Fig. 5.

One example of a separation cascade is the liquid/liquid extraction system where the desired isotopic enrichment takes place in phase A. Using systems with crown ethers or cryptands, these ligands should have a good solubility in one phase, whereas the same compounds should be insoluble or difficult to dissolve in the other phase. This also applies to chemical reactions in solid/liquid systems, e.g. in the system ion

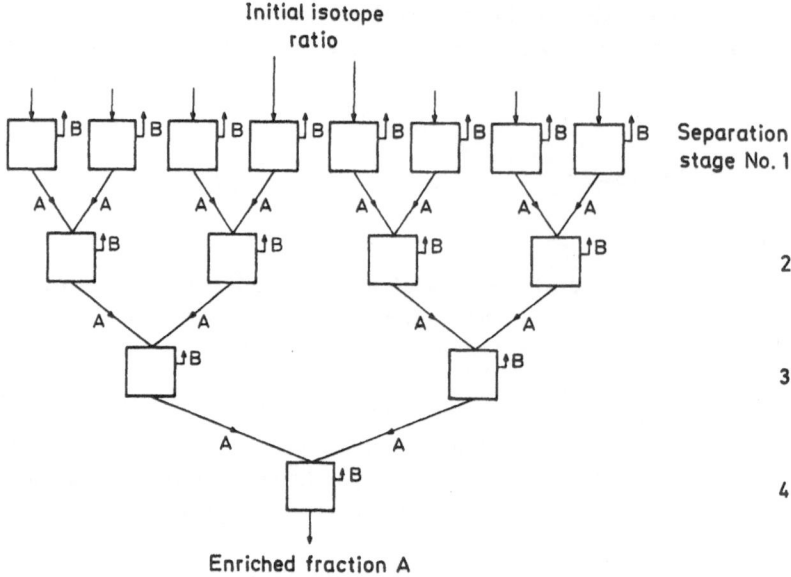

Fig. 5. Schematic figure of a separation cascade (enrichment of the desired isotope in phase A)

exchanger/polyether solution. If the establishment of equilibrium takes place in every separation stage, the total isotopic separation corresponds to Eq. (20).

In principle, one separation unit can also consist of a chromatographic column. If the desired isotopic enrichment takes place in the first fractions of the elution band, the further treatment of the first part of the elution band has to be done for phase A as shown in Fig. 5, whereas the second part of the elution band has to be treated as shown for phase B.

2.5.3 Chromatographic Systems

The advantage of column chromatography is that many equilibrium stages are connected directly in series. Using ion exchange chromatography, N-values of some hundreds up to some thousands are available. In systems with crown ethers and cryptands, these complex-forming compounds could either be used in the dissolved phase or as an anchor group at the exchanger resin. In the first case, the solid matrix can be a substance without ionic groups, e.g. silica gel, or one can use exchangers such as Dowex 50. Because the neutral crown ethers and cryptands form cationic metal complexes, the absorption of the metal complex at a cationic exchanger has to be taken into account [80]. Up to now, various crown ethers and cryptands have been fixed as anchor groups at different matrices. Some significant exchangers of this type are summarized in Table 3, whereas examples for the matrix No. 1, 2, 4 and 5 are given in Fig. 6.

In contrast to batch experiments, the isotopic separation factor α and the ε-value are not determinable directly by chromatographic methods. However, the ε-value and the

Table 3. Different types of exchangers with crown ether and cryptand anchor groups

Matrix	Anchor groups	Type of synthesis	Literature
1. Methylene and phenol groups	Monobenzo/dibenzo crown and cryptand compounds	Condensation with formaldehyde and phenol	Blasius et al. [25,26, 81-84]
2. Polystyrene	Crown and cryptand compounds	Copolymerization of vinylbenzo crown/ cryptand compounds with divinylbenzene	Blasius et al. [26] Kopolow et al. [85]
3. Polystyrene with ether bridges	Monobenzo crown and cryptand compounds	Substitution reaction of chlormethylated poly- sterene	Blasius et al. [26]
4. Silica gel	Monobenzo crown compounds	Substitution reaction of chlormethylated silica gel	Waddell, Leyden [28]
5. Cellulose	Diaminodibenzo[18]- crown-6, amino- [2B.2.2] cryptand	Coupling of diazotized aminophenol cellulose with amino crown/ cryptand compound	Lieser et al. [27] Heumann, Seewald [86]

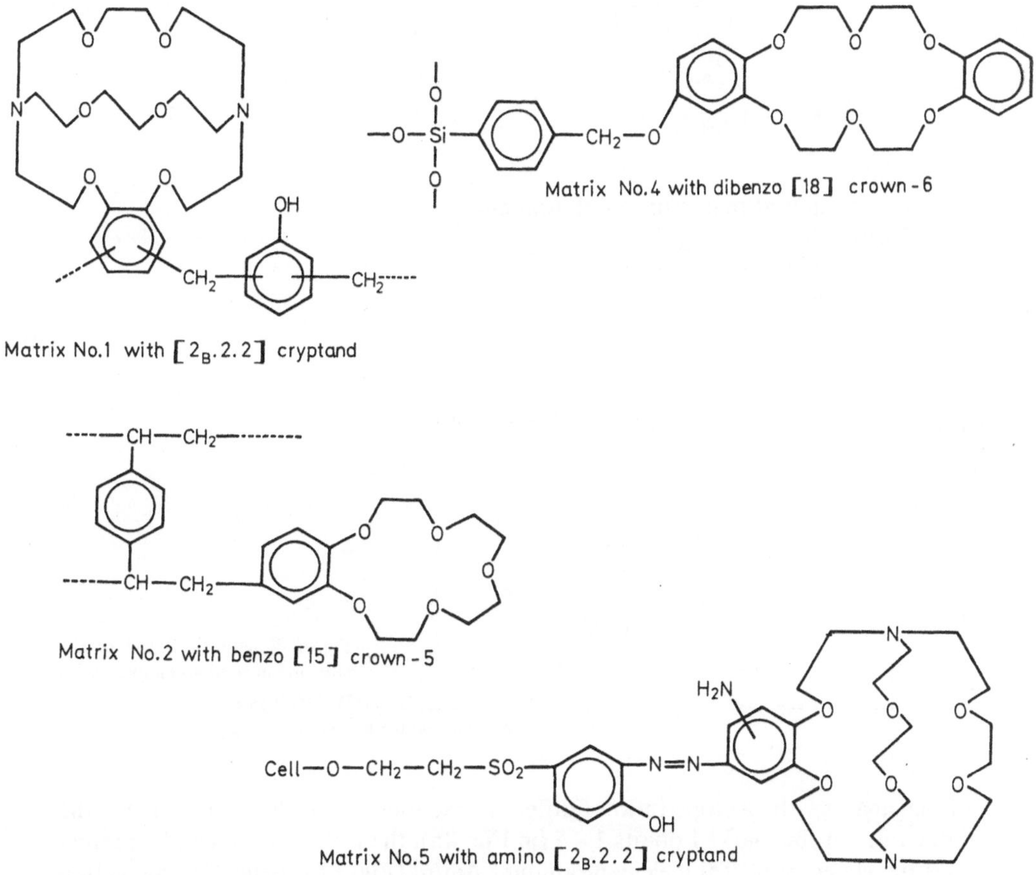

Matrix No.1 with [2B.2.2] cryptand

Matrix No.4 with dibenzo [18] crown-6

Matrix No.2 with benzo [15] crown-5

Matrix No.5 with amino [2B.2.2] cryptand

Fig. 6. Examples of different types of exchangers with crown ether and cryptand anchor groups

number of equilibrium stages N in a column can be calculated with an approximation method measuring the isotopic enrichment dependent on the eluted amount of substance and evaluating the elution curve. Usually, in isotopic chemistry a method developed by Glueckauf [87] is applied. The essential characteristics of this method are described in the following.

If one eluates an element containing the isotopes 1 and 2 out of a chromatographic column, an elution curve is obtained as plotted in Fig. 7 by determining the element concentration dependent on the effluent volume. Under the assumption that the heavy isotope 2 is enriched in the first fractions of the elution band — and the light isotope 1 is enriched in the final fractions — the element elution curve in reality consists of two curves of the isotopes. Because of the slight difference in the effluent volume for the isotopes, the isotope elution curves are not well separated. By evaluating the elution curve with Glueckauf's method [88], one can calculate the number of equilibrium stages N in the chromatographic column with Eq. (21) and (22). The results of these two equations are approximations; therefore, the average of both results is usually used for calculating ε.

$$N = 2\pi \times \left(\frac{c_{max} \times \bar{v}}{m}\right)^2 \tag{21}$$

$$N = 8 \times \left(\frac{\bar{v}}{\beta}\right)^2 \tag{22}$$

c_{max} concentration at maximum of elution curve
$\bar{v}$ effluent volume at c_{max}
m sum of eluted element amount
β width of elution curve at c_{max}/e

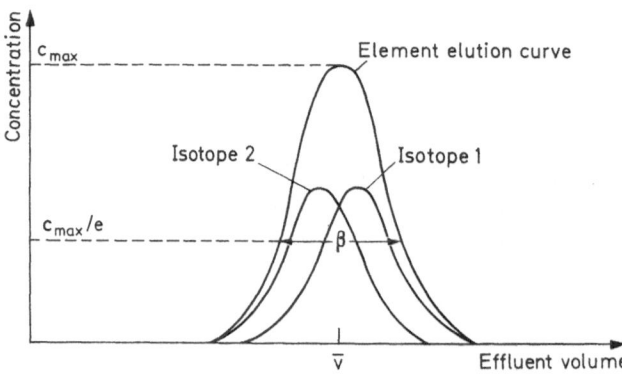

Fig. 7. Schematic figure of the elution curve of an element with two isotopes

If one analyzes the isotope ratio of different fractions of the elution curve, e.g. the isotope ratio of the marked points 1—5 (in Fig. 8a), the isotope ratio can be plotted versus the eluted part of the element amount $\Delta m/m$ (Fig. 8b). After the calculation of the local separation coefficient R (= isotope ratio at any point of the elution curve/

initial isotope ratio), one can plot ln R versus $\Delta m/m$ on probability paper (Fig. 8c). The curve in Fig. 8b corresponds to a Gaussian distribution $A_{(t)}$:

$$A_{(t)} = \frac{1}{\sqrt{2\pi}} \times \int e^{-t^2/2} \times dt \,, \tag{23}$$

if $\Delta m/m$ is equivalent to $A_{(t)}$ and the result of the integrated Eq. (23) from $-\infty$ to $+\infty$ is standardized to one ($= 100\%$). On the basis of the chromatographic experiments, Glueckauf's approximation results in the following equation for ε [87]:

$$\ln R \approx \varepsilon \times \sqrt{N} \times t \,. \tag{24}$$

The plot given in Fig. 8c shows a linear relation between ln R and t, the same as it is in the comparable Eq. (24) where the distance from $\Delta m/m = 0.02\%$ to

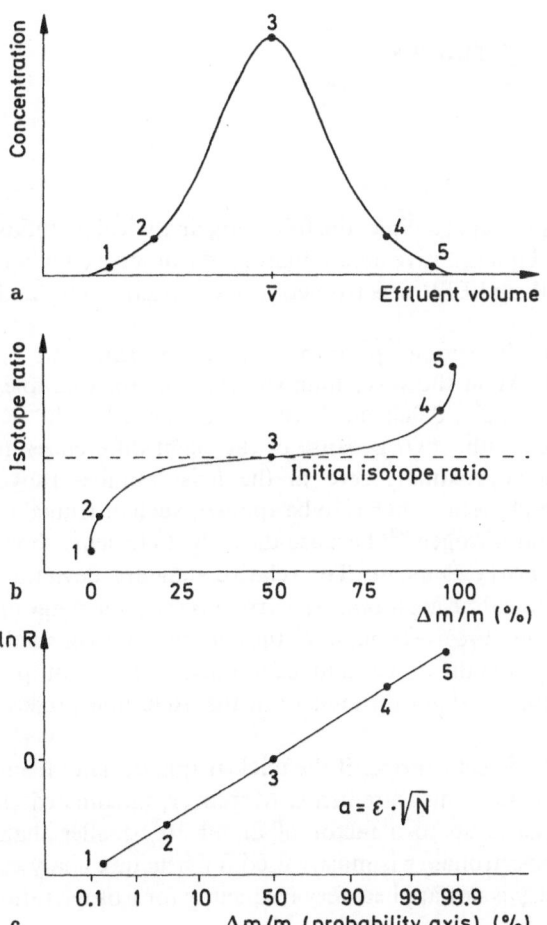

Fig. 8. The graphic determination of ε with the results of a chromatographic elution

$\Delta m/m = 99.98 \%$ (x-axis in Fig. 8c) corresponds to 7 t-units [89]. For the slope a of the straight line in Fig. 8c it follows:

$$a = \frac{\Delta \ln R}{\Delta t} \approx \varepsilon \times \sqrt{N} . \tag{25}$$

From the graphically determined a-value and from the number of equilibrium stages N obtained by Eq. (21) and (22), ε and α can be calculated for an isotopic exchange reaction.

Because the described method is an approximation, the following conditions have to be fulfilled:
— The elution band should be small compared with the column length,
— The dynamic volume of the column (volume not occupied by the resin) should be small compared with the effluent volume,
— The width of the elution curve should not be too wide.

3 Measurement of Isotopic Separations

3.1 General View

Besides a number of special techniques, in principle, the following analytical methods are applicable for the qualitative and quantitative determination of isotopes: emission and absorption spectrometry, NMR and ESR spectrometry, mass spectrometry and radioanalysis.

Using emission spectrometry or absorption spectrometry, the measurement of isotopes is due to the isotopic shift Δv in the wave number which is, for example, $+0.35 \text{ cm}^{-1}$ for ^{6}Li compared with ^{7}Li, $+0.008 \text{ cm}^{-1}$ for ^{39}K compared with ^{41}K and -0.28 cm^{-1} for ^{235}U compared with ^{238}U. Because of the slight differences in the isotopic wave numbers which approximate zero in the mass number range 40–140 [90], optical equipment with high resolution has to be applied. Such instruments are especially used for lithium [91] and nitrogen [92] because these light elements show a larger isotopic shift Δv than the heavy elements. The relative standard deviation of the isotope ratio measurement of metals using atomic spectroscopy is in the range of 1–10% [93]. Therefore, this method is not precise enough for the determination of small isotopic separations (see Chap. 2.5.1). On the other hand, experiments on the isotopic separation of metals with a laser use the slight differences in the excitation process of isotopes [94].

NMR spectrometry is only able to detect isotopes if the nuclear spin deviates from zero. This already limits the isotope ratio measurements. Moreover, in some cases the detection sensitivity for isotopes is up to a factor of about 10^3 smaller than for ^{1}H atoms [93]. Therefore, NMR spectrometry is mostly used for structure analyses of molecules. Also, ESR spectrometry is not applied very frequently for isotope ratio determinations.

3.2 Mass Spectrometry

In principle, all known ionization methods are suitable for mass spectrometric isotope ratio determinations. Today, those methods listed in Table 4 are applied for the isotope measurement of the elements in particular. With high precision thermal ionization instruments as well as with gas isotope mass spectrometers using electron impact ionization, relative standard deviations of the isotope ratio determination in the range of 0.1–0.001 % are available [101, 113, 114]. Using such types of mass spectrometers, very slight isotopic separations are also detectable. First measurements with a quadrupole thermal ionization mass spectrometer, which is a low-cost instrument and simpler to handle than a magnetic field high precision mass spectrometer, have resulted in relative standard deviations of 0.1–0.5 % for the isotope ratio determination [115]. This should be sufficient for the measurement of a number of isotopic separations. Usually, the relative standard deviations are a little higher when using field desorption mass spectrometry instead of the thermal ionization technique. However, relative standard deviations of 0.1–0.5 % are also obtained for the determination of high abundance isotopes, e.g. for isotopes of Ca, Sr and Ba [109]. Using spark source mass spectrometry one has to take standard deviations of some per cent into account [116]. The comparable values for a mass spectrometer with an inductively coupled plasma as ion source are in the range of 0.1–1 % [117].

Table 4. Ionization methods in a mass spectrometer for isotope ratio measurements of the elements

Ionization method	Sample	Preferred elements analyzed
Thermal ionization	Inorganic salts	Metals with not too high ionization energy (positive thermal ions) [95,96]; non-metals with high electron affinity (negative thermal ions) [97−103]
Electron impact	Low-molecular gases	Noble gases, H, C, N, O, S
	Metal chelate after GC-separation	All metals forming stable and volatile chelate complexes [104,105]
Field desorption	Inorganic salts	Alkalis, alkaline earths [106−109]
Spark source	Inorganic solids	All elements [110]
Inductively coupled plasma	Solution of inorganic salts	All metals [111,112]

Today, thermal ionization mass spectrometry is preferably used for the isotope analyses of inorganic solid samples, and electron impact instruments are preferably applied for the analyses of low-molecular gases. Therefore, Fig. 9 gives a summary of the possibilities for the isotope ratio determination in the periodic table of the elements corresponding to these two ionization methods. A position which is not marked in Fig. 9 represents an element with not more than one stable or long-lived radioactive isotope. In these cases an isotope ratio measurement by mass spectrometry is usually not possible. For the isotope ratio determination of those elements which are marked by a black bar, at least one long-lived radioactive isotope has to be used. Except uranium, these elements are only monoisotopic in nature (Be, Al, Mn, Nb, I, Cs, Bi, Th), or they are synthetic elements (Tc, Np, Pu, Am, Cm).

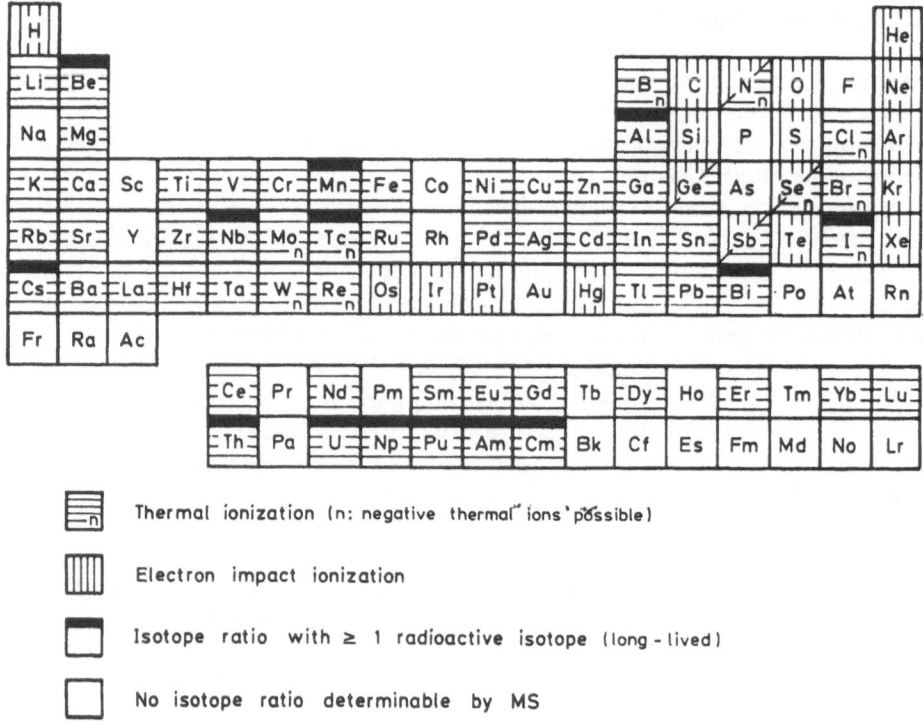

Thermal ionization (n: negative thermal ions possible)

Electron impact ionization

Isotope ratio with ≥ 1 radioactive isotope (long - lived)

No isotope ratio determinable by MS

Fig. 9. Possibilities for mass spectrometric isotope ratio measurements in the periodic table of the elements

3.3 Radioanalysis

A direct isotope ratio determination by radioactivity is possible with all isotopic radionuclides. In principle, the analysis of stable'isotopes in that way can be done after an activation process, e.g. with an (n, γ)-reaction in a nuclear reactor. Because of the high expenditure, this type of activation analysis is only applied in a few exceptional cases. The best type of radioactivity for precise isotope ratio determinations is γ-radiation, whereas the strong absorption of α-activity and low-energetic β-activity strongly influences the reproducibility of the measurement. Moreover, the half-life of the radionuclide should be long enough compared with the duration of the experiment. In contrast, the half-life should not be too long for a sufficient activity which is needed for a good statistical error of the measurements. Under these conditions the natural radionuclides are not suitable for the radioanalytical determination of isotopic separations. This means that synthetic mixtures have to be applied for these investigations.

A survey on those elements in the periodic table of the elements where at least two suitable radionuclides are available to investigate isotopic separations is given in Fig. 10. Only·such elements are marked where the radionuclides emit γ-radiation — or as an alternative, in one case β⁻-particles with an energy ≥0.1 MeV — and where the half-life of the isotopes is ≥ 10 h. All these radionuclides are commercially available or can be easily produced by an activation process in a nuclear reactor. There are also known suitable radioisotopes of a number of other elements which are not marked in Fig. 10 because the production of these radioisotopes is more difficult. A comparison of Fig. 9 and Fig. 10 shows that, on the one hand, a larger number of elements can be investigated by mass spectrometry. On the other hand, fortunately radioanalysis is able to determine isotopic separations for those elements where mass spectrometric detection is not possible because of a lack of stable isotopes. This is valid for the monoisotopic elements Na, Sc, Co, As, Y, Pr, Tb, Tm, Au and for the radioactive elements Pm and Ra.

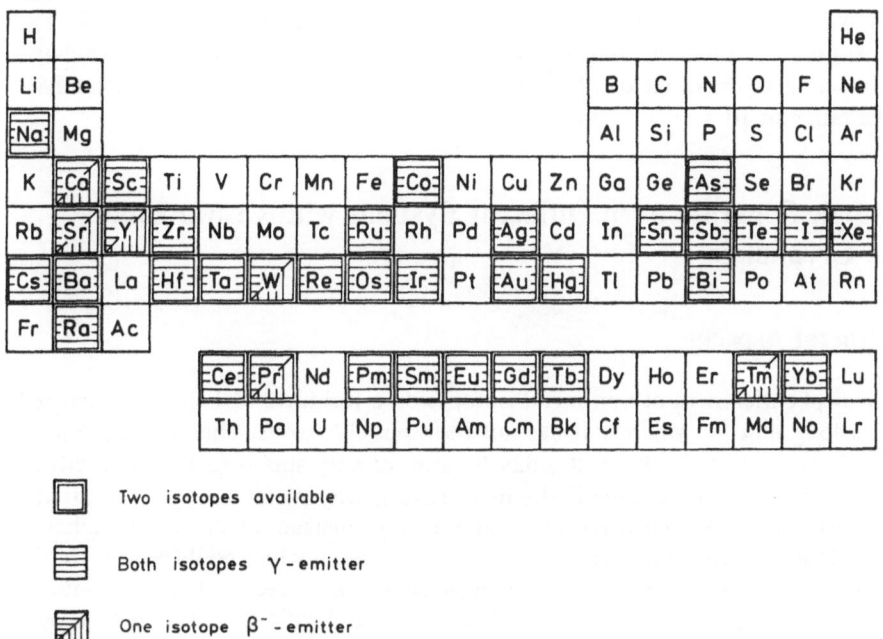

Two isotopes available

Both isotopes γ-emitter

One isotope β⁻-emitter

Fig. 10. Possibilities for radioanalytical isotope ratio measurements in the periodic table of the elements

Until today, radioanalytical measurements for the determination of isotopic separations have been carried out by Betts and coworkers [118], by Räde and Wagener [65,66] as well as by Knöchel and Wilken [64,119] for ^{22}Na/^{24}Na, by Aaltonen for ^{56}Co/^{60}Co and ^{88}Y/^{91}Y [68], by Räde for ^{85}Sr/^{89}Sr and ^{133}Ba/^{139}Ba [65]. It is also possible to investigate isotopic separations by analyzing the total amount of the (stable) element except for one radionuclide. This procedure is particularly suitable for monoisotopic elements or for those elements with one stable isotope of more than 90% natural abundance. This method was applied by Lindner for ^{45}Ca/total Ca [120], by Lee for ^{47}Ca/total Ca, ^{7}Be/total Be and ^{60}Co/total Co [69].

The standard deviations of radioanalytical isotope ratio measurements are significantly higher than those obtained by high precision mass spectrometers. For example, Räde obtained a relative standard deviation of 0.5% for the ratio ^{85}Sr/^{89}Sr [65]. For the isotope ratio $^{22}Na/^{24}Na$ a precision of less than 1% can only be achieved if the activity of the sodium isotopic mixture is followed until the complete decay of of ^{24}Na [118]. Knöchel and Wilken [119] have shown that the isotopic separation factor α is infected by an error of 1% if the statistical error of the radioactivity determination is 0.1%. To obtain such a statistical error, 10^6 decays have to be measured. Additionally, the radioanalysis can be influenced by the following sources of error: radiochemical purity of the radioisotopes, background radiation, interferences of the different radionuclides and drift problems during the detection. It is obvious that mass spectrometry should usually be preferred for precise isotope ratio measurements in investigations on isotopic separations. The radioanalytical method has to be applied if the element to be determined does not consist of two stable or long-lived radioactive isotopes or if the isotope amount to be analyzed is extremely small. The latter is due to the fact that the detection limit for the mass spectrometric measurement is in the range of 10^{-10}–10^{-14} mol, whereas this limit for radioanalysis is in the range of 10^{-15}–10^{-19} mol.

4 Isotopic Separations in Different Systems with Crown Ethers and Cryptands

4.1 General Aspects

The chemical enrichment of non-metal isotopes on a technical scale has been carried out for a long time [37], whereas it has not been possible to use chemical separation of metal isotopes technically until today because of very small fractionation effects in those chemical systems. This is the main reason why metal isotopes are usually enriched by means of expensive physical methods instead of chemical exchange reactions. However, for a number of years chemical systems which could be used for the separation of metal isotopes in a technical scale have been intensively sought. Therefore, it is not surprising that in this connexion chemical systems with crown ethers and cryptands became of interest after an experiment by Jepson and DeWitt in 1976 [13]. They found a significantly higher isotopic separation factor for calcium in an extraction system with crown ethers than had been achieved by other chemical reactions up to that time.

Today, a great number of different crown ethers and cryptands is commercially available. In the meanwhile, a few resins with cryptand anchor groups are also on the market, or one can synthesize them according to instructions given in the literature (see Table 3). From that, there is no problem in applying polyether systems in investigations on isotopic separations where the polyether is used as a complex-forming compound either in a solution phase or in a solid phase. Hence, it follows that the required heterogeneous exchange reaction for the isotopic separation could take place in a liquid/liquid or in a liquid/solid system.

Because of the particular properties of crown ethers and cryptands in forming complexes with alkali and alkaline earth elements, it is understandable that these elements have been exclusively investigated in respect to isotopic separations. Among these elements the enrichment of heavy calcium isotopes for medical investigations and the enrichment of ^{6}Li for the production of tritium is of special interest. Calcium consists of six stable isotopes with the mass numbers 40, 42, 43, 44, 46 and 48, whereas lithium consists of the two stable isotopes ^{6}Li and ^{7}Li. The relative mass difference between the lightest and the heaviest calcium isotope, ^{40}Ca and ^{48}Ca, as well as between the two lithium isotopes has the highest magnitude of all metals. That means that for these two elements a comparably large isotopic separation could be expected. Besides calcium and lithium, sodium, which consists of the stable isotope ^{23}Na, has been investigated in systems with polyethers. Corresponding to the different stable isotopes, calcium and lithium isotopic separations have been measured by mass spectrometry using the thermal ionization technique (see Chap. 3.2), whereas for sodium the isotope ratio ^{22}Na/^{24}Na was investigated by radioanalysis (see Chap. 3.3).

For an exact characterization of the investigated systems, besides the isotope ratio measurement, one has to carry out the quantitative analysis of polyethers, and of the corresponding metal complexes as well. The analysis method which is most commonly described in the literature for polyethers is NMR spectrometry [121]. However, this spectroscopic method is not very precise for a quantitative determination. Therefore, other types of analysis have been developed, e.g. a gravimetric procedure [122] and a titrimetric method using the standard addition technique with a potentiometric indication [123].

4.2 Liquid/Liquid Extraction Systems

4.2.1 Introduction

Up to now, isotopic separations in extraction systems have been investigated for calcium by Jepson and DeWitt [13] as well as by Heumann and Schiefer [124], for lithium by Jepson and Cairns [125], and for sodium by Knöchel and Wilken [64]. In these investigations the system chloroform/water was mainly used. Jepson and DeWitt have also carried out experiments with methylene chloride but no detailed results have been submitted about this system.

The heterogeneous isotopic exchange reaction in an extraction system is analogous to Eq. (18). If one assumes that the crown ether or cryptand complex of the metal ion only exists in the organic phase, then the isotopic exchange reaction, for example between ^{40}Ca^{2+} and ^{44}Ca^{2+}, can be described as:

$$^{40}\text{Ca}^{2+}(\text{aq}) + {}^{44}\text{CaL}^{2+}(\text{org}) \rightleftharpoons {}^{40}\text{CaL}^{2+}(\text{org}) + {}^{44}\text{Ca}^{2+}(\text{aq}) \,. \qquad (26)$$

For the isotopic separation factor of Eq. (26) it follows:

$$\alpha = 1 + \varepsilon = \frac{(^{44}\text{Ca}/^{40}\text{Ca})\ \text{aq}}{(^{44}\text{Ca}/^{40}\text{Ca})\ \text{org}} \,. \qquad (27)$$

4.2.2 Calcium Isotopic Separation

The results of calcium isotopic separations in extraction systems are summarized in Table 5. The chemical structure of the crown ethers and cryptands are shown in Fig. 11.

Table 5. Extraction systems with crown ethers and cryptands for calcium isotopic separations

Investigated system	Crown ether/ cryptand[a]	ε-value $\times 10^3$		Direction of enrichment	Literature
		$(^{44}Ca/^{40}Ca)$	$(^{48}Ca/^{40}Ca)$		
1. $CHCl_3/H_2O$; $CaCl_2$ 2–4 stages, 25 °C	Dicyclohexano[18]-crown-6 (two isomers)	4.0	n. d.[b]	^{44}Ca in H_2O	Jepson, DeWitt [13]
	Dibenzo[18]crown-6	no precise determination	n. d.		
2. $CHCl_3/H_2O$; $CaCl_2$ single-stage 45 °C	[2.2.1] cryptand	5.6	10.8	^{44}Ca and ^{48}Ca in H_2O	Heumann, Schiefer [124]
	[2.2.2] cryptand	8.0	15.1		

[a] Structure of compounds see Fig. 11
[b] n. d. = not determined

To get better precision in the determination of the isotopic separation factor of one equilibrium stage, Jepson and DeWitt carried out experiments with two to four stages [13]. The two-phase system consisted of a concentrated aqueous $CaCl_2$ solution and a crown ether dissolved in chloroform. According to the scheme in Fig. 5, in every multi-stage experiment several primary separation stages were used. The concentration of the aqueous $CaCl_2$ solution was established to be in the range of 5–5.5 M after the complexation. After an intensive shaking period one has to wait for some hours until the separation of the two phases takes place. This is required because of the similar densities of chloroform and the concentrated aqueous solution. Additionally, it is possible that the experimental value could be influenced by small emulsified droplets of the concentrated aqueous phase in chloroform. The experiments were carried out with the two isomeres A and B of dicyclohexano[18]crown-6 and with dibenzo[18]crown-6. The isomers were characterized by their melting points of 61–62 °C and 69–70 °C [126–128]. For comparison, an experiment was also carried out with the isomer mixture in the system $CHCl_3/D_2O$. The relative standard deviation of the mass spectrometric $^{44}Ca/^{40}Ca$-ratio measurement was $\leq 0.2\%$.

Table 6 contains the individual results of the different experiments. The listed

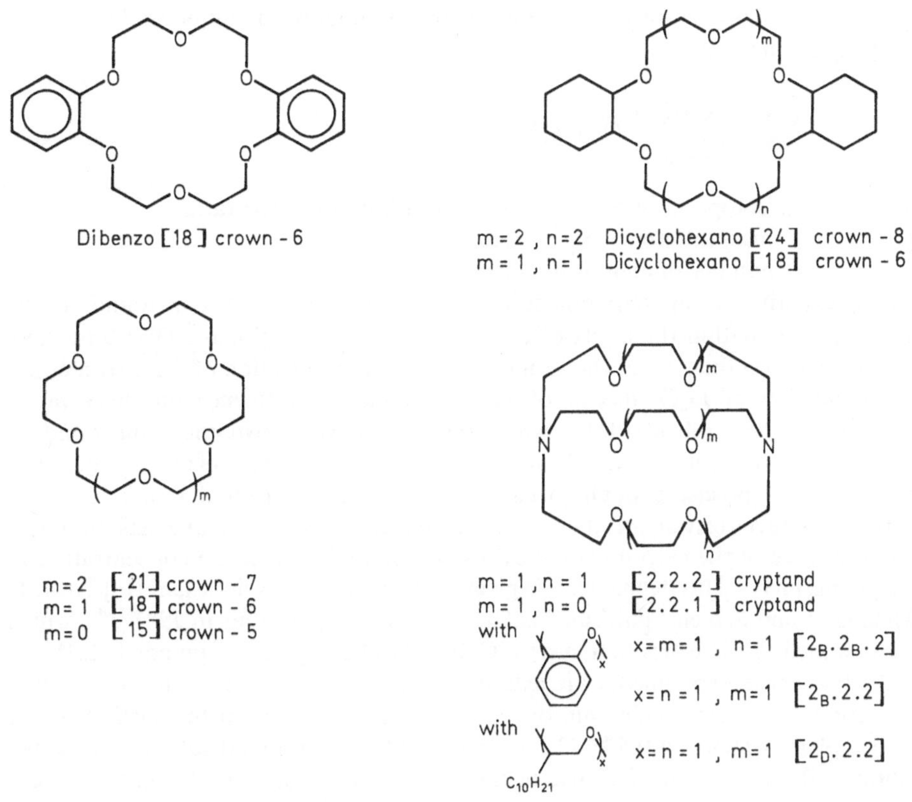

Dibenzo [18] crown - 6

m = 2 , n = 2 Dicyclohexano [24] crown - 8
m = 1 , n = 1 Dicyclohexano [18] crown - 6

m = 2 [21] crown - 7
m = 1 [18] crown - 6
m = 0 [15] crown - 5

m = 1 , n = 1 [2.2.2] cryptand
m = 1 , n = 0 [2.2.1] cryptand
with
x = m = 1 , n = 1 [2_B.2_B.2]
x = n = 1 , m = 1 [2_B.2.2]
with
x = n = 1 , m = 1 [2_D.2.2]
$C_{10}H_{21}$

Fig. 11. Structures of crown ethers and cryptands used for isotopic separation

Table 6. Single-stage ε-values of calcium for a system with dicyclohexano[18]crown-6 [13]

System	ε-value $\times 10^3$/mass unit
1. $CHCl_3/H_2O$ Isomer mixture, 0.07 M	1.0 ± 0.2
2. $CHCl_3/H_2O$ Isomer B, 0.07 M	1.0 ± 0.4
3. $CHCl_3/H_2O$ Isomer A, 0.07 M	1.0^a
4. $CHCl_3/H_2O$ Isomer A, 0.2 M	0.9 ± 0.1
5. $CHCl_3/D_2O$ Isomer mixture, 0.07 M	0.9 ± 0.2
Weighted average 1.–5.	1.0 ± 0.2

[a] Insufficient number of samples for calculation of standard deviation; other experiments: five independent determinations

single-stage ε-values per mass unit were calculated under the assumption that the following equation is valid:

$$\frac{(^{44}Ca/^{40}Ca)_N}{(^{44}Ca/^{40}Ca)_0} = (1 + \varepsilon)^{\Delta m \times N}.$$

(28)

$(^{44}Ca/^{40}Ca)_{N,0}$ isotope ratio of stage No. N and initial isotope ratio
Δm mass difference of isotopes
N   number of equilibrium stages

In all experiments an enrichment of the heavy isotope ^{44}Ca is found in the aqueous phase. Within the limits of error, the experiments No. 1–5 (Table 6) show no dependence on the type of the isomer or on the concentration of the crown ether. The application of D_2O instead of H_2O also has no influence on the ε-value. Experiments where dibenzo[18]crown-6 was used have shown the same trend of enrichment as was found with dicyclohexano[18]crown-6. However, the measurements were not precise enough to calculate the ε-values from this data. For the complete characterization of the discussed extraction systems, one has to know the dependence of the calcium complex distribution of the calcium concentration in the aqueous phase. This dependence is given in Table 7 for dicyclohexano[18]crown-6.

Heumann and Schiefer published calcium isotopic separations in 1981 [124], which were obtained in the extraction system $CHCl_3/H_2O$ using the cryptands [2.2.1] and [2.2.2]. For these experiments an extraction apparatus as shown in Fig. 12 was used. The chloroform condensate drops from the condensor to the surface of the aqueous phase which is a 0.5 M $CaCl_2$ and 4×10^{-2} M cryptand solution in the beginning of the experiment. Because of the higher density of chloroform, the $CHCl_3$ droplets sink down through the aqueous phase and reach the distillating flask through an overflow tube. The aqueous phase is heated up to a temperature of about 45 °C by the warm chloroform condensate. Compared with shaking experiments, this arrangement as described has the advantage that a good separation of the phases can be achieved. For all experiments an extraction period of seven days was chosen to get a sufficient amount of calcium in the distillating flask. Short extraction times would result in a small amount of calcium in the distillating flask because the distribution of calcium ions as well as of cryptated calcium ions completely

Table 7. Phase compositions for dicyclohexano[18]crown-6 complex formation (initial chloroform phase concentration = 0.069 mol/l) [13]

Aqueous phase calcium concentration [mol/l]	Chloroform phase complex concentration [mol/l]	Distribution ratio
0.166	0.0017	98
5.11	0.041	125
5.36	0.041	131
5.79	0.044	132
6.5	0.050	130
7.1	0.066	108

lies on the side of the aqueous solution. The isotope ratios $^{44}Ca/^{40}Ca$ and $^{48}Ca/^{40}Ca$ were determined by mass spectrometry with relative standard deviations in the range of 0.1–0.8%.

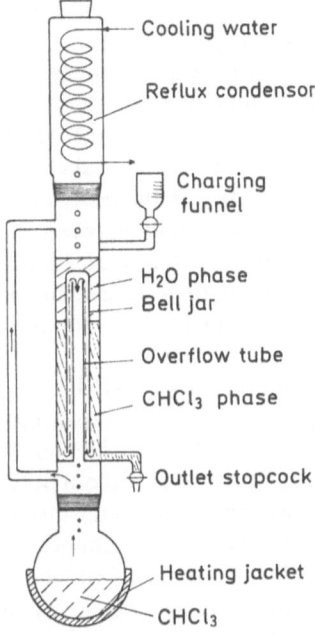

Cooling water

Reflux condensor

Charging funnel

H_2O phase

Bell jar

Overflow tube

$CHCl_3$ phase

Outlet stopcock

Heating jacket

$CHCl_3$

Fig. 12. Extraction apparatus for calcium isotopic separations with cryptands in the system $CHCl_3/H_2O$

The results of the described experiments are summarized in Table 8 where the totally extracted amounts of calcium and cryptand are also listed. Additionally, a blank result is given which was obtained under the same conditions, however without a cryptand compound. The ε-values were calculated under the assumption that during the passage of the chloroform droplets through the aqueous solution a distribution equilibrium has developed. Only under this condition the measured isotope ratio of calcium in the distillating flask is equal to that of one extraction equilibrium stage. Because of the fact that the establishment of the equilibrium is not completely

Table 8. Extracted calcium amounts, cryptand amounts and ε-values for the isotopic separation in the system $CHCl_3/H_2O$ [124]

Cryptand	Aqueous phase initial amount [mmol]		Chloroform phase extracted amount [mmol]		ε-value × 10^3	
					$(^{44}Ca/^{40}Ca)$	$(^{48}Ca/^{40}Ca)$
	calcium	cryptand	calcium	cryptand		
None	25	0	0.350	0	4.3	10.8
[2.2.1]	25	2	0.369	0.093	5.6	10.8
[2.2.2]	25	2	0.325	0.708	8.0	15.1

fulfilled in this experiment, the real ε-values should be somewhat higher than the listed values in Table 8.

In both systems with the different cryptand compounds an enrichment of the heavier isotopes ^{44}Ca and ^{48}Ca compared with ^{40}Ca is found in the aqueous phase. Within the limits of error for the mass spectrometric measurement, the ε-values for $^{48}Ca/^{40}Ca$ are a factor of two higher compared with the ratio $^{44}Ca/^{40}Ca$. Using [2.2.2] cryptand, the ε-values are significantly higher than those for experiments where [2.2.1] cryptand had been used. It is remarkable that the ε-value of the system with [2.2.1] is very similar to or identical ($^{44}Ca/^{40}Ca$ and $^{48}Ca/^{40}Ca$) with that which was obtained without a cryptand compound. Therefore, one has to take into account that a significant portion of the calcium isotopic separation in the experiment with [2.2.1] is also due to the different distribution of the isotopes only between the H_2O phase and the $CHCl_3$ phase. This assumption is confirmed by the result that in the case of [2.2.1], an essentially smaller amount of cryptand than of calcium is extracted, whereas with [2.2.2] the extracted cryptand is about twice the amount of calcium.

Summarizing the calcium isotopic separations in the extraction system $CHCl_3/H_2O$, one has to point out that for all polyethers an enrichment of the heavy isotopes compared with the light isotopes is found in the aqueous solution. Because Heumann and Schiefer [124] also measured an isotopic separation in the same system without polyethers, in all further experiments of this type is should be proved whether an isotopic fraction is already established by the pure phases or not. Therefore, it cannot be excluded that the ε-values determined by Jepson and DeWitt [13] are partly reducible to such an effect. The different solvatation of the isotopic ions was discussed in Chap. 2.3. Hence, it follows that differences in the distribution coefficient of the isotopes can appear between two immiscible phases of solvents. In any case, such a solvatation effect has to be taken into account for the aqueous phase of all described experiments because of the high calcium concentration compared with the low polyether concentration used in the aqueous solutions. For the organic phase a possible difference in the complex stability of the calcium isotopes corresponding to Eq. (19) could also be of considerable influence. The different isotopic shifts for the two cryptands (Fig. 4) might explain the higher ε-values in the extraction system with [2.2.2] compared with that where [2.2.1] was applied (Table 8).

The application of the described extraction systems for a practical enrichment of calcium isotopes is not too attractive because ^{40}Ca instead of the more important ^{48}Ca is enriched in the chloroform phase. A mass balance can show that no usable enrichment of ^{48}Ca is obtained in the aqueous phase because of the high distribution ratio of calcium ions between the aqueous solution and chloroform (see Chap. 2.4).

4.2.3 Lithium Isotopic Separation

Jepson and Cairns [125] have investigated lithium isotopic separations in the extraction system $H_2O/CHCl_3$ using the cryptands [2.2.1] and [2.2.2] as well as the crown ethers dicyclohexano[18]crown-6 and tert-butylcyclohexano[15]crown-5 (see Fig. 11). Recently, a Chinese investigation was published [129] where the isotopic separation of lithium was established by a crown ether called "C_{401}". In this experiment LiSCN was distributed between an aqueous and an organic solution phase; the polyether "C_{401}"

Table 9. Single-stage ε-values of lithium for the extraction system $H_2O/CHCl_3$ [125]

Polyether	Chemical species	Single/multistage experiment	ε-value × 10^3 [a]
[2.2.1]	LiTFA, excess HTFA	single	41 ± 6
	LiTFA	single	35 ± 3
	LiBr	single	26 ± 6
[2.2.2]	LiBr	multi	11 ± 1
Dicyclohexano-[18]crown-6	LiBr	multi	9 ± 2
tert-Butylcyclohexano-[15]crown-5	LiBr	single	20 ± 12
	LiTFA	single	20 ± 9
"C_{401}" [b]	LiSCN		30

[a] The listed confidence limits are ±2 s   [b] Ref. [129]

was only very slightly soluble in the aqueous phase (1.9×10^{-3} g/100 g H_2O). The ε-value which was obtained in this system was 30×10^{-3}. This result as well as the results of Jepson and Cairns' work are summarized in Table 9. The newest results in this field of another chinese group which investigated the anion effect of the lithium isotopic separation in an extraction system with benzo[15]crown-5 showed that the isotopic separation is higher the larger and softer the anions are [149].

The first column of Table 9 contains the used polyethers, whereas in the second column the lithium compounds are listed. During the first experiment with lithium trifluoroacetate (LiTFA), a little excess of trifluoroacetic acid (HTFA), which causes a protonation of the cryptand [2.2.1], was applied. As far as single-stage experiments were carried out, equal portions of an aqueous solution containing the lithium salt, and the polyether-containing chloroform phase were shaken 18–24 h at a temperature of 25 °C. The distribution of lithium between the two phases can be quoted from Table 10. So far as multi-stage experiments were investigated, a countercurrent extraction column was used. In this case it was assumed that Eq. (20) is valid for the calculation of ε (see Chap. 2.5.2). To obtain a smallest possible column height for one equilibrium stage, the exchange reaction rate must be high. This proved true for the system with [2.2.1] and LiTFA with a half-life of 15–20 s for the establishment of the equilibrium.

Table 10. Distribution ratio of lithium between the aqueous and the chloroform phase using [2.2.1] cryptand (initial concentration of [2.2.1] in the chloroform phase 0.15 M) [125]

Chemical species	Distribution ratio $(Li^+)org/(Li^+)aq$
LiTFA, excess HTFA	0.047
LiTFA	0.080
LiBr	0.057

If one formulates an isotopic exchange reaction for lithium in analogy to Eq. (26), the ε-values for this reaction correspond to Eq. (27). From this the ε-values were calculated which are listed in the last column of Table 9 (average of 3–6 parallel experiments). The relative standard deviation of the mass spectrometric $^6Li/^7Li$-measurement is $\pm 0.2\%$.

For all systems, the lighter isotope ^6Li is enriched in the organic phase compared with ^7Li. Obviously, this also proved true for the Chinese work [129] where the obtained ε-value of 30×10^{-3} is comparable with the results of Jepson and Cairns. On the whole, a significant dependence of the isotopic separation on the polyether as well as on the anion of the lithium salt was found. This can be understood by means of the different interaction of the cryptated cation with the anion in chloroform. According to Jepson and Cairns, the enrichment of ^6Li in the organic phase is attributed to a more stable ^6Li polyether complex compared with the ^7Li compound.

In contrast to the investigations with calcium where ^{40}Ca, which is not the preferable calcium isotope for application purposes, is enriched in the chloroform phase (see Chap. 4.2.2), ^6Li which is of importance for the production of tritium is enriched in that phase. Certainly, corresponding to the disadvantageous distribution coefficient (see Table 10) a small amount of enriched ^6Li is only obtainable within one equilibrium stage. Therefore, a cascade experiment as described in Chap. 2.5.2 must be applied.

4.2.4 Sodium Isotopic Separation

Sodium is monoisotopic so that investigations on isotopic separations of this element have to be carried out by means of synthetic mixtures of radioactive sodium isotopes (see Chap. 3.3, Fig. 10). The most suitable isotopes are ^{22}Na and ^{24}Na due to their good detectability with γ-radiation.

Investigations on isotopic separations for $^{22}Na/^{24}Na$ have been carried out by Knöchel and Wilken in the extraction system $H_2O/CHCl_3$ with polyethers [64]. The different crown ethers and cryptands used in the single-stage experiments are listed together with the results in Table 11 (structures of the compounds see in Fig. 11).

Table 11. α-values of sodium for the extraction system $H_2O/CHCl_3$ [64]

Polyether	α-value	Number of determinations	log K_s
[21]crown-7	0.938 ± 0.050	8	
[18]crown-6	0.970 ± 0.004	7	0.3
Dicyclohexano[18]crown-6	0.993 ± 0.011	14	1.6
Dicyclohexano[24]crown-8	0.995 ± 0.024	7	
[15]crown-5	1.025 ± 0.028	7	0.3
[2.2.1]	0.951 ± 0.016	21	5.4
[2$_B$.2.2]	0.971 ± 0.004	12	4.0
[2.1.1]	0.971 ± 0.021	7	2.8
[2$_B$.1$_B$.2]	0.998 ± 0.015	5	3.3
[2$_D$.2.2]	1.011 ± 0.022	8	
[2.2.2]	1.057 ± 0.005	14	3.9

The α-values in Table 11 correspond to an isotopic exchange reaction between $^{22}Na^+$ and $^{24}Na^+$ which is analogous to Eqs. (26) and (27). That means, an enrichment of ^{24}Na takes place in the aqueous solution for all values α > 1 but an enrichment of this isotope in the organic phase, which also contains the main polyether portion, is true for values α < 1. In the third column of Table 11 the number of parallel experiments is given. From the results of the individual experiments, the average and the standard deviation for α, which are listed in the second column, were calculated. To obtain an as-precise-as-possible value for the isotope distribution, a high precision measuring system with a well-typed scintillation detector for the γ-radiation of the nuclides was applied. Possible long-term instabilities of the detector system were corrected by calibration samples. Additionally, the method described by Betts and coworkers [118] was used (see Chap. 3.3) where the γ-activity of the isotopic mixture is followed until the complete decay of ^{24}Na ($t_{1/2}(^{24}Na) = 15.0$ h; $t_{1/2}(^{22}Na) = 2.602$ a).

It was possible to measure an isotopic separation out of the standard deviation for the α-values in the system $H_2O/CHCl_3$ using the polyethers [21]-crown-7, [18]-crown-6, [2.2.1], [2$_B$.2.2], [2.1.1] and [2.2.2]. This effect is particularly high for the polyethers [21]crown-7, [18]crown-6, [2.2.1], [2$_B$.2.2] and [2.2.2]. The heavy isotope ^{24}Na is enriched in the aqueous phase only in the systems where [15]crown-5, [2$_D$.2.2.] and [2.2.2] is used. All other experiments result in an enrichment of ^{24}Na in the organic phase and, therefore, in the complex with the polyether ligand. The latter direction of enrichment is inverse compared with the results of the extraction experiments on calcium and lithium where the lighter isotope is always enriched in the chloroform phase (see Chap. 4.2.2 and Chap. 4.2.3). The α-values in Table 11 were discussed by Knöchel and Wilken under the aspect that there appears to be a correlation between the direction of the effect and decreasing ring size and number of oxygen donors in the macrocyclic compound. Using cryptands, a certain correlation between the measured isotopic separation and the stability constant K_s for the sodium complex is found. To make this obvious, the last column of Table 11 contains the log K_s values [3, 130], so far as they are known for aqueous solutions.

Table 12. Distribution ratio of sodium between the aqueous and the chloroform phase using different polyethers [64]

Polyether	Distribution ratio $(Na^+)aq/(Na^+)org$
[2.2.1]	1.7
[2.2.2]	49
[2$_B$.2.2]	60
[2$_B$.2$_B$.2]	62
[2.1.1]	230
[2$_D$.2.2]	320
Dicyclohexano[18]crown-6	100
[18]crown-6	7000
[15]crown-5	17 200
Dicyclohexano[24]crown-8	40 000
[21]crown-7	61 500

In Table 12 the distribution of sodium ions between water and chloroform (referred to g H_2O/g $CHCl_3$) is presented in dependence on the different polyethers. All results were obtained under analogous conditions with 0.1 mmol Na^+ and 0.1 mmol polyether in the system where the pH-value was established to be ≥ 8 by adding 10 mmol of tetraethylammonium chloride. The establishment of the equilibrium requires less than 60 min in all systems and was followed by the γ-activity of the sodium isotopes and the β-activity of ^{14}C-labeled polyethers. The enrichment of one of the sodium isotopes in a practical scale from a $^{22}Na/^{24}Na$-mixture can only be achieved in a system where the distribution ratio $(Na^+)_{aq}/(Na^+)_{org}$ is not too high. However, in contrast to the enrichment of stable isotopes from a sample with natural isotope abundance, the enrichment of ^{22}Na or of ^{24}Na from an isotopic mixture is not of great importance because these two isotopes can be produced by nuclear reactions. On the other hand, the investigations on sodium isotopic separations are of common interest in respect to further knowledge about isotopic effects.

4.3 Solid/Liquid Systems

4.3.1 Cation Exchanger/Polyether Solution

4.3.1.1 Fundamentals

Systems with a strongly acidic cation exchanger resin, e. g. Dowex 50, and a solution containing polyether are especially applied for investigations with the sodium isotopes ^{22}Na and ^{24}Na. Knöchel and Wilken have investigated sodium isotopic separations with the cryptands [2.1.1], [2.2.1] and [2.2.2] in batch experiments [119]; Delphin and Horwitz have determined such separations with the crown ethers [18]-crown-6 and dicyclohexano[18]crown-6 by column chromatography [131]. Because of the similar structure of the antibiotic monactin compared with the polyethers discussed in this review (see Fig. 13), the isotopic separation during the formation of the sodium monactin complex, which was investigated by Räde and Wagener [66], will be also described in the following. Calcium isotopic separations have been investigated by Schiefer in the system Dowex 50/[2.2.1] solution and Dowex 50/[2.2.2] solution, respectively [80].

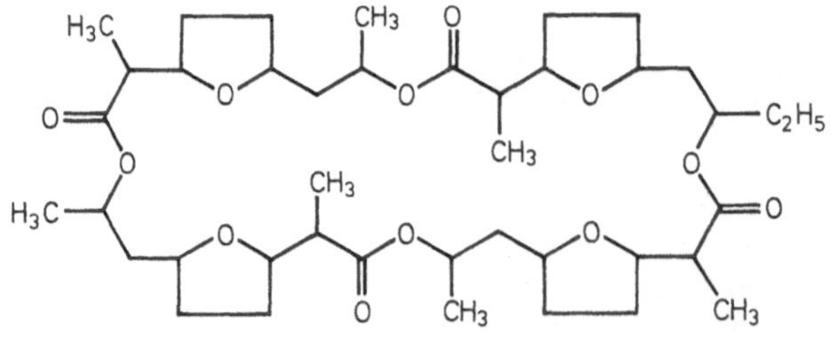

Fig. 13. Structure of monactin

4.3.1.2 Determination Method for Complex Stability Constants of Isotopes

As mentioned in Chap. 2.4, it is possible to determine the complex stability constants of isotopes corresponding to Eqs. (19a) and (19b) in an indirect way by distributing the complexed and the non-complexed species between two phases and then measuring the isotopic distribution. A suitable system for that is the exchange between an ion exchanger resin and a solution with and without a complex-forming ligand.

For example, the exchange reaction between the isotopic ion ^{24}Na$^+$ and the cryptand [2.2.1] corresponds to the following Eq.:

$$^{24}\text{Na}^+ + [2.2.1] \rightleftharpoons \{[2.2.1]\,^{24}\text{Na}^+\}\,. \tag{29}$$

An analogous equilibrium reaction is valid for ^{22}Na$^+$. If the neutral complex-forming ligand is characterized by L, from Eq. (29) it follows for the complex stability constant:

$$K^{24} = \frac{[^{24}\text{NaL}^+]}{[^{24}\text{Na}^+] \times [\text{L}]}\,, \tag{30}$$

and for the isotopic separation factor α_K between ^{24}Na$^+$ and ^{22}Na$^+$:

$$\alpha_K = \frac{K^{24}}{K^{22}}\,. \tag{31}$$

A procedure for the determination of stability constants of elements, which was described by Schubert [132], can be transferred for the determination of α_K-values of isotopes [65,66,119]. In particular, the following two distributions of isotopes have to be analyzed between the ion exchanger and the solution:
1) Distribution of the isotopes without a complexing agent, e. g. for ^{24}Na$^+$:

$$K_R^{24} = \frac{[^{24}\text{Na}^+]_R}{[^{24}\text{Na}^+]_S}\,. \tag{32}$$

Index R resin, S solution
$[^{24}\text{Na}^+]_R$ = mmol ^{24}Na$^+$ at resin per g resin
$[^{24}\text{Na}^+]_S$ = mmol ^{24}Na$^+$ in solution per ml solution
From the K_R-values of different isotopes, one can calculate the α_R-value.
2) Distribution of the isotopes in the presence of a complexing agent, e. g. for ^{24}Na$^+$:

$$K_{RL}^{24} = \frac{[^{24}\text{Na}^+]_R + [^{24}\text{NaL}^+]_R}{[^{24}\text{Na}^+]_S + [^{24}\text{NaL}^+]_S}\,. \tag{33}$$

The term $[\text{NaL}^+]_R$ in Eq. (33) has to be taken into consideration because Schiefer [80] was able to show that calcium cryptate complexes were fixed at a cation exchanger resin.

If one uses the distribution ratio K_L for the complex compound

$$K_L = \frac{[\text{NaL}^+]_R}{[\text{NaL}^+]_S}\,, \tag{34}$$

which is not directly determinable, and substitutes Eqs. (32) and (34) in Eq. (33), one can obtain the following expression for K_{RL} if Eq. (30) is taken into account:

$$K_{RL} = \frac{K_R + K_L \times K \times [L]_S}{1 + K \times [L]_S}. \tag{35}$$

After resolving Eq. (35) for K it follows:

$$K = \frac{K_R - K_{RL}}{(K_{RL} - K_L) \times [L]_S}. \tag{36}$$

From a transformation of Eq. (36), one obtains:

$$\frac{K_R - K_{RL}}{[L]_S} = K \times K_{RL} - K \times K_L. \tag{37}$$

If one plots the term $(K_R - K_{RL})/[L]_S$, which is experimentally determinable dependent on the ligand concentration $[L]_S$, versus K_{RL}, the value for K can be calculated from the slope of the straight line and K_L can be obtained by an extrapolation for the condition $K_{RL} = 0$. The described determination of K_L is not very precise for complex compounds with high stability constants [119].

If one substitutes Eq. (36) for the corresponding isotopes ^{22}Na and ^{24}Na in Eq. (31), the following expression is obtained for the isotopic separation factor:

$$\alpha_K = \frac{(K_R^{24} - K_{RL}^{24})}{(K_{RL}^{24} - K_L^{24})} \times \frac{(K_{RL}^{22} - K_L^{22})}{(K_R^{22} - K_{RL}^{22})}. \tag{38}$$

The values for K_R and K_{RL} in Eq. (38) are directly analyzable whereas the value for K_L has to be determined by the procedure described above.

4.3.1.3 *Sodium Isotopic Separation Using Polyethers*

The results for the isotopic separation of $^{22}Na/^{24}Na$ which were obtained by Knöchel and Wilken [119] in the system Dowex 50/aqueous or methanolic solution of cryptands are summarized in Table 13 (explanation for K_{RL} and K_R see Chap. 4.3.1.2). To reach a high total enrichment compared with one equilibrium stage, the batch experiments were carried out as a cascade (Chap. 2.5.2). Then Eq. (20) was used for the calculation of α-values. To determine the isotopic separation factor α_K for the complex formation as well, the K_R-values were analyzed in the same system without cryptands [119,133] (see Chap. 4.3.1.2). In all experiments 30 mg cation exchanger resin (Li^+- or Cs^+-form) were equilibrated with a 10^{-3} M Na^+-solution where a lithium or cesium salt, which corresponds to the counterion of the resin, was added up to a total cation concentration of 10^{-2} M. If one has used a complexing agent, the initial cryptand concentration has been established to be 10^{-2} M (pH = 8). For most of the systems, the standard deviations given in Table 13 correspond to seven parallel experiments. The measurement of the radionuclides ^{22}Na and ^{24}Na was carried out as described in Chap. 4.2.4.

Table 13. Isotopic separation of sodium in the system cation exchanger resin/solution with and without cryptand [119]

Cryptand	Resin form	Solvent	$K_{RL}^{22} K_{RL}^{24}$	K_R^{22}/K_R^{24}	α_K-value
[2.1.1]	Cs^+	H_2O	1.004 ± 0.005	—	1.001 ± 0.025
[2.2.1]	Cs^+	H_2O	0.997 ± 0.007	—	1.000 ± 0.025
[2.2.1]	Li^+	CH_3OH	1.003 ± 0.015	—	1.046 ± 0.032
[2.2.1]	Cs^+	CH_3OH	1.108 ± 0.030	—	1.108 ± 0.053
[2.2.2]	Li^+	CH_3OH	1.002 ± 0.007	—	1.087 ± 0.071
—	Cs^+	H_2O	—	0.996 ± 0.017	—
—	Cs^+	CH_3OH	—	1.013 ± 0.014	—
—	Li^+	CH_3OH	—	0.939 ± 0.020	—

All experiments with an aqueous solution show no significant isotopic separation. In the methanolic systems without cryptand, an effect was only found with a resin in the Li^+-form. In this case the heavy isotopic ion $^{24}Na^+$ was enriched in the resin phase which corresponds to a value $K_R^{22}/K_R^{24} < 1$. Up to now, the known isotopic exchange reactions between a cation exchanger resin and a diluted aqueous solution have resulted in an enrichment of the light isotope in the resin phase [49-52]. However, corresponding to the different solvatation of isotopic ions (Chap. 2.3), an inversion of the separation effect should be possible when using non-aqueous solutions. An evident enrichment of ^{22}Na was found in the resin phase ($K_{RL}^{22}/K_{RL}^{24} > 1$) for the system cation exchanger (Cs^+-form)/methanolic [2.2.1] cryptand solution (Table 13). A conversion corresponding to Eq. (38) results in $\alpha_K = 1.108$. That means that ^{24}Na is enriched in the $\{[2.2.1]Na^+\}$ cryptate complex. The assumption for this result is that the solution, and not the resin, contains the main portion of the cryptate complex. If the results of Schiefer (see Chap. 4.3.1.5), who found a nearly total absorption of $\{[2.2.1]Ca^{2+}\}$ cryptate at Dowex 50, can be transferred to $\{[2.2.1]Na^+\}$, Eq. (40a) or (40b), and not the assumed Eq. (39), must be responsible for the sodium isotopic separation:

$$RSO_3^{24}Na + {}^{22}NaL_{sol}^+ \rightleftharpoons RSO_3^{22}Na + {}^{24}NaL_{sol}^+ , \tag{39}$$

$$RSO_3^{24}NaL + {}^{22}NaL_{sol}^+ \rightleftharpoons RSO_3^{22}NaL + {}^{24}NaL_{sol}^+ , \tag{40a}$$

$$RSO_3^{24}NaL + {}^{22}Na_{sol}^+ \rightleftharpoons RSO_3^{22}NaL + {}^{24}Na_{sol}^+ . \tag{40b}$$

The methanolic system with [2.2.1] and a resin in the Li^+-form shows no significant isotopic separation for sodium. Because an enrichment was obtained in the same system without cryptand, one can assume that the addition of [2.2.1] has an inverse influence which neutralizes the separation effect without cryptand.

Delphin and Horwitz [131] investigated sodium isotopic separations in chromatographic experiments using dicyclohexano[18]crown-6 and [18]crown-6 (see Fig. 11). A steel column was applied which was 50 cm in height and 0.3 cm in diameter and filled with the strongly acidic polystyrene resin Aminex A 7. A mixture of ^{22}Na and ^{24}Na which had a nearly equal decay rate for both isotopes was fixed at the top of the column and was then eluted by a solution of methanol/water (80%/20%, v/v) which contained nitric acid (1 M) and a crown ether (0.025 M). The eluted fractions were measured with a γ-spectrometer in a direct comparison with the initial isotopic mixture, using the 0.51 McV energy of ^{22}Na and the 2.75 MeV energy of ^{24}Na. From these measurements the local separation coefficients could be calculated and then the ε-value could be evaluated by Glueckauf's method (see Chap. 2.5.3). The results for two temperatures are listed in Table 14. The statistical error of the results corresponds to 95% certainty.

Table 14. ε-values of sodium isotopes in the system cation exchanger resin/solution of 1:0 M HNO$_3$, 0.025 M crown ether, 80% CH$_3$OH (v/v) [131]

Crown ether	T [°C]	ε-value × 10^4/mass unit
Dicyclohexano[18]crown-6 (Isomer A)	25	3.65 ± 0.85
	10	7.85 ± 0.85
[18]crown-6	25	3.55 ± 0.08
	10	3.8 ± 1.5

The experiments by Delphin and Horwitz are based upon the following exchange reaction:

$$^{24}Na^+(crown)_{sol} + {}^{22}Na^+_{resin} \rightleftharpoons {}^{24}Na^+_{resin} + {}^{22}Na^+(crown)_{sol} . \tag{41}$$

The results in Table 14 show that the heavy isotope is enriched in the resin phase. The direction of this enrichment is contrary to the results of Knöchel and Wilken [119] who found an enrichment of ^{22}Na$^+$ in the exchanger using a methanolic system with [2.2.1] cryptand (Table 13). Betts and coworkers [118] also obtained an enrichment of ^{22}Na$^+$ compared with ^{24}Na$^+$ at a strongly acidic cation exchanger in an aqueous system without complexing agent which is in agreement with results of calcium isotopes [49-52]. The inverse isotopic separation in the system with crown ethers [131] is possibly due to the different solvents used in the experiments or to an absorption of the coronate cation in analogy to Eq. (40).

The α- and ε-values in systems with cation exchangers and a polyether solution are significantly lower than in extraction systems (see Chap. 4.2.4). However, a multiplication of the elementary isotopic fractionation can be easily achieved by column chromatography. The increase of the ε-value with decreasing temperature (Table 14) is in accordance with the theory (see Chap. 2.2). Comparable trends of the temperature dependence were also found in systems cation exchanger/aqueous solution for sodium isotopes by Betts and coworkers [118], for lithium isotopes by Lee [54] and for calcium isotopes by Heumann and Klöppel [51]. Heumann and Hoffmann [134] were able to obtain a similar effect for chloride with a strongly basic anion exchanger. From the results of Betts and coworkers, Glueckauf has calculated the enthalpy for the sodium isotopic reaction to be $\Delta H = -1.7$ J/mol [87] (exchange reaction corresponding to equation (41) but without crown ether), whereas for analogous reactions between ^{6}Li$^+$ and ^{7}Li$^+$ $\Delta H = -9.5$ J/mol [54] and, between ^{40}Ca^{2+} and ^{48}Ca^{2+}, $\Delta H = -5.8$ J/mol [51] were determined.

4.3.1.4 Sodium Isotopic Separation Using Monactin

Räde and Wagener have investigated the isotopic separation $^{22}Na/^{24}Na$ executing experiments with the antibiotic monactin [65,66] (structure see Fig. 13). The conditions of these experiments can be best compared with the investigations of Knöchel and Wilken [119]: Some mg of the strongly acidic cation exchanger Dowex 50W-X8 were added to a 10^{-2} M $LiClO_4$ methanolic solution (pH = 8, 25 °C) which contained the radioisotopes ^{22}Na and ^{24}Na. For a precise determination of these radionuclides, the γ-activity was followed until the total decay of ^{24}Na (see Chap. 3.3). During the measurements, a radioactive standard sample was analyzed to correct long term drift-effects in the counting equipment. To determine the α_K-value for the monactin complex with the sodium isotopes, the experiments and the evaluation of the results were carried out as described in Chap. 4.3.1.2.

Single-stage batch experiments without monactin showed no isotopic separation, which does not agree with the comparable investigations of Knöchel and Wilken [119] (Table 13). In contrast to that, Räde and Wagener found an enrichment of ^{24}Na in the monactin complex. The average of two independent investigations was $\alpha_K(^{24}Na/^{22}Na) = 1.038$.

4.3.1.5 Calcium Isotopic Separation Using Cryptands

It is well known that the distribution coefficient of cations at a cation exchanger decreases if one adds to the system a complex compound which forms negative charged species [135]. For example, this is true for calcium in the system Dowex 50/EDTA solution where complex anions $[EDTA, Ca]^{2-}$ are formed. By adding the neutral crown ether or cryptand ligands to such a heterogeneous cation exchange system, a similar decrease in the distribution coefficient should be expected if the complex cations are only present in the solution phase and if they are not fixed at the resin. In a number of investigations it was assumed that the coronates and cryptates are mainly present in the solution and not in the resin phase, or that the complex ion causes no isotopic separation in equilibrium with the exchanger [119,131].

Schiefer has shown that this assumption cannot be accepted for cryptates of [2.2.1] and [2.2.2] [80]. This has very extensive consequences for the interpretation of isotopic separations at cation exchangers if polyethers are used. In batch experiments with Dowex 50-X12 (Ca^{2+}-form) and a $CaCl_2$ solution, it was measured

Table 15. Absorption of the cryptand [2.2.1] and [2.2.2] with a strongly acidic cation resin (Ca^{2+}-form, total Ca^{2+}-capacity = 6,25 mmol) [80]

Cryptand	Amount in solution [mmol]			
	initial:		equilibrium:	
	cryptand	Ca^{2+}	cryptand	Ca^2
[2.2.1]	1.25	1.23	0	1.21
[2.2.2]	1.25	1.23	0.007	1.23

that after adding a cryptand to the system, this ligand was completely absorbed by the resin (Table 15). On the other hand, if one first establishes a calcium cryptate solution and adds this solution to a cation exchanger resin in the Li$^+$-form, the total amount of the calcium cryptate is absorbed by the exchanger. An indication for the assumption that the cryptate ion and not the free ligand is absorbed at the resin can be concluded from the fact that a surplus amount of cryptand compared with the exchanger capacity is found in the solution. The protonated cryptands can also be fixed at a cation exchanger. For example, the protonated cryptand ion $\{[2.2.2]2H^{2+}\}$ can only be eluted from a column filled with Dowex 50–X8 if a a 6 M HCl solution is applied, whereas the cryptate $\{[2.2.2]Na^+\}$ already leaves the column with a 1 M HCl solution [136]. Therefore, the elution otherwise used with acids cannot be applied in the case of cryptate systems, in contrast to coronate systems.

In Table 16 are listed the $^{44}Ca/^{40}Ca$ and $^{48}Ca/^{40}Ca$ isotope ratios which were measured in the solution after equilibrium was established between Dowex 50-X12 in the Ca^{2+}-form and a CaCl$_2$ solution using an excess of cryptand compared with the total calcium amount in the system [80]. In correspondence to the above mentioned results, the exchange reaction between $^{40}Ca^{2+}$ and $^{48}Ca^{2+}$ is as follows for the [2.2.1] cryptand:

$$(RSO_3)_2 \{[2.2.1]^{48}Ca\} + \{[2.2.1]^{40}Ca^{2+}\} \rightleftharpoons (RSO_3)_2 \{[2.2.1]^{40}Ca\} + $$
$$+ \{[2.2.1]^{48}Ca^{2+}\} \tag{42}$$

The increase of the isotope ratio $^{44}Ca/^{40}Ca$ and $^{48}Ca/^{40}Ca$ in the solution compared with the intial isotope ratio using [2.2.1] means that the $^{40}Ca^{2+}$-cryptate complex is enriched in the exchanger resin. As one can see from the results in Table 16, the effect is inverse for [2.2.2]. Of course, one has to take into account that the analyzed isotopic separations are in the range of the measuring errors.

Table 16. Calcium isotopic separation in the system Dowex 50/CaCl$_2$ solution with an excess of cryptand [80]

Cryptand	Isotopic separation in solution		K_L^{40}/K_L^{44}	K_L^{40}/K_L^{48}
	$^{44}Ca/^{40}Ca$	$^{48}Ca/^{40}Ca$		
[2.2.1]	0.02157 ± 0.00007	0.001951 ± 0.000011	1.003	1.009
[2.2.2]	0.02144 ± 0.00003	0.001924 ± 0.000004	0.997	0.995
Initial isotope ratio	0.02150 ± 0.00005	0.001933 ± 0.000009		

In these experiments the calcium distribution ratio between the Dowex 50 resin and the solution was established to be about ten by using the corresponding amounts of exchanger and CaCl$_2$ salt. With such a distribution ratio, the isotopic fractionation effect is almost completely shifted into the solution, whereas the initial isotope ratio is present in the resin phase (see Chap. 2.4). Under this condition and under the assumption that Eq. (42) is valid, the distribution ratios K_L^{40}/K_L^{44} and K_L^{40}/K_L^{48} can be calculated (Eq. (34) in Chap. 4.3.1.2) by the isotope ratios given in Table 16. As experience has shown up to now, no isotopic separation of calcium can be expected without complexing agents in the system Dowex 50/aqueous $CaCl_2$ solution by single-stage batch experiments [49]. Because cation exchangers are able to absorb metal cryptates, and because protonated cryptands are formed during the elution with acids, it is not very suitable to use cryptand solutions as eluants for chromatographic separations of isotopes in systems with cation exchanger resins. Therefore, it is obvious that resins are preferred for chromatographic investigations where polyethers are fixed as anchor groups at the resin.

4.3.2 Ion Exchanger Resins with Polyether Anchor Groups

4.3.2.1 General Aspects

The matrices which have been used to fix neutral polyethers as anchor groups are presented in Table 3 (Chap. 2.5.3). A selection of structures of those exchanger resins is given in Fig. 6. Up to now, investigations on isotopic separations have been carried out for calcium using a cellulose exchanger [137] and a condensation resin [79, 138], for lithium [140, 141] and calcium [79, 139, 142] using a polymerization resin. The essential advantage of the application of resins with neutral anchor groups — compared with those resins where ionic groups are fixed — is that one only needs pure solvents or mixtures of solvents instead of electrolyte solutions for the chromatographic elution of the metal ions [82]. For example, the isotopic exchange reaction between $^{40}Ca^{2+}$ and $^{48}Ca^{2+}$ using a resin with $[2_B.2.2]$ anchor groups is as follows (R = organic matrix, Ca_{sol}^{2+} = solvated calcium ion in the free solution):

$$R\text{-}\{[2_B.2.2]^{48}Ca^{2+}\} + {}^{40}Ca_{sol}^{2+} \rightleftharpoons R\text{-}\{[2_B.2.2]^{40}Ca^{2+}\} + {}^{48}Ca_{sol}^{2+} . \qquad (43)$$

4.3.2.2 Cellulose Matrix with Diaminodibenzo[18]Crown-6 for Calcium Isotopic Separation

Diaminodibenzo[18]crown-6 was coupled with diazonium cellulose using Lieser and coworkers' specification [27]. The structure of the reaction product is shown in Fig. 14. The capacity of the exchanger was determined to be 0.21 mmol/g in the nitrogen content analyzed.

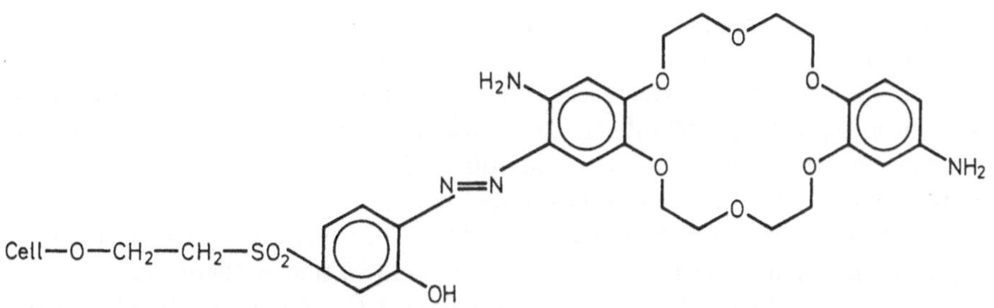

Fig. 14. Structure of a cellulose resin with diaminodibenzo[18]crown-6 anchor groups

115

A plexiglass column, which was used to prevent calcium contaminations, was filled up to 80 cm with 15 g of the exchanger (column diameter = 1.2 cm). The distribution coefficient of calcium between this exchanger and a mixture of 90% CH_3OH/10% H_2O (v/v) was analyzed to be 4 ml/g which was sufficiently high for an elution experiment. The half-life of the calcium ion exchange reaction with this resin is about 20 s [27] which means that not too high elution speeds must be applied (about 20 ml/h). Different fractions of the calcium elution were analyzed for their $^{44}Ca/^{40}Ca$ and $^{48}Ca/^{40}Ca$ isotope ratios by thermal ionization mass spectrometry (see Chap. 3.2) [137].

An exact evaluation of all fractions was not possible because of the small fractionation effect compared with the standard deviation of the measurements. However, a trend was found in the direction of a depletion of the heavy isotopes in the first fractions and of an enrichment of these isotopes in the last fractions. This means that the heavy calcium isotopes were enriched in the cellulose exchanger with the polyether anchor group. If one uses Glueckauf's method (Chap. 2.5.3) to evaluate the $^{48}Ca/^{40}Ca$ ratios for the isotopic fractionation effect of one equilibrium stage, $\varepsilon = -4 \times 10^{-3}$ is obtained. This ε-value can only be used for orientation as an order of magnitude. The absolute value of ε is comparable with those isotopic fractionation effects obtained in extraction systems (Chap. 4.2.2–4.2.4) as well as in the chromatographic elution of sodium isotopes by a polyether solution from a column filled with a strongly acidic cation exchanger (Chap. 4.3.1.3).

In addition, it is of interest to point out that the elution of calcium by acetic acid from a column filled with the exchanger resin cellulose hyphan [143] causes no isotopic separation out of the limits of error [86]. The constitution of the cellulose hyphan exchanger is the same as given for the resin in Fig. 14, except that in the hyphan exchanger 2-naphthol is fixed at the diazonium group instead of diaminodibenzo-[18]crown-6.

4.3.2.3 Condensation Resin with Dibenzo[18]Crown-6 for Calcium Isotopic Separation

In accordance with a description by Blasius and coworkers [25], a condensation resin was synthesized from phenol and formaldehyde which contains dibenzo[18]crown-6 anchor groups [144]. The structure of this resin can be seen in Fig. 6 in connexion with Fig. 11. The capacity of this resin was determined to be 0.85 mmol/g for the uptake of calcium ions from a methanolic solution of $CaCl_2$. 0.052 mmol CaI_2 were eluted from a column with a mixture of 30% methanol/70% tetrahydrofuran (v/v) (2.4 g resin, particle size = 45–90 μm). The elution curve is given in Fig. 15 [79].

The maximum calcium concentration is reached after an effluent volume of about 10 ml. Then, the concentration decreases to a level of about 3.5 μg Ca^{2+} per ml which continues over more than 100 ml. At the marked fraction No. 25 the amount of eluted calcium is not more than 36% of the total calcium which was deposited at the top of the column. This unusual elution curve can be attributed to kinetic effects. The mass spectrometric $^{44}Ca/^{40}Ca$ and $^{48}Ca/^{40}Ca$ ratio measurements of fractions No. 3 and 5, and of the combined fractions 18–25 are listed in Table 17. The last two columns contain the local separation coefficient R (see Chap. 2.5.3). The results show a significant enrichment of the heavy calcium isotopes in the first fractions

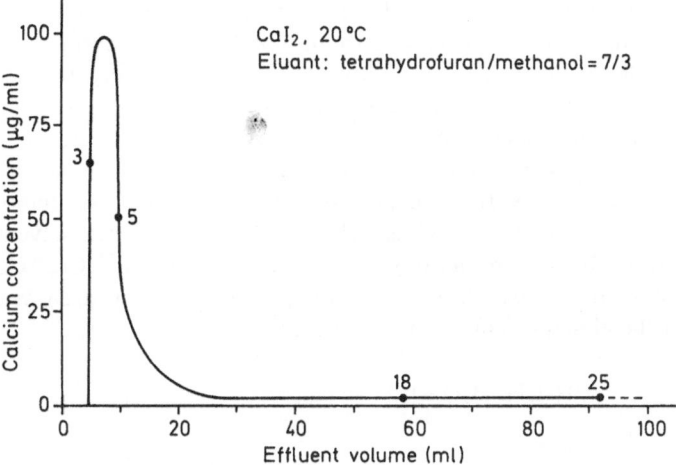

Fig. 15. Calcium elution curve using a condensation resin with dibenzo[18]crown-6 anchor groups

compared with the initial isotope ratio. This means that the light isotopes have been concentrated in the complex compound at the resin. Because of the strong deviation of the elution curve from a Gaussian distribution, the calculation of ε with Glueckauf's plot (see Chap. 2.5.3) is not possible in this case. Although the tailing of the elution curve is not very suitable for a practical application of this resin, it is of principal interest that this type of resin is able to separate isotopes.

Table 17. Calcium isotopic separation using a condensation resin with dibenzo[18]crown-6 anchor groups [79]

Fraction No.	Eluted calcium/ total calcium [%]	Isotope ratio		R	
		$^{44}Ca/^{40}Ca$	$^{48}Ca/^{40}Ca$	$^{44}Ca/^{40}Ca$	$^{48}Ca/^{40}Ca$
3	1.6	0.02186 ± 0.00013	0.001989 ± 0.000019	1.016	1.027
5	21.7	0.02180 ± 0.00009	0.001980 ± 0.000022	1.013	1.022
18–25	35.6	0.02168 ± 0.00003	0.001956 ± 0.000004	1.008	1.010
Initial material		0.02151	0.001937		

4.3.2.4 Condensation Resin with [2$_B$.2.2] Cryptand for Calcium Isotopic Separation

In accordance with the literature [83], a condensation resin with [2$_B$.2.2] anchor groups was synthesized from phenol and formaldehyde [138] (see Fig. 6). The capacity of the resin corresponding to the number of anchor groups was 0.567 ± 0.003 mmol/g. This is the average of determinations with the calcium uptake in methanol, the proton uptake in water, and the elementary analysis of nitrogen. 4 g of the resin were applied in a thermostatically controlled column (diameter = 0.5 cm, height = 12 cm). A solution containing 10 mg CaCl$_2$ was deposited at the top of the column and then calcium was eluted with methanol and a mixture of 50% methanol/50% water (v/v), respectively, using an elution speed of 0.15 ml/min. The slow elution speed was chosen because the establishment of the equilibrium requires time, e.g. for calcium ions in methanol 60–80 min for the temperature range from $+20$ °C to -20 °C. The distribution coefficient for calcium ions between the resin and an aqueous solution is too low for practical work so that only a methanol/water mixture or pure methanol could be applied as eluants. Experiments were carried out with methanol at temperatures between -20 °C and $+20$ °C as well as with the mentioned solvent mixture at a temperature of 10 °C. The mass spectrometric determined $^{44}Ca/^{40}Ca$ and $^{48}Ca/^{40}Ca$ isotope ratios for the experiments at a temperature of 10 °C are given in Fig. 16 dependent on the eluted calcium amount $\Delta m/m$.

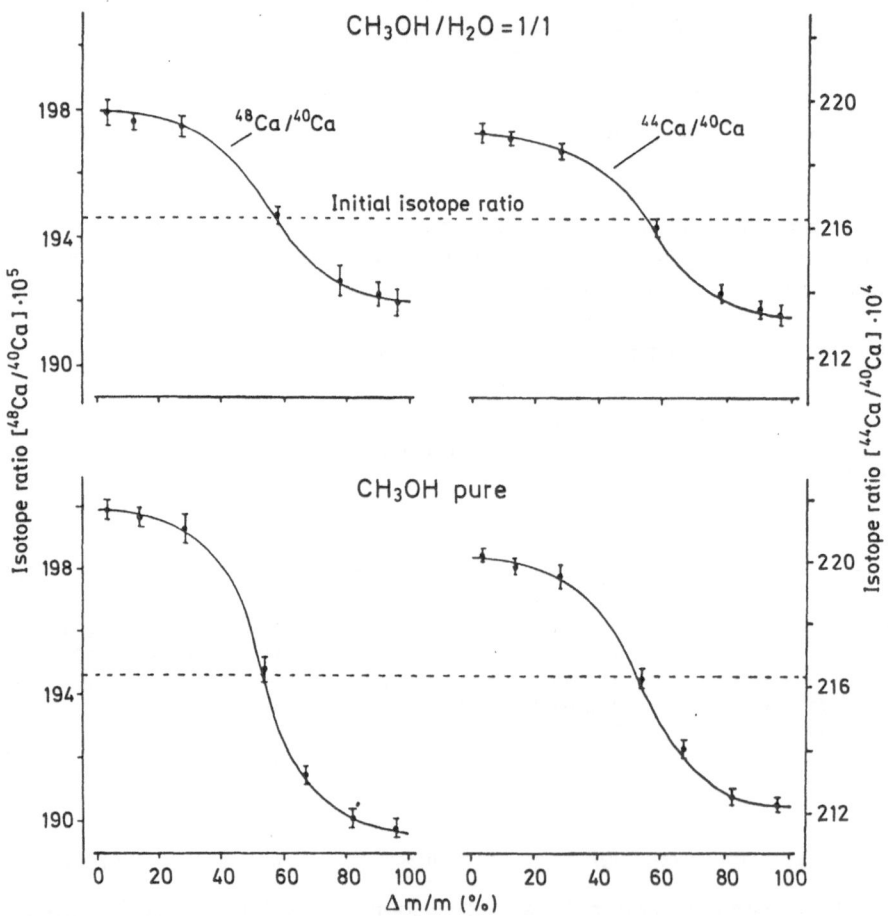

Fig. 16. Calcium isotopic separation dependent on the eluted calcium amount using a condensation resin with [2$_B$.2.2] anchor groups [138]

A high isotopic separation of calcium was found with the CH_3OH/H_2O eluant as well as with CH_3OH. The effect is a little bit higher in the case of pure methanol which was the reason for analyzing the temperature dependence with this eluant. In both cases the enrichment of the heavy isotopes took place in the first fractions, that is in the solution phase of this system. This direction of the enrichment agrees with that obtained for calcium isotopes at strongly acidic cation exchangers [49, 51, 52]. In contrast, the shape of the curves in Fig. 16 is different from those obtained with a strongly acidic resin (see Fig. 3, Chap. 2.3). Whereas the curve using the strongly acidic exchanger has a ⌐ shape, that of the cryptand resin is ⌐. It follows that a significantly higher amount of enriched material can be obtained with the cryptand resin than with the cation exchanger with sulfonic acid groups if equal amounts of calcium are employed.

On the one hand, the enrichment of the heavy calcium isotopes in the solution corresponding to Eq. (43) can be attributed to a stronger solvation of $^{48}Ca^{2+}$ compared with $^{40}Ca^{2+}$ (see Fig. 2, Chap. 2.3). During the incorporation of Ca^{2+} into the cage of the cryptand, the solvent molecules must be stripped off the calcium ion. As Kakihana and coworkers [76, 145] have suggested, one can only expect a particularly high isotopic separation if the chemical states of the metal in the two phases are as different as possible. On the other hand, the preferable incorporation of $^{40}Ca^{2+}$ into the cryptand rather than the heavier isotopes is possibly due to the cavity effect characterized in Fig. 4. The concentration of the lighter isotopes in the phase containing polyether also corresponds to the result for calcium and lithium in extraction systems (Chap. 4.2.2 and Chap. 4.2.3).

If one plots the calcium ratios of all experiments with pure methanol in the temperature range of $-20\ °C$ to $+20\ °C$ dependent on the $\Delta m/m$-values, curves similar to those shown in Fig. 16 are obtained. The exact temperature dependence of the isotope ratios can be seen from the results listed in Table 18. Because the isotope ratios could not be measured exactly at the same amount of eluted calcium, rounded off mean values are given in the first column of Table 18 for experiments where the same temperature was applied.

Table 18. Calcium isotopic separation dependent on the temperature using a condensation resin with [2$_B$.2.2] anchor groups [138]

$\Delta m/m$	$(^{44}Ca/^{40}Ca) \times 10^5$					$(^{48}Ca/^{40}Ca) \times 10^6$				
[%]	T [°C]= -20	. -10	0	$+10$	$+20$	-20	-10	0	$+10$	$+20$
4	2190	2200	2205	2202	2193	1988	1997	2004	1999	1991
15	2187	2197	2199	2198	2190	1986	1993	1998	1996	1987
34	2177	2179	2181	2195	2178	1972	1973	1979	1993	1972
54	2158	2157	2162	2162	2164	1946	1944	1946	1948	1950
65	2149	2147	2140	2139	2148	1930	1928	1919	1914	1925
86	2131	2129	2122	2125	2130	1911	1905	1901	1901	1906
96	2129	2125	2119	2123	2128	1908	1901	1895	1898	1905

The isotopic separation is stronger the higher the isotope ratio is for low $\Delta m/m$-values, and the lower the isotope ratio is for $\Delta m/m$-values $>50\%$. For all temperatures, a decrease of the isotope ratios is found with increasing amounts of eluted calcium. The maximum of the isotopic separation is achieved at a temperature of

0 °C whereas the total isotopic effect decreases for higher as well as for lower temperatures. The decrease of the isotopic separation with increasing temperature can be explained by thermodynamics (see Chap. 2.2). This explanation is not possible for the inverse tendency in the temperature range from 0 °C to —20 °C. For this tendency, kinetic effects have to be taken into account. This assumption is based on the result that the establishment of equilibrium between calcium ions in methanol and the [2$_B$.2.2] condensation resin requires about 60 min at a temperature of +20 °C and about 80 min at —20 °C [138].

The great advantage of the described condensation resin with [2$_B$.2.2] anchor groups lies in the very favourable isotopic separation dependent on the eluted amount of substance (see Fig. 16) and by the fact that a pure solvent can be used as eluant which simplifies the isolation of the isotopic-enriched fractions. Now, additional experiments with larger columns and the application of the cascade principle (Chap. 2.5.2) will have to prove whether or not an enrichment of the heavy isotopes on a technical scale is possible with this exchange system.

4.3.2.5 Polymerization Resin with [2$_B$.2.1] and [2$_B$.2.2] Cryptand for Lithium and Calcium Isotopic Separation

For a couple of years, polymer resins with [2$_B$.2.1] and [2$_B$.2.2] anchor groups have been commercially available (trade name Merck-Schuchardt: "Kryptofix 221B polymer" and "Kryptofix 222B polymer"; for their structures compare Fig. 6). In the meanwhile, a number of lithium [140a, 141] and calcium [79, 139, 142] isotopic separations have been investigated with these resins.

Fujine and coworkers [140] have carried out chromatographic displacement experiments with lithium; Nishizawa and coworkers [141] have investigated the same resin with [2$_B$.2.1] cryptand anchor groups in batch and chromatographic breakthrough experiments for lithium isotopic separations. The displacement experiments were carried out in a column of 2 m length (diameter = 3 mm) using methanol as an eluant (40 °C). The column was first loaded with CsCl and, afterwards, lithium acetate was deposited on the top of the column. Then, lithium was displaced by a 0.12 M methanolic sodium acetate solution (band speed = 1.48 m/h) which corresponds to the following sequence of the stability constants of the alkalis with

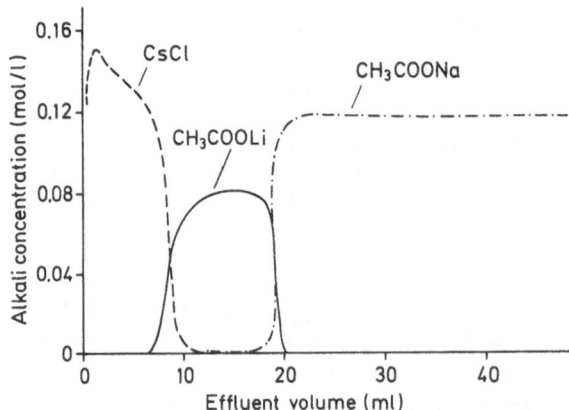

Fig. 17. Chromatogram obtained by displacement chromatography using a resin with [2$_B$.2.1] anchor groups [140]

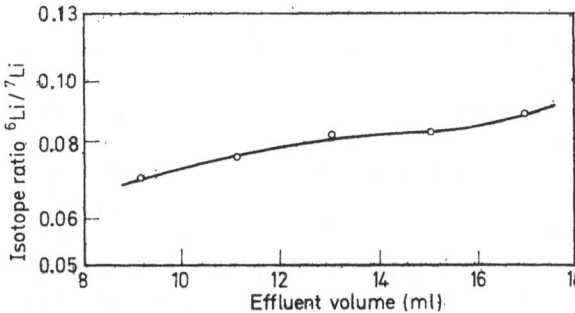

Fig. 18. $^6Li/^7Li$ isotope ratio dependent on the effluent volume by displacement chromatography [140]

[2.2.1]: $Cs^+ < Li^+ < Na^+$ [5]. Some of the lithium fractions were analyzed for their $^6Li/^7Li$ isotope ratio by atomic absorption spectrometry (see Chap. 3.1).

The alkali elution curves of the displacement chromatography are shown in Fig. 17, the ratio $^6Li/^7Li$ dependent on the effluent volume is given in Fig. 18. As one can see from Fig. 18, an increase of the $^6Li/^7Li$ ratio from 0.07 to 0.09 is found within the lithium elution band which corresponds to a column length of 91 cm. The relative enrichment of the heavy lithium isotope ^7Li in the first fractions — that is in the methanolic phase — agrees with isotopic separations of calcium using a condensation resin with dibenzo[18]crown-6 and [2_B.2.2], respectively (Chap. 4.3.2.3 and Chap. 4.3.2.4). Fujine and coworkers have also carried out one breakthrough experiment with methanolic solutions of cesium chloride and lithium acetate [140]. The evaluation of the front analysis with Spedding and coworkers' method [146] resulted in an isotopic separation factor of $\alpha = 1.014$.

The results which were obtained by Nishizawa and coworkers [141] in batch experiments for the system cryptand resin [2_B.2.1]/methanolic solution dependent on the anion of the lithium salt and on the temperature are summarized in Table 19 (mass spectrometric measurement, see Chap. 3.2). Additionally, recent results of the same research group for the extraction system aqueous solution/organic solution with benzo[15]crown-5 (see Fig. 11) are also listed in Table 19 for comparison [147].

Table 19. Single-stage isotopic separation factors for $^6Li/^7Li$ with a [2_B.2.1] resin [141] and an extraction system with benzo[15]crown-5 [147]

Polyether	Temp. [°C]	α-value		
		LiCl	LiBr	LiI
[2_B.2.1] polymer	0	1.047 ± 0.003[a]	1.045 ± 0.003	1.045 ± 0.002
	20	1.039 ± 0.002	1.041 ± 0.002	1.041 ± 0.002
	40	1.034 ± 0.002	1.036 ± 0.003	1.036 ± 0.002
Benzo[15]crown-5	1	—	—	1.042 ± 0.002
(extraction system)	25	1.002 ± 0.002	1.014 ± 0.002	1.026 ± 0.002
	40			1.012 ± 0.003

[a] Confidence limit $= \pm 2 s$

The isotopic separation factors in the system with the $[2_B.2.1]$ resin are the highest values which have been obtained up to now with the use of ion exchangers. Only the α-values for the extraction systems are of comparable magnitude (see Chap. 4.2). In all experiments, the light isotope ^6Li is enriched in the resin phase, which agrees in direction with the above-described calcium and lithium investigations at other types of ion exchangers or with other eluants. Within the limits of error, no dependence of the isotopic separation on the used halide ion is found for the $[2_B.2.1]$ resin, whereas a very strong dependence can be seen from the extraction experiments with benzo[15]crown-5. As predicted by theory (Chap. 2.2), in all cases a decrease in the isotopic separation is measured with increasing temperature. From the temperature dependence of the α-values ($K_c \equiv \alpha$, see Eq. (12)), one can calculate the enthalpies for the lithium isotopic exchange reaction between the $[2_B.2.1]$ resin and a methanolic solution to be $\Delta H(LiCl) = -211$ J/mol, $\Delta H(LiBr) = -153$ J/mol and $\Delta H(LiI) = -147$ J/mol. The enthalpy for the extraction system is $\Delta H(LiI) = -469$ J/mol. The absolute values of these enthalpies are much higher than all other ΔH-values which have been determined for isotopic ion exchange reactions up to the present day (see Chap. 4.3.1.3).

From a breakthrough experiment in the system $[2_B.2.1]$ resin/0.116 M methanolic LiCl solution (column: height = 22 cm, diameter = 0.9 cm), Nishizawa and co-workers [141] were able to calculate the following isotopic fractionation factors: $\alpha = 1.047$ (0 °C), $\alpha = 1.037$ (30 °C). From the results of the batch and breakthrough experiments it follows that there is a promising possibility for an enrichment of lithium isotopes on a technical scale with the use of the $[2_B.2.1]$ cryptand resin exchanger.

Calcium isotopic investigations were carried out with the resin "Kryptofix 222B

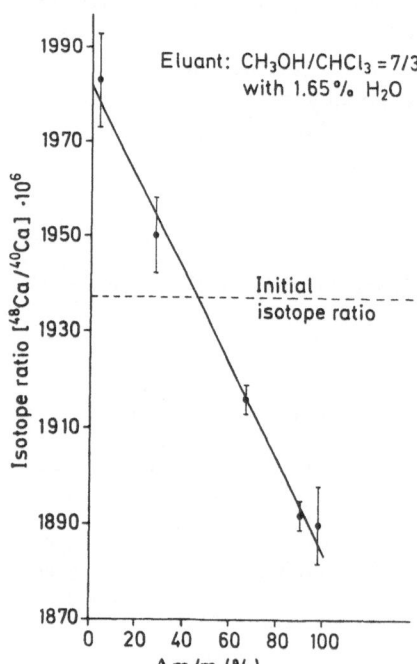

Fig. 19. Calcium isotopic separation by elution with $CH_3OH/CHCl_3/H_2O$ using a copolymerization resin with $[2_B.2.2]$ anchor groups (0 °C) [79]

polymer" (capacity = 0,315 mmol Ca^{2+}/g) by Heumann and Schiefer, using a mixture of methanol and chloroform (70%/30%, v/v) as eluant which contained 1.65% water [79,142]. A thermostatically controlled column (diameter = 2.1 cm, height = 60 cm) and 30 mg $CaCl_2$ were used for the chromatographic experiments (elution speed = 0.5 ml/min) at temperatures of −21, 0 and +20 °C. The isotope ratios $^{44}Ca/^{40}Ca$ and $^{48}Ca/^{40}Ca$ were analyzed mass spectrometrically (see Chap. 3.2) dependent on the eluted calcium amount $\Delta m/m$. The result obtained for the $^{48}Ca/^{40}Ca$ ratio at a temperature of 0 °C is given in Fig. 19.

The measured values for the $^{44}Ca/^{40}Ca$ ratio follow the same trend as for $^{48}Ca/^{40}Ca$ but the isotopic separation is smaller due to the lower mass difference of the isotopes. Whereas an enrichment of the heavy isotopes is found in the first half of the fractions, a depletion of these isotopes is found in the second half compared with the initial isotopic composition. This means that the light isotopes favour the resin phase with the cryptand anchor group (see Eq. (43)). An analogous enrichment was found for lithium isotopes at the same type of resin with $[2_B.2.1]$ anchor groups, as well as for calcium using a condensation resin with different polyether anchor groups (see Chap. 4.3.2.3 and Chap. 4.3.2.4). It is of importance for the application of this isotopic separation system that the isotope ratio decreases in an approximately linear manner with an increasing amount of eluted substance. This is less favourable than the shape of the elution curve with condensation resins (see Chap. 4.3.2.4) but essentially more advantageous compared with the strongly acidic cation exchanger (see Fig. 3, Chap. 2.3). If one carries out a second elution, using the first half of the eluted calcium amount of the first elution, one obtains another straight line with the same slope as in Fig. 19 but with higher isotope ratios. In an experiment where 210 mg $CaCl_2$ of natural isotopic composition were used as starting material, it was possible to isolate 30 mg $CaCl_2$ after a three-stage experiment of this type which showed relative enrichments of 1.9% for $^{44}Ca/^{40}Ca$ and of 3.3% for $^{48}Ca/^{40}Ca$, respectively [79].

By evaluating the results described above with Glueckauf's method (see Chap. 2.5.3) — neglecting possible kinetic effects during these experiments — one obtains the ε-values listed in Table 20. On the average, the $\varepsilon(^{48}Ca/^{40}Ca)$-value is by a factor 1.8 higher than the $\varepsilon(^{44}Ca/^{40}Ca)$-value which corresponds to the relative mass differences $[(8/48):(4/44) = 1.83]$. The magnitude of the ε-values is comparable to those of the extraction systems for calcium (see Chap. 4.2.2) but smaller than those for lithium in the system "221B polymer"/methanol (see Table 19).

Table 20. ε-values of calcium as a function of temperature using a copolymerization resin with $[2_B.2.2]$ anchor groups ($CH_3OH/CHCl_3/H_2O$) [142]

Temperature [°C]	$\varepsilon(^{44}Ca/^{40}Ca) \times 10^3$	$\varepsilon(^{48}Ca/^{40}Ca) \times 10^3$
−21	2.6	4.5
0	4.2	7.7
+20	5.7	10.4

For a practical application of the enrichment of calcium isotopes, the cryptand resin system with the eluant mixture $CH_3OH/CHCl_3/H_2O$ is not ideal because there is a strong tailing during the elution of calcium (up to an effluent volume of about 2000 ml under the above described conditions). Therefore, Heumann and Schrödl [139] used water/methanol mixtures (95%/5% and 90%/10%, v/v) and pure water as eluants for the calcium isotopic separation with the resin "Kryptofix 222B polymer" (capacity of this fraction of resin = 0.389 mmol Ca^{2+}/g). If water is only used as eluant, relatively sharp elution bands and small effluent volumes are obtained which continuously increase with the addition of methanol. At temperatures in the range of 0–30 °C the calcium isotopic separation was investigated by eluting 56 mg $CaCl_2$ from a column (diameter = 1.5 cm, elution speed = 1 ml/min) which was filled with the resin up to 84 cm (84 g resin). The mass spectrometrically measured isotope ratios dependent on $\Delta m/m$ are shown in Fig. 20 using water as an eluant and a temperature of 7 °C

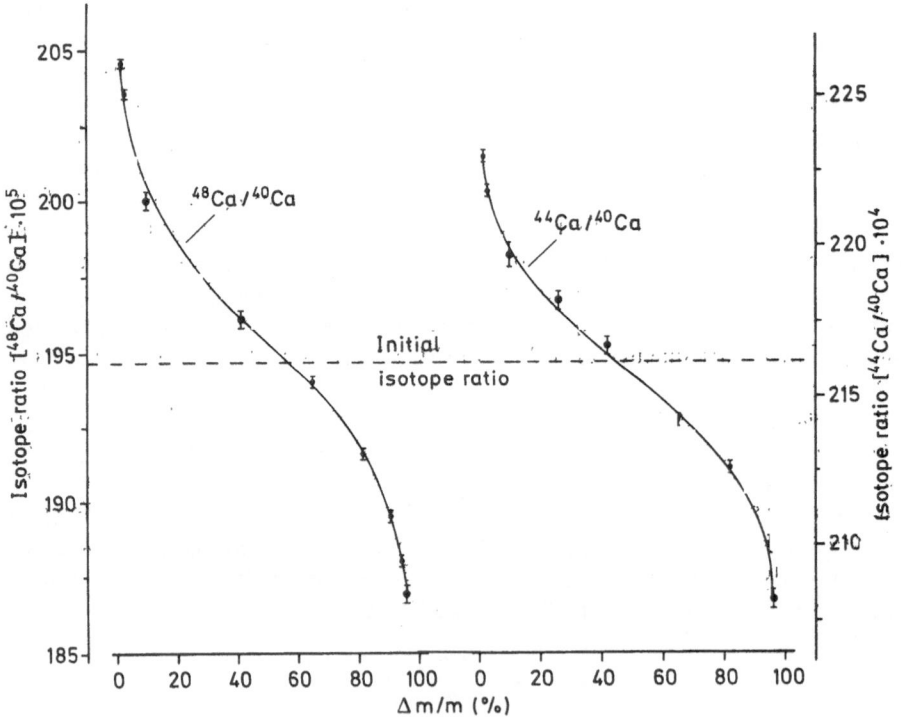

Fig. 20. Calcium isotopic separation by elution with H_2O using a copolymerization resin with [2_B.2.2] anchor groups (7 °C) [139]

In analogy to the elution with the mixture $CH_3OH/CHCl_3/H_2O$, one again finds an enrichment of the heavy calcium isotopes in the first half of the fractions whereas the light isotopes are concentrated in the resin with the complex-forming ligand. Although the plot of the isotope ratio versus $\Delta m/m$ is not linear as with the $CH_3OH/CHCl_3/H_2O$ eluant, the total isotopic separation nearly reaches the same magnitude with significantly lower effluent volumes of about 120 ml. In the investigated temperature range, the isotopic separation is greatest for 7 °C, whereas a decrease in the isotopic fractionation effect up to higher as well as down to lower temperatures is found. Such a temperature dependence cannot be explained thermodynamically. Therefore, this effect should be due to kinetic influences similar to those discussed for calcium isotopic separations at the condensation resin with $[2_B.2.2]$ anchor groups (see Chap. 4.3.2.4). If one calculates the isotopic fractionation of one equilibrium stage for 7 °C with Glueckauf's method (Chap. 2.5.3) — neglecting the possible kinetic effect — one obtains $\varepsilon(^{44}Ca/^{40}Ca) = 3.04 \times 10^{-3}$ and $\varepsilon(^{48}Ca/^{40}Ca) = 5.34 \times 10^{-3}$. These isotopic fractionation effects are comparable to those of the same system but elute with $CH_3OH/CHCl_3/H_2O$. The experiments with a 5% and 10% methanol addition in the eluant show no significantly measurable difference in the calcium isotopic separation compared with the elution where pure water is used. Because of the advantage that water results in smaller effluent volumes and is less expensive than methanol/water mixtures, water should be applied for experiments to enrich calcium isotopes on a technical scale.

The first investigations on a semitechnical scale were carried out in the system "Kryptofix 222B polymer"/water for the calcium isotopic enrichment by Heumann and Schrödl [139]. First of all, the isotopic separation was investigated in preliminary tests where the collected first half of the calcium elution was again deposited at the top of the same column. Afterwards, a second elution cycle (second stage) was carried out, etc. For such a four-stage experiment, the $^{44}Ca/^{40}Ca$ and $^{48}Ca/^{40}Ca$ ratios have been measured. The same shape of curves as given in Fig. 20 was obtained for all stages, but the curves were shifted parallel up to higher isotope ratios. Additionally, such an experiment was combined with a cascade arrangement as described in Chap. 2.5.2. The schematic diagram of a ten-stage experiment is given in Fig. 21 [139]. The same column as for the previous investigation has been used at a temperature of 7 °C. The first separation stage consisted of 16 parallel column units. Because of the high costs for the resin, the same column was used for all elutions. Therefore, in the first separation stage 16 successive elutions were carried out, etc. The numerical values given in Fig. 21 are the mg-amounts of $CaCl_2$ which were applied in the beginning of the corresponding separation unit. In the first separation stage, calcium was of natural abundance. 45% to 48% of the calcium amount of the first half of one elution unit was combined with the same portion of the next unit. These collected fractions were concentrated by evaporation of the solvent. Afterwards, they were applied in one of the chromatographic units of the second separation stage etc. Using 16 initial separation units, the $CaCl_2$ amount per unit is approximately the same up to the 5th separation stage if the number of units is halved from stage to stage. Then, the calcium amount is halved from stage to stage so that it was possible to isolate at least 2.6 mg enriched $CaCl_2$ after the 10th separation stage. The eluted calcium chloride was determined by a conductivity detector which has the advantage of being an on-line detection system. The condition for the application of this type of detection system is a non-conducting eluant. The multi-stage experiment could be automatically controlled by the conductivity detector in connexion with the limiting value detector which regulated the substance currents by the different magnetic valves MV1–MV4 (Fig. 22) [139].

Figure 23 shows the measured calcium isotope ratios dependent on the number of

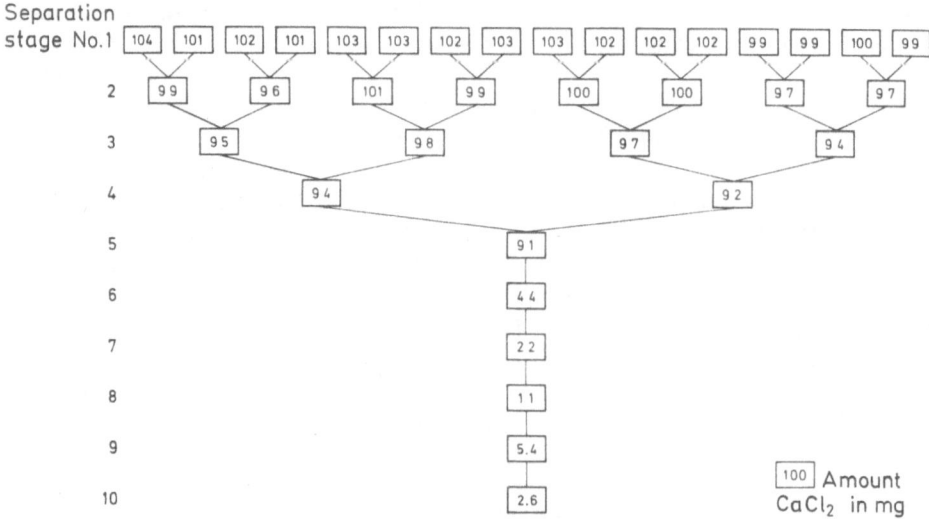

Fig. 21. Schematic diagram of a multi-stage calcium isotopic separation experiment using a co-polymerization resin with [2$_B$.2.2] anchor groups [139]

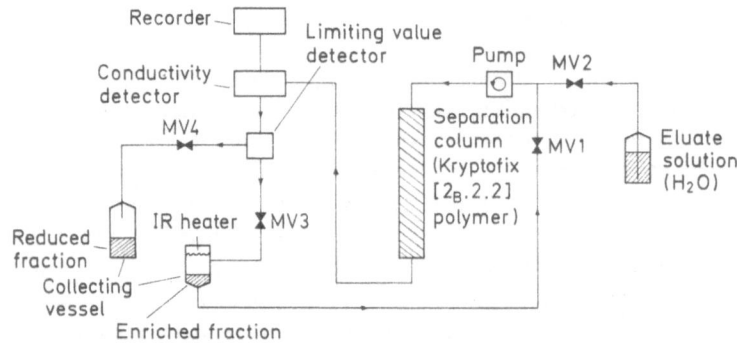

Fig. 22. Schematic diagram of the automatically controlled multi-stage isotopic separation system [139]

separation stages. It follows that the relative enrichment after ten stages is 17% for ^{44}Ca/^{40}Ca and 22.9% for ^{48}Ca/^{40}Ca compared with the initial isotope ratio (see also Table 21).

To demonstrate the capability of this system for an enrichment of heavy calcium isotopes on a scale which is of technical interest, a column (diameter = 1.5 cm) filled up to a height of 195 cm with the resin "Kryptofix 222B polymer" (189 g resin) was used for an additional multi-stage experiment (eluant = H$_2$O, elution speed = 1 ml/min, 7 °C). One elution cycle with this column has shown that the total isotopic separation is higher compared with the smaller column (Fig. 20). The increase of the total isotopic fractionation is directly proportional to the percentage increase of the resin amount. Furthermore, it was found that the isotopic separation under these conditions is independent of the calcium amount up to 450 mg CaCl$_2$

126

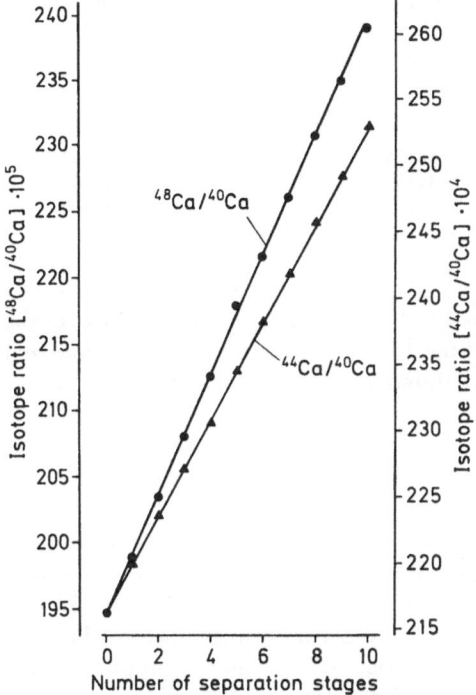

Fig. 23. $^{44}Ca/^{40}Ca$ and $^{48}Ca/^{40}Ca$ isotope ratios dependent on the number of separation stages [139]

which is used for the elution. On the other hand, the isotopic separation decreases if a greater amount of substance is applied. Taking this result into account, an experiment with 13 stages was carried out analogous to the investigation with 10 stages [139]. The results of this experiment together with those of the ten-stage cascade are listed in Table 21. One obtains after the 10th stage in the 13-stage experiment a relative enrichment of the heavy calcium isotopes ^{44}Ca and ^{48}Ca which is more than a factor of two higher compared with the ten-stage experiment using a smaller column. Additionally, the amount of enriched calcium is also higher. After the 13th stage, a relative enrichment of 48.5 % for ^{44}Ca and of 60.7 % for ^{48}Ca is reached. This is the most eminent isotopic enrichment of calcium in a chemical system which has been

Table 21. Calcium isotopic separation of two multi-stage experiments [139]

Experiment	Resin [g]	CaCl$_2$ [mg]		Isotope ratio		Rel. enrichment [%]	
		initial	enriched	$^{44}Ca/^{40}Ca$	$^{48}Ca/^{40}Ca$	$^{44}Ca/^{40}Ca$	$^{48}Ca/^{40}Ca$
Initial isotope ratio[a]				0.02163	0.001946		
10 separation stages	84	1625	2.6	0.02530	0.002391	17.0	22.9
13 separation stages:							
after 10 stages	189	7200	8.1	0.02971	0.002857	37.4	46.8
after 13 stages	189	7200	0.8	0.03213	0.003128	48.5	60.7

[a] Measured values, not calculated with listed isotope abundances [148]

Table 22. Calcium isotope abundances of two multi-stage isotopic separation experiments [139]

Experiment	Isotope abundance [%]					
	^{40}Ca	^{42}Ca	^{43}Ca	^{44}Ca	^{46}Ca	^{48}Ca
Natural abundance [148]	96.941	0.647	0.135	2.086	0.004	0.187
10 separation stages	96.477	0.696	0.151	2.441	0.004	0.231
13 separation stages	95.701	0.750	0.168	3.075	0.006	0.300

obtained up to now. The mass spectrometrically determined isotope abundances of the enriched end products of the ten-stage and the 13-stage experiments are summarized in Table 22 in comparison with the natural abundances of calcium [148].

4.4 Conclusion

The investigation of isotopic separations in systems with cyclic polyethers has been carried out up to now for the elements lithium, calcium and sodium, in particular. Among these elements, the enrichment of ^6Li is of essential importance for the production of tritium and that of the heavy calcium isotopes for medical labeling experiments. An enrichment aspect does not exist for the monoisotopic element sodium. Investigations with the radioactive nuclides ^{22}Na and ^{24}Na are obviously of interest for fundamental investigations because these isotopes can be easily and precisely measured by their γ-activity. Except for uranium, most of the investigations on other chemical exchange systems with metal ions are also based on measurements with lithium and calcium, respectively.

From the isotopic separations with cyclic polyethers it follows that the extraction systems and the resins with polyether anchor groups are the most successful chemical systems for a practical enrichment of metal isotopes. In the extraction system $CHCl_3/H_2O$, ε-values in the range of $10^{-2} - 4 \times 10^{-2}$ have been found for $^6Li/^7Li$, of $0.4 \times 10^{-2} - 0.8 \times 10^{-2}$ and of $10^{-2} - 1.5 \times 10^{-2}$ for $^{44}Ca/^{40}Ca$ and $^{48}Ca/^{40}Ca$, respectively. Usually, the ε-values for these isotopes are slightly smaller in resin exchanger systems with polyether anchor groups than in comparable extraction systems. But the highest elementary isotopic fractionation effect was found to be $\varepsilon = 4.7 \times 10^{-2}$ for $^6Li/^7Li$ in the system resin with $[2_B.2.1]$ anchor groups/ methanolic solution [141]. The ε-values of polyether systems are usually higher by a factor of ≥ 10 compared with systems where a strongly acidic cation exchanger is used, e.g. for lithium [53] and calcium [79]

The investigated extraction systems with lithium and calcium have shown that the lighter isotopes are always enriched in the organic phase where the cyclic polyether is present. This is advantageous for the production of ^6Li but not for the production of the heavy calcium isotopes. The different distribution of the element between the two phases, which one needs for as high as possible isotopic fractionations in one phase, causes a practical isotopic separation only in the organic phase (see Chap. 2.4). In most of the chromatographic experiments with cyclic polyethers, the heavier isotopes were enriched in the first fractions of the elution band. Here, it is of minor importance whether the first or the last fractions of the elution band are enriched by the isotope, which is of particular interest. Besides this advantage of column chromatography, a high number of equilibrium stages can be obtained more easily than with an extraction system. Furthermore, the chromatographic experiments with cryptand resin exchangers have shown that for the elution of the isotopes the use of pure solvents and not the application of electrolyte solutions is sufficient. This reduces the expenditure for the isolation of the enriched material and, it is an inexpensive procedure, especially if water is used as eluant. Therefore, this type of enrichment systems has the best conditions for a practical application.

The highest relative isotopic enrichment of metals which has been obtained in a chemical system up to now, is probably the enrichment of ^{44}Ca and ^{48}Ca by the described multi-stage chromatographic experiment using a resin with [2$_B$.2.2] anchor groups [139] (see Chap. 4.3.2.5). In principle, this type of a cascade experiment has shown that an isotopic separation of metal ions is possible on a practical scale. However, it has to be pointed out that a very high enrichment of metal isotopes by chemical systems seems to be very improbable. But this does not reduce the advantage of the production of low-enriched isotopes by chemical systems because these systems are much cheaper than physical methods for isotopic separations. In many cases, low-enriched isotopes can be applied for labeling experiments or they can be used as a starting material for high-enriched isotopes produced by physical methods. For example, a calculation based on the results of the 13-stage investigation described in Chap. 4.3.2.5 has shown that one needs a column filled with about 10 kg of the resin "Kryptofix 222B polymer" to obtain 100 mg calcium chloride with an ^{48}Ca abundance of 3.5% instead of the natural abundance of 0.187%. Except for the costs for investment, the operating expenses are relatively low for such a chemical system because pure water is the only chemical which is consumed during the process.

As a summary, one can confirm that the application of systems containing cyclic polyethers for isotopic enrichments is of great interest, especially under economic aspects. Therefore, more and more investigations have been carried out in this field during the last few years, and they will gain increasing importance in the future.

5 References

1. a) Lehn, J.-M.: Design of Organic Complexing Agents. Strategies towards Properties, in: Structure and Bonding, Vol. 16, Dunitz, J. D. (ed.), Springer-Verlag, Berlin, Heidelberg, New York 1973;
   b) Dietrich, B., Lehn, J.-M., Sauvage, J.-P.: Chem. uns. Zeit 7, 120 (1973);
   c) Lehn, J.-M., Sauvage, J.-P.: J. Am. Chem. Soc. 97, 6700 (1975)
2. Kappenstein, C.: Bull. Soc. Chim. Fr., 89 (1974)
3. a) Christensen, J. J., Eatough, D. J., Izatt, R. M.: Chem. Rev. 74, 251 (1974);
   b) Izatt, R. M., Christensen, J. J. (eds.): Progress in Macrocyclic Chemistry —, Vol. 1 and 2, John Wiley and Sons, New York 1979 and 1981
4. Pedersen, C. J., Frensdorff, H. K.: Angew. Chem. 84, 16 (1972)
5. a) Vögtle, F., Weber, E. in: The chemistry of functional groups, Suppl. E: The chemistry of ethers, crown ethers, hydroxyl groups and their sulphur analogues, Part 1, Patai S. (ed.), John Wiley and Sons, London 1980;
   b) Weber, E.: Kontakte (Merck) 1984 (1), 26 and preceding contributions of this series
6. Vögtle, F., Weber, E. (eds.): Host Guest Complex Chemistry- I-III, Topics in Current Chemistry, Vol. 98 (1981), 101 (1982), 121 (1984), Springer Verlag, Berlin, Heidelberg, New York

7. Aaltonen, J.: Suomen Kemistilehti *B44*, 1 (1970)
8. Klinskii, G. D., Knyazev, D. A., Vlasova, G. I.: Russ. J. Phys. Chem. *48*, 380 (1974)
9. Heumann, K. G.: Z. Naturforsch. *27b*, 492 (172)
10. Lee, D. A.: Adv. Chem. Ser. *89*, 57 (1969)
11. Horwitz, E. P. et al.: J. Chromatogr. *125*, 203 (1976)
12. Knop, J., Reichstein, K.-H., Montz, R.: Eur. J. Nucl. Med. *2*, 35 (1977)
13. Jepson, B. E., DeWitt, R.: J. Inorg. Nucl. Chem. *38*, 1175 (1976)
14. Weber, E., Vögtle, F.: Kontakte (Merck) *1978* (2), 16
15. Frankfurter Allgemeine Zeitung, May 18, 1977
16. Demande de brevet d' invention, Institut National de la Propriété Industrielle, No. de publication 2.214.509, Paris 1974
17. Krumbiegel, P.: Isotopieeffekte, Berlin, Akademie-Verlag 1970
18. Simon, H., Palm, D.: Angew. Chem. *78*, 993 (1966)
19. Ceraso, J. M., Dye, J. L.: J. Am. Chem. Soc. *95*, 4432 (1973)
20. Cahen, Y. H., Dye, J. L., Popov, A. L.: J. Phys. Chem. *79*, 1289 (1975)
21. Weber, E., Vögtle, F.: Crown-Type Compounds — An Introductory Overview, in: Topics in Current Chemistry, Vol. 98, Vögtle, F. (ed.), Springer-Verlag, Berlin, Heidelberg, New York 1981
22. Graczyk, D. G. et al.: J. Am. Chem. Soc. *97*, 7382 (1975)
23. Cox, B. G., Schneider, H., Stroka, J.: ibid. *100*, 4746 (1978)
24. Loyola, V. M., Pizer, R., Wilkins, R. G.: ibid. *99*, 7185 (1977)
25. Blasius, E. et al.: J. Chromatogr. *96*, 89 (1974)
26. Blasius, E. et al.: Z. Anal. Chem. *284*, 337 (1977)
27. Djamali, M. G., Burba, P., Lieser, K. H.: Angew. Makromol. Chem. *92*, 145 (1980)
28. Wadell, T. G., Leyden, D. E.: J. Organ. Chem. *46*, 2406 (1981)
29. Bigeleisen, J. Mayer, M. G.: J. Chem. Phys. *15*, 261 (1947)
30. Urey, H. C.: J. Chem. Soc. (London), 562 (1947)
31. Silverman, J., Cohen, K.: Ann. Rev. Phys. Chem. *7*, 335 (1956)
32. Wolfsberg, M.: ibid. *20*, 449 (1969)
33. Bigeleisen, J., Lee, M. W., Mandel, F.: ibid. *24* 407 (1973)
34. Bigeleisen, J., Wolfsberg, M.: Isotope Effects in Chemical Kinetics, in: Advances in Chemical Physics, Vol. 1, Prigogine, I. (ed.), Interscience Publ., London 1958
35. Brodsky, A. E.: Isotopenchemie, Akademie-Verlag, Berlin 1961
36. Roginski, S. S.: Theoretische Grundlagen der Isotopenchemie, VEB Deutscher Verlag der Wissenschaften, Berlin 1962
37. Lieser, K. H.: Einführung in die Kernchemie, p. 53 ff, Verlag Chemie, Weinheim 1980²
38. Stahl, W., Wendt, I.: Earth Planet. Sci. Lett. *11*, 192 (1971)
39. Haissinsky, M.: Nuclear Chemistry and its Application, p. 250, Addison-Wesley Publ. Comp., Reading/Massachusetts 1964
40. Stern, M. J., Wolfsberg, M.: J. Chem. Phys. *45*, 2618 (1966)
41. Glueckauf, E.: J. Am. Chem. Soc. *81*, 5262 (1959)
42. Dickel, G.: Z. Phys. Chem. (Frankfurt/M.) *25*, 233 (1960)
43. Dickel, G., Richter, K.: Z. Naturforsch. *19a*, 111 (1964)
44. Ohtaki, H.: Naturwissenschaften *44*, 417 (1957)
45. Wicke, E., Eigen, M.: Z. Elektrochem. *57*, 319 (1953)
46. Glueckauf, E.: Trans. Faraday Soc. *51*, 1235 (1955)
47. Thewlis, J.: Acta Crystallogr. *8*, 36 (1955)
48. Knyazev, D. N.: Russ. J. Phys. Chem. *35*, 298 (1961)
49. Heumann, K. G., Lieser, K. H.: Z. Naturforsch. *27b*, 126 (1972)
50. Heumann, K. G., Gindner, F., Klöppel, H.: Angew. Chem. Int. Ed. Engl. *16*, 719 (1977)
51. Heumann, K. G., Klöppel, H.: Z. Anorg. Allg. Chem. *472*, 83 (1981)
52. Russell, W. A., Papanastassiou, D. A.: Anal. Chem. *50*, 1151 (1978)
53. Lee, D. A., Begun, G. M.: J. Am. Chem. Soc. *81*, 2332 (1959)
54. Lee, D. A.: J. Phys. Chem. *64*, 187 (1960)
55. Lee, D. A.: J. Am. Chem. Soc. *83*, 1801 (1961)
56. Lee, D. A., Drury, J. S.: J. Inorg. Nucl. Chem. *27*, 1405 (1965)
57. Heumann, K. G., Hoffmann, R.: Angew. Chem. Int. Ed. Engl. *15*, 55 (1976)

58. Heumann, K. G. et al.: Z. Naturforsch. *34b*, 406 (1979) ·
59. Heumann, K. G., Baier, K.: Z. Anorg. Allg. Chem. *453*, 23 (1979)
60. Heumann, K. G., Baier, K.: Z. Naturforsch. *35b*, 1538 (1980)
61. Heumann, K. G., Klöppel, H.: ibid. *34b*, 1044 (1979)
62. Heumann, K. G., Baier, K., Wibmer, G.: ibid. *35b*, 642 (1980)
63. Lehn, J.-M., Sauvage, J.-P.: Chem. Comm., 440 (1971)
64. Knöchel, A., Wilken, R.-D.: J. Am. Chem. Soc. *103*, 5707 (1981)
65. Räde, H.-S.: Isotopieeffekte bei Chelatkomplexbildung, Promotion, Technische Hochschule Aachen 1971
66. Räde, H.-S., Wagener, K.: Radiochim. Acta *18*, 141 (1972)
67. Aaltonen, J.: Suomen Kemistilehti *B45*, 141 (1972)
68. Aaltonen, J.: Annales Academiae Scientarium Fennicae AII *137*, 1 (1967)
69. Lee, D. A.: J. Inorg. Nucl. Chem. *38*, 161 (1976)
70. Kobayashi, N. et al.: Bull. Res. Lab. Nucl. Reactors *5*, 19 (1980)
71. Sabau, C., Calusaru, A.: A Literature Survey on the Separation of Isotopes by Ion Exchange, Report R. C. 6, Institutul de Fizica Atomica, Bukarest, 1970
72. Aaltonen, J., Heumann, K. G., Pietilä, P.: Z. Naturforsch. *29b*, 190 (1973)
73. Ponta, A., Calusaru, A.: Isotopenpraxis *11*, 422 (1975)
74. Okamoto, M. et al.: ibid. *16*, 293 (1980)
75. Oi, T. et al.: J. Chromatogr. *248*, 281 (1982)
76. Kakihana, H., Nomura, T., Mori, Y.: J. Inorg. Nucl. Chem. *24*, 1145 (1962)
77. Knyazev, D. A.: Russ. J. Phys. Chem. *37*, 885 (1963)
78. Nandan, D., Gupta, A. R.: Indian J. Chem. *16A*, 256 (1978)
79. Heumann, K. G., Schiefer, H.-P., Spiess, W.: Enrichment of Heavy Calcium Isotopes by Ion Exchanger Resins with Cyclopolyethers as Anchor Groups, in: Stable Isotopes, Schmidt, H.-L., Förstel, H., Heinzinger, K. (eds.), Elsevier, Amsterdam 1982
80. Schiefer, H.-P.: Calcium-Isotopieeffekte bei heterogenen Austauschgleichgewichten mit Kryptanden, Promotion, p. 129, Universität Regenburg 1979
81. Blasius, E., Maurer, P.-G.: J. Chromatogr. *125*, 511 (1976)
82. Blasius, E., Janzen, K.-P., Neumann, W.: Mikrochim. Acta (Wien), 279 (1977)
83. Blasius, E., Maurer, P.-G.: Makromol. Chem. *178*, 649 (1977)
84. Blasius, E., Janzen, K.-P.: Analytical Applications of Crown Compounds and Cryptands, in: Topics in Current Chemistry, Vol. 98, Vögtle, F. (ed.), Springer-Verlag, Berlin, Heidelberg, New York 1981
85. Kopolow, S., Hogen Esch, T. E., Smid, J.: Macromolecules *6*, 133 (1973)
86. Heumann, K. G., Seewald, H.: in preparation (Seewald, H., Diplomarbeit, Universität Regensburg 1982)
87. Glueckauf, E.: Trans. Faraday Soc. *54*, 1203 (1958)
88. Glueckauf, E.: ibid *51*, 34 (1955)
89. Hodgman, C. D. (ed.): Handbook of Chemistry and Physics, p. 212, The Chemical Rubber Publishing Comp., Cleveland 1961/62[43]
90. Stern, R. C., Snavely, B. B.: Annals New York Acad. Sci. *267*, 71 (1976)
91. Artaud, J., Blaise, J., Gerstenkorn, S.: Spectrochim. Acta *10*, 110 (1957)
92. Guiraud, G., Fardeau, J. C.: Analysis *8*, 148 (1980)
93. Müller, G., Mauersberger, K., Sprinz, H.: Analyse stabiler Isotope durch spezielle Methoden, Akademie-Verlag, Berlin 1969
94. Güsten, H.: Chem. uns. Zeit *11*, 33 (1977)
95. Turnbull, A. H.: Surface Ionisation Techniques in Mass Spectrometry, AERE Report 4295, Harwell 1963
96. Crouch, E. A. C.: Adv. Mass Spectrom. *2*, 157 (1963)
97. Kaminsky, M.: Atomic and ionic impact phenomena on metal surfaces, p. 136, Springer-Verlag, Berlin, Heidelberg, New York 1965
98. Shields, W. R. et al.: J. Am. Chem. Soc. *84*, 1519 (1962)
99. Catanzaro, E. J. et al.: J. Res. Nat. Bur. Stand. *68A*, 593 (1964)
100. Heumann, K. G., Schindlmeier, W.: Z. Anal. Chem. *312*, 595 (1982)
101. Heumann, K. G.: Int. J. Mass Spectrom. Ion Phys. *45*, 87 (1982)
102. Unger, M., Heumann, K. G.: ibid. *48*, 373 (1983)

103. Zeininger, H., Heumann, K. G.: ibid. *48*, 377 (1983)
104. Kownatzki, R. et al.: Biomed. Mass Spectrom. *7*, 540 (1980)
105. Schäfer, W., Ballschmiter, K.: Z. Anal. Chem. *315*, 475 (1983)
106. Schulten, H.-R.: Int. J. Mass Spectrom. Ion Phys. *32*, 97 (1979)
107. Lehmann, W. D., Schulten, H.-R.: Anal. Chem. *49*, 1744 (1977)
108. Schulten, H.-R., Lehmann, W. D., Ziskoven, R.: Z. Naturforsch. *33c*, 484 (1978)
109. Bahr, U. et al.: Z. Anal. Chem. *312*, 307 (1982)
110. Ahearn, A. J.: Trace Analysis by Mass Spectrometry, Academic Press, New York, London 1972
111. Douglas, D. J., French, J. B.: Anal. Chem. *53*, 37 (1981)
112. Date, A. R., Gray, A. L.: Analyst *106*, 1255 (1981)
113. Mook, W. G., Grootes, P. M.: Int. J. Mass Spectrom. Ion Phys. *12*, 273 (1973)
114. Brunnée, C.: Chemie-Technik *7*, 111 (1978)
115. Schmidt, M.: Technical Report No. 406, Finnigan MAT, Bremen 1984
116. Van Puymbroeck, J., Gijbels, R.: Z. Anal. Chem. *309*, 312 (1981)
117. Date, A. R., Gray, A. L.: Int. J. Mass Spectrom. Ion Phys. *48*, 357 (1983)
118. Betts, R. H., Harris, W. E., Stevenson, M. D.: Can. J. Chem. *34*, 65 (1956)
119. Knöchel, A., Wilken, R.-D.: J. Radioanal. Chem. *32*, 345 (1976)
120. Lindner, R.: Z. Naturforsch. *9a*, 798 (1954)
121. Dietrich, B., Lehn, J.-M., Sauvage, J.-P.: Tetrahedron *29*, 1647 (1973)
122. Müller, W. H., Beaumation, J.: Anal. Chem. *46*, 2218 (1974)
123. Heumann, K. G., Schiefer, H.-P.: Z. Anal. Chem. *298*, 358 (1979)
124. Heumann, K. G., Schiefer, H.-P.: Z. Naturforsch. *36b*, 566 (1981)
125. Jepson, B. E., Cairns, G. A.: Report MLM-2622, UC-22 for the U.S. Department of Energy, Monsanto Res. Corp., Miamisburg, 1979
126. Izatt, R. M. et al., J. Am. Chem. Soc. *93*, 1619 (1971)
127. Christensen, J. J., Hill, J. O., Izatt, R. M.: Science *174*, 459 (1971)
128. Frensdorff, H. K.: J. Am. Chem. Soc. *93*, 600 (1971)
129. Fang, S. et al.: Lanzhou Daxue Xuebao, Ziran Kexueban *18*, 187 (1982); ref. in: Chem. Abstracts *99*, 504 (1983)
130. Pocnia, N. S., Bajaj, A. V.: Chem. Rev. *79*, 389 (1979)
131. Delphin, W. H., Horwitz, E. P.: Anal. Chem. *50*, 843 (1978)
132. Schubert, J.: J. Phys. Colloid Chem. *52*, 340 (1948)
133. Knöchel, A., Wilken, R.-D.: Chem. Comm., 968 (1981)
134. Heumann, K. G., Hoffmann, R.: Angew. Chem. Int. Ed. Engl. *16*, 114 (1977)
135. Schubert, J.: Ann. Rev Phys. Chem. *5*, 473 (1954)
136. Shih, J. S., Liu, L., Popov, A. I.: J. Inorg. Nucl. Chem. *39*, 552 (1977)
137. Beck, E. W.: Diplomarbeit, Technische Hochschule Darmstadt 1981
138. Heumann, K. G., Schrödl, W.: in preparation (Schrödl, W.: Promotion, p. 144, Universität Regensburg 1983)
139. Heumann, K. G., Schrödl, W.: in preparation (Schrödl, W.: Promotion, p. 150, Universität Regensburg 1983)
140. Fujine, S., Saito, K., Shiba, K.: J. Nucl. Sci. Technol. *20*, 439 (1983)
141. Nishizawa, K. et al.: ibid *21*, 133 (1984)
142. Heumann, K. G., Schiefer, H.-P: Angew. Chem. Int. Ed. Engl. *19*, 406 (1980)
143. Burba, P., Lieser, K. H.: Z. Anal. Chem. *279*, 17 (1976)
144. Spiess, W.: Diplomarbeit, Universität Regensburg 1980
145. Kakihana, H.: J. Chim. Phys. *60*, 81 (1983)
146. Spedding, F. H., Powell, J. E., Svec, H. J.: J. Am. Chem. Soc. *77*, 6125 (1955)
147. Ishino, S., Watanabe, H., Nishizawa, K.: J. Nucl. Sci. Technol., in press
148. Seelmann-Eggebert et al.: Chart of Nuclides, Kernforschungszentrum Karlsruhe, Gersbach und Sohn Verlag, München 1981[5]
149. Sheng, H. et al.: Youji Huaxue, 217 (1984), ref. in: Chem. Abstracts 101, 62198 (1984)

Chiral Pyrrolidine Diamines
as Efficient Ligands in Asymmetric Synthesis

Teruaki Mukaiyama and Masatoshi Asami

Department of Chemistry, Faculty of Science
The University of Tokyo, Hongo, Bunkyo-ku, Tokyo 113, Japan

Table of Contents

Teruaki Mukaiyama, Masatoshi Asami

Complex biologically active substances possessing an array of chiral centers within the same molecule continue to be challenging synthetic targets in organic chemistry. The synthesis of such complex compounds in optically active form has led to the concept of "the asymmetric reaction", a potentially direct and effective method for the construction of chiral framework. Herein, a historical review of new and useful highly selective asymmetric reactions developed in our laboratory, employing chiral diamines and amino alcohols derived easily from (S)-proline, is described. The high selectivity observed in these reactions is postulated to result from efficient non-covalent bonded interactions between organometallic reagent and chiral auxiliary, offering a conceptually new dimension to asymmetric synthesis. Application to the successful synthesis of several natural products is also illustrated.

1 Introduction

Although asymmetric reactions have been studied for over 50 years, dramatic progress has only been made in recent years [1,2]. We undertook an investigation of asymmetric reactions around 1976, as an integral part of our studies on the concept of "Synthetic Control", that is, utilization of common metal chelates for inter- or intramolecular interactions leading to highly selective or entropically advantageous reactions. At the outset of this fruitful study we considered the following criteria as paramount for efficient asymmetric synthesis;

(I) a chiral compound, containing heteroatoms, which interacts with organometallic compound(s) generating structurally rigid "chelate complexes" as intermediates, and

(II) such intermediates should necessarily be of conformationally restricted cis-fused 5-membered ring structure.

Chiral auxiliaries derived from (S)-proline appeared to be particularly attractive since they possess conformationally rigid pyrrolidine ring(s), a prerequisite to the above specified criteria. In this chapter highly stereoselective asymmetric reactions, employing the chiral diamines *1* and chiral diamino alcohols *2* and *3* (Fig. 1),

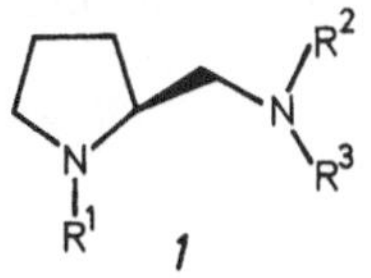

	R^1	R^2	R^3
a	H	Ph	H
b	H	2,6-Xylyl	H
c	H	Isopropyl	H
d	H	Hexyl	H
e	H	Cyclohexyl	H
f	H	(R)-1-phenylethyl	H
g	H	(S)-1-phenylethyl	H
h	H	1-Naphtyl	H
i	H	2-Methoxyphenyl	H
j	H	4-Methoxyphenyl	H
k	H	2-Pyridyl	H
l	H	4-Pyridyl	H
m	H	3,4-Dichlorophenyl	H
n	H	Et	Et
o	H	Ph	Me
p	H	$-(CH_2)_4-$	
q	H	$-(CH_2)_2-O-(CH_2)_2-$	
r	H	$-(CH_2)_2-NMe-(CH_2)_2-$	
s	H	$-(CH_2)_5-$	
t	Me	$-(CH_2)_4-$	
u	Me	$-(CH_2)_5-$	
v	Me	$-(CH_2)_2-O-(CH_2)_2-$	
w	Me	Cyclohexyl	H
x	Me	Ph	H
y	Me	1-Naphthyl	H
z	Me	2-Naphthyl	H

Fig. 1.

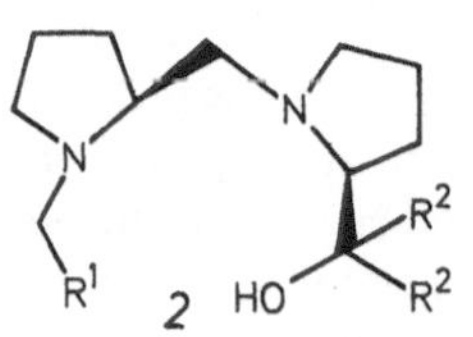

	R^1	R^2
a	H	H
b	H	Me
c	H	Et
d	H	Propyl
e	H	Butyl
f	Me	H
g	t-Bu	H

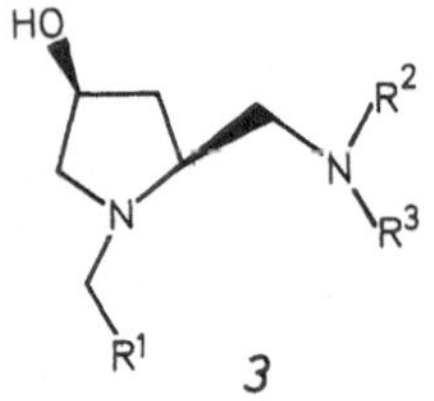

	R^1	R^2	R^3
a	Me	Ph	H
b	Me	Cyclohexyl	H
c	Me	2,6-Xylyl	H
d	Me	1-Naphthyl	H
e	Me	Ph	Me
f	t-Bu	Ph	H

easily prepared from commercially available (S)-proline (or L-4-hydroxypyroline), and their application to the syntheses of a series of optically active natural products will be described.

2 Preparation of Chiral Diamines and Diamino Alcohols

2.1 Preparation of Chiral Diamines *1*

The diamines *1a–s* were readily prepared from N-benzyloxycarbonyl-(S)-proline [N-Z-(S)-proline] according to the following procedures:

I) coupling with the corresponding amines using dicyclohexylcarbodiimide (DCC) or ethyl chlorocarbonate/N-methylmorpholine as condensing reagent,

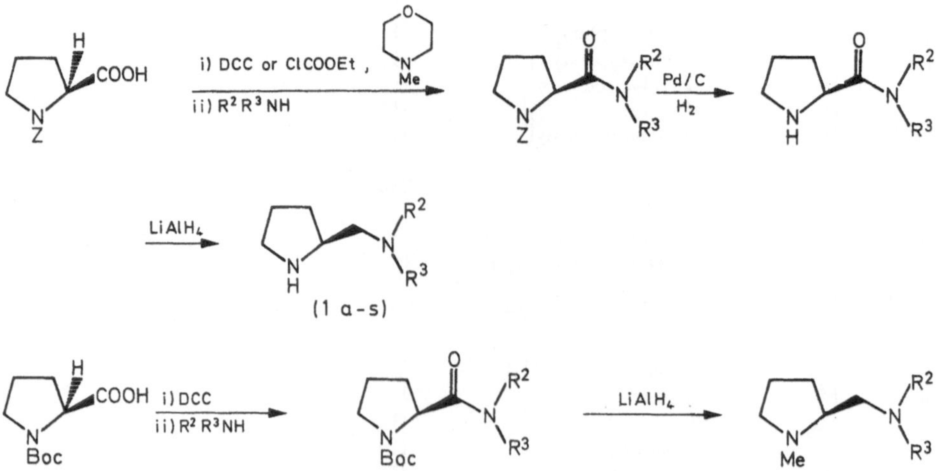

Scheme 1.

136

II) catalytic hydrogenolysis of the benzyloxycarbonyl group, and

III) reduction with LiAlH$_4$ (Scheme 1) [3,4].

The diamines *1 t–z* were prepared from N-*tert*-butoxycarbonyl-(S)-proline [N-BOC-(S)-proline] by analogous procedures (Scheme 1) [5].

2.2 Preparation of Diamino Alcohols 2

All diamino alcohols (*2a–g*) were derived from N-Z-(S)-proline via the intermediate N-[(N-benzyloxycarbonyl)prolyl]proline methyl ester (*4*). The diamino alcohol *2a* was obtained by the reduction of *4* with LiAlH$_4$, and *2b–e* were obtained by the reaction of *4* with the corresponding Grignard reagents. Diamino alcohols *2f–g* and triamino alcohol *5* were obtained by the following sequence of reactions:

I) removal of benzyloxycarbonyl group by HBr-AcOH,

II) reduction with LiAlH$_4$,

III) N-acylation with corresponding acyl chlorides, and

IV) reduction with LiAlH$_4$, as illustrated in Scheme 2 [6].

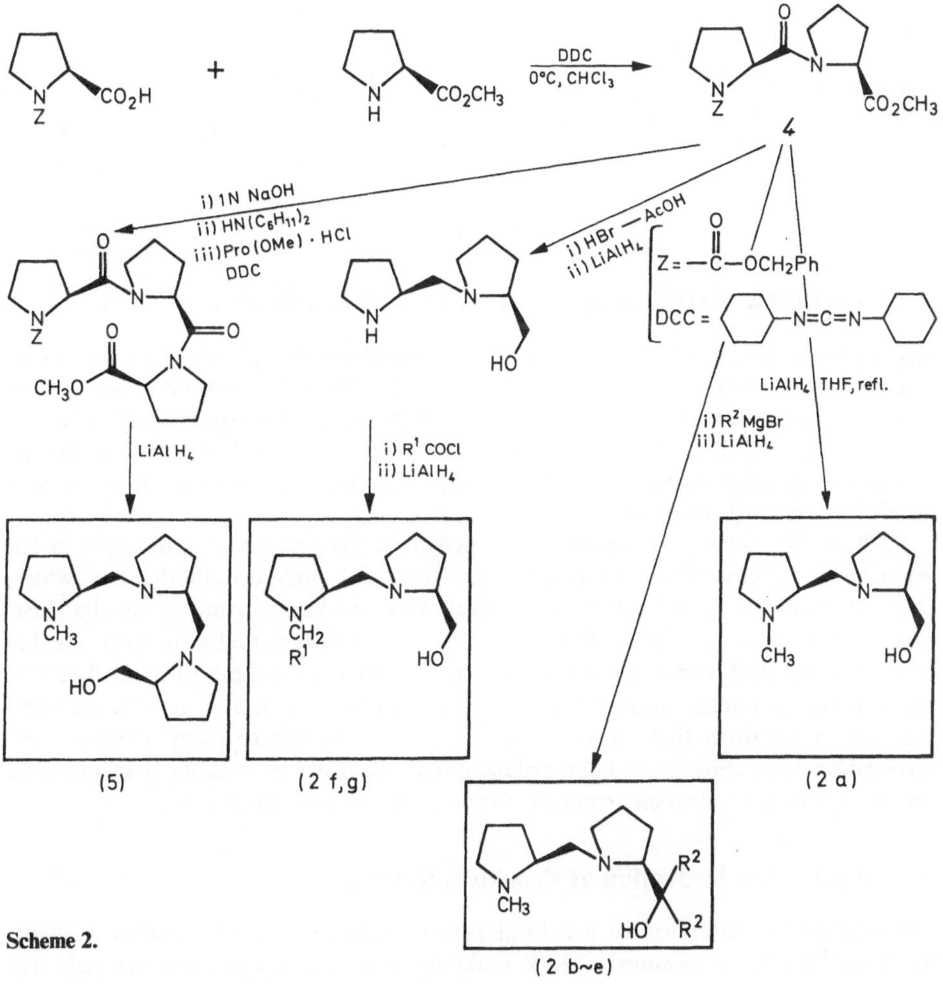

Scheme 2.

137

2.3 Preparation of Diamino Alcohols 3

The diamino alcohols *3 a–f* were prepared from the natural amino acid, L-4-hydroxy-proline according to the following reaction sequences:

I) conversion to (2S, 4S)-1-benzyloxycarbonyl-4-hydroxyproline by known procedure,

II) acetylation,

III) condensation with corresponding amines,

IV) catalytic hydrogenolysis of the benzyloxycarbonyl group, followed by

V) acylation and reduction with LiAlH₄ (Scheme 3) [7].

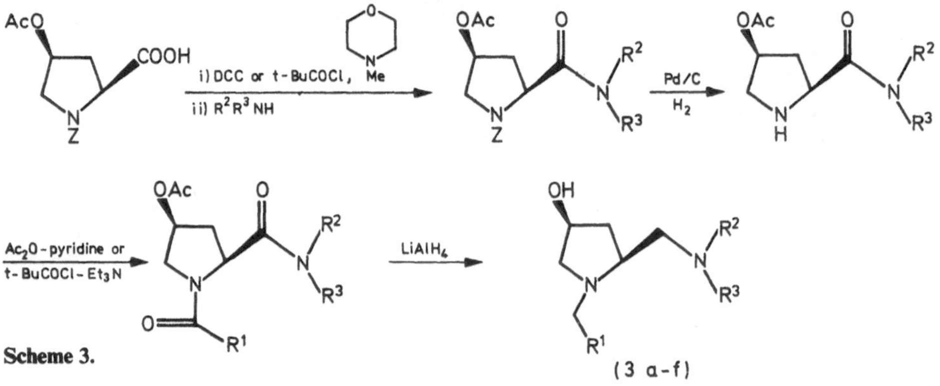

Scheme 3.

3 Asymmetric Syntheses of Optically Active Alcohols

Among various optically active compounds, optically active alcohols are considered one of the most important classes of substrates. As versatile synthetic intermediates, optically active alcohols can readily be converted stereospecifically into the corresponding halides, thiols, and amines via a variety of methods, such as the use of onium salts of azaaromatics. [8] Therefore, the preparation of alcohols in optically active form is most desirable.

Both 1) the asymmetric reduction of prochiral ketones and, 2) the asymmetric addition of organometallic reagents to aldehydes have proved to be the most widely investigated routes to optically active alcohols. Over the years, several studies had been investigated; however, no satisfactory method has been reported until 1977. At that time, we embarked on a detailed study of the asymmetric synthesis of optically active alcohols by both of the aforementioned methods. Our assumption was that an intermediate, a conformationally restricted *cis*-fused 5-membered ring chelate, constructed between chiral ligand and reducing reagent or organometallic reagent would create an efficient chiral environment for such asymmetric inductions.

3.1 Asymmetric Reduction of Prochiral Ketones

The asymmetric reduction of prochiral ketones employing chiral hydride reagents has been the subject of extensive work, and a number of methods have been reported.

In general, the chiral hydride reagent is generated in situ by reaction of a suitable metal hydride with chiral ligands such as alkaloids [9], sugar derivatives [10], amino alcohol [11], chiral oxazolines [12], tartaric acid derivatives [13], chiral amines [14] and chiral diols [15].

Relatively high optical yields were achieved in the asymmetric reduction of acetophenone by these chiral hydride reagents; however, the optimum enantiomeric excess (e.e.) achievable was 83% at that time. Two effective methods have been reported since then [16]. Thus, we initiated a study on the exploration of a new and efficient chiral ligand suitable for the asymmetric reduction of prochiral ketones, and found that a chiral hydride reagent formed in situ from $LiAlH_4$ and the chiral diamine (S)-2-(anilinomethyl)pyrrolidine (1a) is efficient for the reduction of acetophenone, affording (S)-1-phenylethanol in 92% e.e. [3a]. Examination of the effect of the N-substituent in the diamine (1a–m) on the enantioselectivity in the asymmetric reduction of acetophenone, revealed that when a phenyl or 2,6-xylyl substituent was employed, 1-phenylethanol was obtained in 95% e.e. (Table 1) [3b,3c]. Other results obtained by the asymmetric reduction of various ketones, in ether, using the chiral hydride reagent prepared from $LiAlH_4$ and 1a or 1b are summarized in Table 2 [3b,3c].

We assume that the high selectivity is accountable for by considering the following factors;

1) the complex 6 formed by the reaction of $LiAlH_4$ with diamine (1a–m) is extremely rigid with cis-fused 5-membered bicyclic structure, and

Table 1. Asymmetric Reduction of Acetophenone with $LiAlH_4$-Chiral Diamine (1) Complex[a]

Entry	Diamine (1)	Yield/%	Opt. Purity/%
1	a	84 (93)[b]	84 (92)[b]
2	b	80 (87)[b]	87 (95)[b]
3	c	65	47
4	d	45	49
5	e	78	59
6	f	69	52
7	g	72	54
8	h	67	77
9	i	18	~ 0
10	j	63	76
11	k	88	2
12	l	64	7
13	m	66	77

a Molar ratio of acetophenone:$LiAlH_4$:(1) = 1:1.3:1.5; reaction temperature −78 °C.

b Molar ratio of acetophenone:$LiAlH_4$:(1) = 1:1.75:2.00; reaction temperature −100 °C.

Table 2. Asymmetric Reduction of Various Ketones with LiAlH$_4$-Chiral Diamine (1b) or (1a) Complex[a]

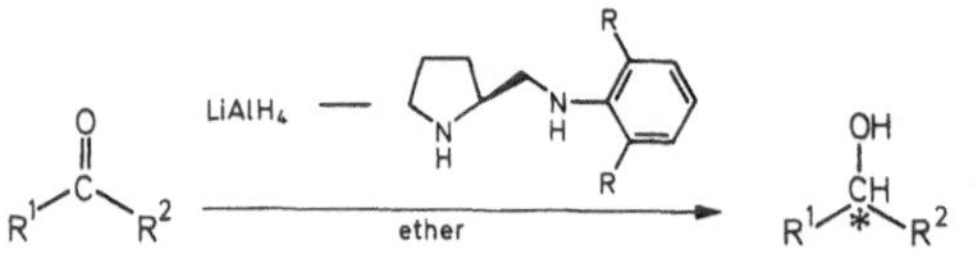

Entry	Ketone	Diamine	Yield/%	Opt. Purity/%	Config.
1	PhCOMe	(1b)	87	95	S
2		(1a)	93	92	S
3	PhCOEt	(1b)	90	96	S
4	PhCOCHMe$_2$	(1b)	97	89	S
5	α-Tetralone	(1b)	85	86	S
6	PhCH$_2$COMe	(1b)	82	11	S
7		(1a)	78	42	S
8	n-C$_6$H$_{13}$COMe	(1b)	83	26	S

a Molar ratio of acetophenone:LiAlH$_4$:(1a) or (1b) = 1.0:1.75:2.0; reaction temperature −100 °C.

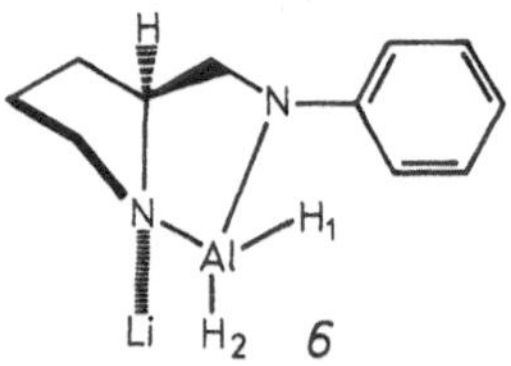

Fig. 2.

2) the reactivity of the two diastereotopic hydrogen atoms contained in this chiral hydride reagent is remarkably different. It is plausible that only H$_2$ is delivered in the reduction step, as H$_1$ is sterically hindered by the pyrrolidine ring and N-substituent of chiral auxiliary;

3) lithium cation, presumably coordinated to the nitrogen atoms on the pyrrolidine ring and/or the side chain in the complex (6), directs the approach of ketone (Fig. 2).

3.2 Asymmetric Addition of Organometallic Reagents to Aldehydes

In contrast to the asymmetric reduction of carbonyl compounds where fairly high e.e. had been realized, the asymmetric addition of organometallic reagents to aldehydes in a chiral solvent, or by using chiral ligands such as (—)-spartein [17a, 17b, 18] and (—)-N-methylephedrine [18], to afford optically active alcohols had met with only moderate success [17−22]. Based on our assumption that conformationally rigid 5-membered ring systems generated by chelation of ligand with organometallic reagent are desirable for enhanced enantioselectivity, we subsequently examined the asymmetric addition of organometallic reagents to aldehydes employing the chiral auxiliaries 2a–g derived from (S)-proline.

140

The influence of molar ratio of reactants, reaction temperature, and solvent on the optical yield in the asymmetric addition of butyllithium to benzaldehyde was examined in detail using *2a* as chiral ligand. The molar ratio of the aldehyde:butyl-lithium:*2a* was found to be an optimum at 1.0:6.7:4.0. The effect of solvent is remarkable, and dimethoxymethane was found to be the solvent of choice when the addition reaction was carried out at −78 °C (entry 11). As it became apparent that lowering the temperature increased the degree of asymmetric induction (entries 2, 3, 11, and 12), in order to carry out the reaction at −123 °C a mixed solvent system of dimethoxymethane-methyl ether or dimethoxymethane-pentane (entries 13 and 14) was employed, and (S)-1-phenyl-1-pentanol was obtained in up to 95% e.e. (Table 3) [6a,6b,6c].

Results of the asymmetric addition of various alkyllithiums to benzaldehyde and an aliphatic aldehyde under the optimized reaction conditions are summarized in Table 4 [6a,6b,6c]. As shown in Table 4, the chiral ligand *2a* is very effective and, in comparison with previously published methods, substantially higher optical yields are achieved. Even in the case of the reaction with an aliphatic aldehyde, the corresponding alcohol is obtained in high optical yield (entry 8). It should be noted

Table 3. Asymmetric Addition of Butyllithium to Benzaldehyde Using the Ligand (2a)[a]

Entry	Solvent	Temp./°C	Yield/%	Opt. Purity[b]/%
1	hexane	− 78	49	20
2	Me$_2$O	− 78	80	53
3	Me$_2$O	−123	80	82
4	Et$_2$O	− 78	76	44[c]
5	Et$_2$O	− 78	60	49[d]
6	Et$_2$O	− 78	57	55
7	Et$_2$O	−123	60	72
8	n-Pr$_2$O	− 78	83	31
9	THF	− 78	80	48
10	DME[e]	− 78	94	53
11	DMM[f]	− 78	67	72
12	DMM	−100	77	87
13	DMM-pentane (1:1)	−123	57	77
14	DMM-Me$_2$O (1:1)	−123	77	95

[a] Molar ratio of benzaldehyde:n-butyllithium:(2a) = 1.0:6.7:4.0, unless otherwise noted.
[b] All of the alcohols had the S configuration.
[c] Molar ratio of benzaldehyde:n-butyllithium:(2a) = 1.0:7.2:3.6.
[d] Molar ratio of benzaldehyde:n-butyllithium:(2a) = 1.0:6.3:3.6.
[e] 1,2-Dimethoxyethane.
[f] Dimethoxymethane.

Table 4. Asymmetric Addition of Alkyllithium to Benzaldehyde Using (2a)[a]

Entry	R^1	R^2	Solvent	Temp./°C	Yield/%	Opt. Purity/%	Config.
1	Me	Ph	Et_2O	0[b]	82	21	R
2	Me	Ph	DMM[c]	0[b]	81	40	R
3	Et	Ph	Et_2O	−123	32	39	R
4	Et	Ph	DMM	−100	59	54	S
5	Et	Ph	DMM-Me_2O (1:1)	−123	70	45	S
6	n-Pr	Ph	DMM-Me_2O (1:1)	−123	64	60	S
7	n-Bu	Ph	DMM-Me_2O (1:1)	−123	77	95	S
8	n-Bu	i-Pr	DMM-Me_2O (1:1)	−123	57	80	S

[a] Molar ratio of aldehyde:alkyllithium:(2a) = 1.0:6.7:4.0.
[b] When the reaction was carried out at lower temperature, the optical purity decreased.
[c] Dimethoxymethane.

Table 5. Addition of Alkylmetals to Benzaldehyde Using the Lithium Salt of (2a)[a]

Entry	Alkylmetal	Solvent	Temp./°C (Time/h)	Yield/%	Opt. Purity/%
1	n-BuCu	Et_2O	−78 (1)	22	0
2	Et_2Zn	Et_2O	−78 → 0 (3)	76	0
3	Et_3Al	Et_2O	r.t. (6)	[b]	
4	n-BuMgBr	Et_2O	−78 (1)	90	47
5	$n\text{-}Bu_2Mg$	Et_2O	−78 (1)	93	68
6	$n\text{-}Bu_2Mg$	Et_2O	−123 (1)	89	73
7	$n\text{-}Bu_2Mg$	Me_2O	−123 (1)	84	43
8	$n\text{-}Bu_2Mg$	THF	−110 (1)	91	59
9	$n\text{-}Bu_2Mg$	DMM[c]	−78 (1)	87	51
10	$n\text{-}Bu_2Mg$	DME[d]	−78 (1)	96	28
11	$n\text{-}Bu_2Mg$	toluene	−78 (1)	93	60
12	$n\text{-}Bu_2Mg$	toluene	−110 (1)	94	88

[a] Molar ratio of benzaldehyde:alkylmetal:lithium salt of (2a) = 1:4:4; this ratio was optimum.
[b] Benzyl alcohol was obtained in 28% yield.
[c] Dimethoxymethane.
[d] Dimethoxyethane.

that the configuration of the alcohol obtained in the addition of ethyllithium to benzaldehyde is dependent on the solvent employed (entries 3 and 4).

Similarly, various other organometallic reagents were examined, such as Grignard reagents, dialkylmagnesium, alkylcopper, dialkylzinc, or trialkylaluminium, which may be applicable to the synthesis of alcohols in optically active form by the asymmetric addition of the organometallic compound to the aldehyde. Furthermore, it was determined that in the presence of the lithium salt of the chiral ligand 2a in the reaction with benzaldehyde, dialkylmagnesium is the organometallic reagent of choice, as shown in Table 5 [6c,6d].

Both effects of solvent and temperature on the reaction of dibutylmagnesium with benzaldehyde in the presence of the lithium salt of 2a was examined and enhanced optical yield was exhibited at low temperature (entries 5, 6, 11, and 12 in Table 5) in toluene, in contrast with the preferred use of ethereal solvents observed in the same reaction using alkyllithiums (see Table 3) [6c,6d].

The results obtained by the reaction of various dialkylmagnesium with benz-aldehyde are listed in Table 6. All of the alcohols thus obtained are of R configuration, whereas the alcohols obtained in the similar reaction with alkyllithiums possess either R or S configuration, depending on the bulkiness of employed alkyllithium [6c,6d].

As shown in Tables 4 and 6, 1-phenyl-1-propanol, 1-phenyl-1-butanol, and 1-phenyl-1-pentanol are obtained in high optical yields by the reaction of dialkyl-magnesiums with benzaldehyde in the presence of the lithium salt of the chiral addend 2a.

On the other hand, 1-phenylethanol was obtained in 86% e.e. (entry 4, Table 7) by the reaction of methyllithium in the presence of the modified chiral ligand, $2d$ ($R^1 = H$, $R^2 = n$-Pr) [6d].

Table 6. Asymmetric Addition of Dialkylmagnesium to Aldehydes Using the Lithium Salt of (2a)[a]

Entry	R^1	R^2	Yield/%	Opt. Purity/%
1	Me	Ph	56	34
2	Et	Ph	74	92
3	n-Pr	Ph	90	70
4	i-Pr	Ph	59	40
5	n-Bu	Ph	94	88
6	i-Bu	Ph	81	42
7	n-Bu	i-Pr	70	22

[a] The reaction was carried out in toluene at -100 °C for 1 h. Molar ratio of aldehyde:dialkylmagnesium:lithium salts of (2a) = 1:4:4. All alcohols possess the R configuration.

Table 7. Effect of Substituents R^1 and R^2 in (2)[a]

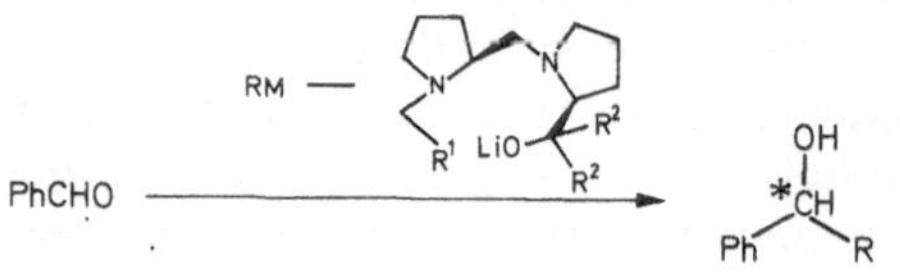

Entry	Ligand	Alkylmetal	Yield/%	Opt. Purity/%	Config.
1	2b ($R^1 =$ H; $R^2 =$ Me)	MeLi	62	5	R
2	2b	Me$_2$Mg	82	43	R
3	2c ($R^1 =$ H; $R^2 =$ Et)	MeLi	79	77	R
4	2d ($R^1 =$ H; $R^2 =$ n-Pr)	MeLi	82	86	R
5	2d	Me$_2$Mg	77	21	R
6	2e ($R^1 =$ H; $R^2 =$ n-Bu)	MeLi	69	70	R
7	2f ($R^1 =$ Me; $R^2 =$ H)	MeLi	61	17	R
8	2g ($R^1 =$ t-Bu; $R^2 =$ H)	MeLi	79	20	R
9	2a ($R^1 =$ H; $R^2 =$ H)	BuLi	60	72	S
10	2b	BuLi	77	0	
11	2c	BuLi	80	11	R
12	2e	BuLi	85	23	R
13	2f	BuLi	78	5	S
14	2g	BuLi	85	54	R
15	5	BuLi	80	68	R

[a] In the case of the reaction using alkyllithium, the molar ratio of benzaldehyde:alkyllithium:(2) = 1.0:6.7:4.0; the reaction was carried out in Et$_2$O at -123 °C for 1 h. In the case of the reaction using dimethylmagnesium, the molar ratio of benzaldehyde:dimethylmagnesium:lithium salt of (2) = 1:4:4; the reaction was carried out in toluene at -110 °C for 1 h.

The ligands *2b–g* were also employed for the reaction of butyllithium with benzaldehyde, and it was observed that formation of alcohol with R configuration is preferred when ligands with larger substituents, R^1 and R^2, were employed (see Table 7). In addition, it is noteworthy that (R)-1-phenyl-1-pentanol is obtained in 68% e.e. by employing the chiral ligand *5* which has three pyrrolidine moieties (entry 15) [6d].

In the above asymmetric addition of organometallic reagents to aldehydes, two pyrrolidine moieties and lithiated hydroxymethyl group are essential for attainment of high enantioselectivity. A rigid complex *7* may be formed by virtue of the coordination of oxygen and the two nitrogen atoms of the ligand to alkyl metal providing an efficient chiral environment for asymmetric induction (Fig. 3).

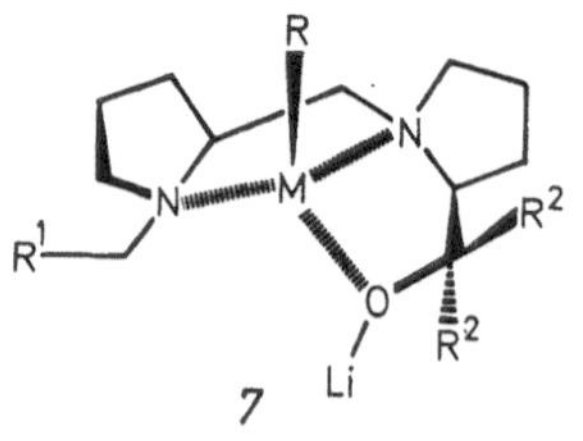

Fig. 3.

The explosive increase in the use of lithium compounds in organic synthesis during recent years prompted us to extend the abovementioned asymmetric reaction to the addition of the lithiated organic compounds of methyl phenyl sulfide, acetonitrile, N-nitrosodimethylamine, and 2-methylthiothiazoline to aldehyde. Various optically active oxiranes, thiiranes, amino alcohols, etc. are thus obtained by suitable derivation (Eq. 1–4) [23].

$$\text{(Eq. 1)}$$

$$\text{(Eq. 2)}$$

$$\text{(Eq. 3)}$$

$$\text{(Eq. 4)}$$

Furthermore, optically active alkynyl alcohols, useful intermediates for the synthesis of several optically active natural products, were obtained by the asymmetric addition of lithium acetylides to aldehyde in the presence of chiral ligand *2a* [24]. Enhanced enantioselectivity in this reaction depends apparently on the substituent group in the acetylene moiety. As shown in Table 8, use of trialkylsilylacetylides gave the best results [24]. Various optically active ethynyl alcohols were obtained by the reaction of lithium trimethylsilylacetylide with aliphatic aldehydes, as summarized in Table 9 [25].

The alkynyl alcohols are successfully converted to γ-ethyl-γ-butyrolactone, an insect pheromone of *Trogoderma*, and (R)-5-octyl-2[5H]-furanone, an important synthetic intermediate for the synthesis of naturally occurring avenaciolide (Scheme 4) [25].

Table 8. Enantioselective Ethynylation to Benzaldehyde

Run	R	Temp./°C	Yield[a]/%	Opt. Purity[b]/%	Config.
1	H-	− 78	76	54	S (+)
2	Me₃Si-	− 78	99	78	S (+)
3	Me₃Si-	−123	87	92	S (+)
4	Et₃Si-	−123	93	80	S (+)
5	t-BuMe₂Si-	−123	67	72	S (+)
6	Ph₂MeSi-	−123	88	80	S (+)
7	Ph₃Si-	−123	83	76	S (+)

[a] The silyl groups were removed by the action of methanolic NaOH (run 2, 3, 4, 5) or KF in two-phase system in the presence of $n\text{-Bu}_4\text{N}^+\text{HSO}_4^-$ (run 6, 7).

[b] The optical yield of the resulting 1-phenyl-2-propyn-1-ol was determined by 100 MHz ¹H-NMR measurement of the (S)-(−)-MTPA ester utilizing benzene-d⁶ as the solvent.

Table 9. Enantioselective Addition of Lithium Trimethylsilylacetylide to Aliphatic Aldehydes

Run	R	Yield/%	Opt. Purity[a]/%	Config.
1	$n\text{-C}_2\text{H}_5\text{-}$	77	68	R
2	$n\text{-C}_5\text{H}_{11}\text{-}$	87	76	R
3	$n\text{-C}_8\text{H}_{17}\text{-}$	83	80	R
4	$n\text{-C}_{11}\text{H}_{23}\text{-}$	82	70	R
5	$n\text{-C}_{13}\text{H}_{27}\text{-}$	76	73	R
6	$(\text{CH}_3)_2\text{CHCH}_2\text{-}$	54	65	R
7	$\text{CH}_3(\text{CH}_2)_2\text{CH}=\text{CH-}$	74	40	R

[a] The optical purity of the resulting alcohol was determined by NMR measurement of its MTPA derivative utilizing benzene-d⁶ as a solvent.

3.3 An Asymmetric Transformation of Cyclohexene Oxide to Optically Active 2-Cyclohexen-1-ol by Chiral Lithium Amides

The transformation of epoxides to allylic alcohols by the use of lithium dialkylamides is well known [26]. Consequently, it should be possible to apply this reaction to the asymmetric synthesis of allylic alcohols by the selective deprotonation of the enantiotopically related protons in symmetrical epoxides.

Since it has been established that the deprotonation of cyclohexene oxide is highly selective for the *syn* proton with pseudoaxial orientation [26], efficient chiral recognition

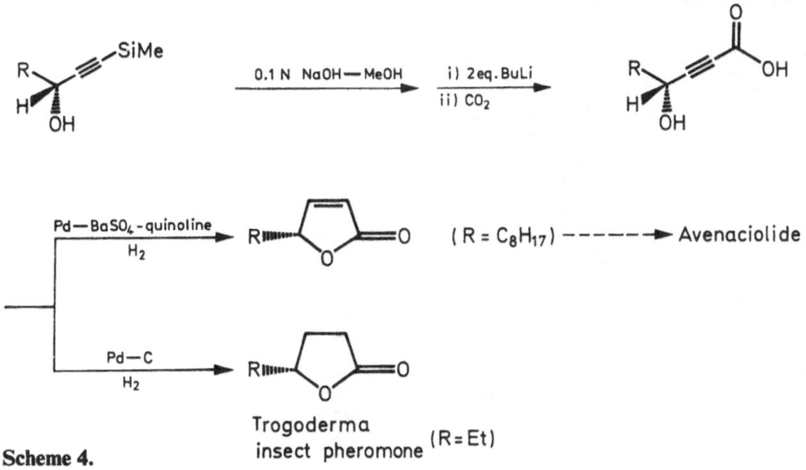

Scheme 4.

of the equilibrating enantiomeric conformations *I* and *II* by chiral lithium amide should generate optically active 2-cyclohexen-1-ol (Fig. 4). Whitesell and Felman reported the first example of such a reaction in 1980 by treating cyclohexene oxide with a number of chiral mono- and dialkyl lithium amides; however, the degree of the asymmetric induction was not high [27]. Thus, we examined the use of the chiral lithium amides derived from butyllithium and the diamines *1n–s* for selective deprotonation of cyclohexene oxide. (S)-2-cyclohexen-1-ol, a useful synthetic intermediate for the synthesis of several sesquiterpenes such as (+)-2-carene and (4S)-trans-β-elemenone, were obtained in as high as 92% e.e. when the diamine *1p* was employed. Results are summarized in Table 10.

The high selectivity realized in the above reaction may be accounted for by assuming the six-membered pericyclic transition state suggested by Thummel and Richborn. Thus, as depicted in the transition state *8* (conformation *I*), no severe steric interactions between cyclohexane ring and diamine is apparent. On the other hand, in the transition state *9* (conformation *II*) severe steric repulsion arises between the cyclohexane ring and chiral diamine (Fig. 5) suggesting that the transition state *8* is preferentially favored, the alcohol having S configuration being obtained.

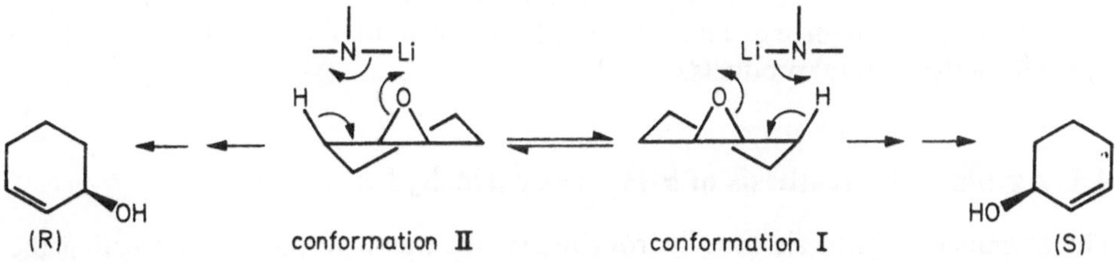

Fig. 4.

Table 10. Asymmetric Transformation of Cyclohexene Oxide

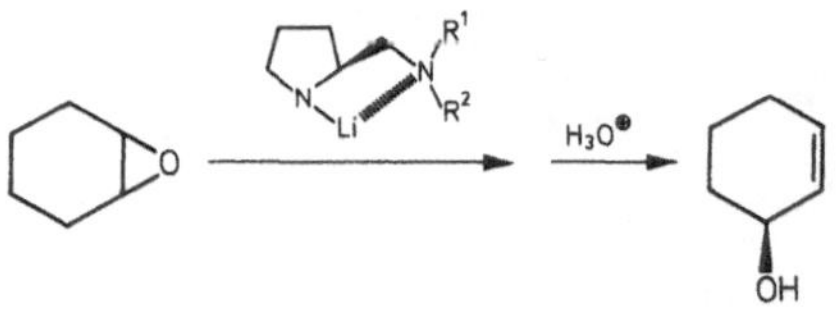

Entry	Diamine	Temp./°C	Yield/%	Opt. Purity/%
1	(1n)	−78	69	83
2	(1o)	−78	55	47
3	(1p)	−78	78	90
4	(1p)	0	77	92
5	(1q)	−78	70	86
6	(1q)	0	72	88
7	(1r)	−78	67	68
8	(1s)	−78	71	81

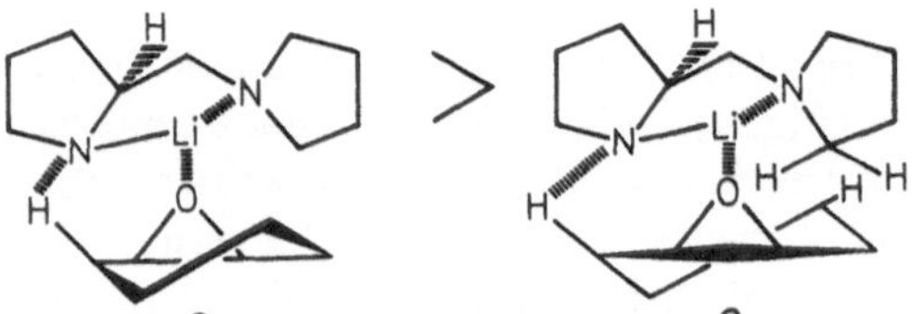

Fig. 5.

4 Asymmetric Syntheses of Optically Active Aldehydes

Optically active aldehydes are desirable synthetic precursors for the construction of chiral carbon skeltones in organic synthesis. Several methods had been devised for the synthesis of optically active aldehydes employing chiral enamines [28], chiral imines [29], or chiral hydrazone [30]; however, little was known about the asymmetric synthesis of chiral aldehyde having a functional group in the same molecule [31].

The fundamental idea of sterically restricted cis-fused 5-membered bicyclic structure as an efficient chiral environment for enhanced asymmetric induction was extended to the synthesis of several synthetically useful, optically active aldehydes. The reaction of aminals (1,3-diazabicyclo[3.3.0] octane derivatives) having a fixed cis-fused 5-membered bicyclic ring structure, obtained from the diamine *1a* and appropriate aldehydes, with nucleophiles was investigated. Several functionalized aldehydes were obtained in high e.e. and successfully applied to the synthesis of selected optically active natural products.

4.1 Asymmetric Synthesis of α-Hydroxy Aldehydes

The asymmetric synthesis of α-hydroxy aldehydes was investigated by treating the derived keto aminal with Grignard reagents. The aminal *10a*, which is easily prepared from the diamine *1a* and phenyl glyoxal monohydrate, was subsequently treated with Grignard reagent, and α-alkyl-α-hydroxyphenylacetaldehyde was obtained in greater than 94% e.e. after mild acidic work-up (Eq. 5) [32].

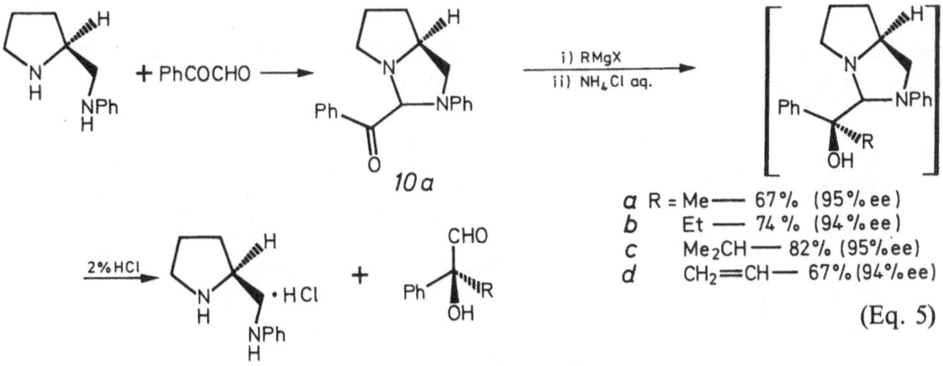

$$
\begin{array}{ll}
a & R = Me \text{——} 67\% \ (95\% \ ee) \\
b & Et \text{——} 74\% \ (94\% \ ee) \\
c & Me_2CH \text{——} 82\% \ (95\% \ ee) \\
d & CH_2{=}CH \text{——} 67\% \ (94\% \ ee)
\end{array}
$$

(Eq. 5)

Various keto aminals were readily obtained by treating aminal *11*, prepared from the diamine *1a* and methyl hydroxymethoxyacetate, with Grignard reagent (R^1MgX) in the presence of $MgCl_2$ in good yields (Eq. 6) [33]. Successive addition of

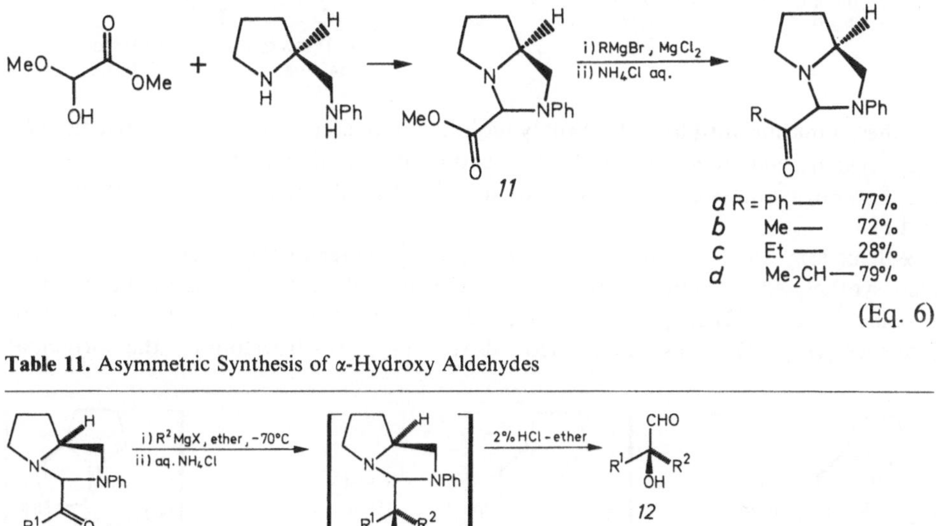

$$
\begin{array}{ll}
a & R = Ph \text{——} 77\% \\
b & Me \text{——} 72\% \\
c & Et \text{——} 28\% \\
d & Me_2CH \text{——} 79\%
\end{array}
$$

(Eq. 6)

Table 11. Asymmetric Synthesis of α-Hydroxy Aldehydes

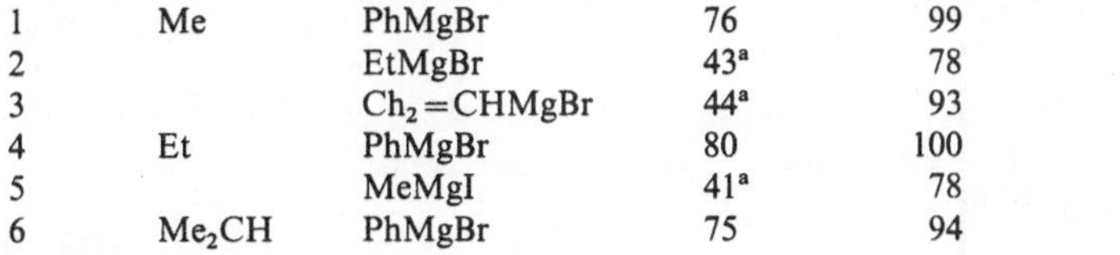

Entry	R^1	R^2MgX	Yield/%	Opt. Purity/%	Config.
1	Me	PhMgBr	76	99	R
2		EtMgBr	43[a]	78	R
3		$Ch_2{=}CHMgBr$	44[a]	93	R
4	Et	PhMgBr	80	100	R
5		MeMgI	41[a]	78	S
6	Me_2CH	PhMgBr	75	94	R

[a] Isolated as benzyl ether.

149

a second Grignard reagent (R²MgX), followed by mild acidic work-up, facilitates a general and versatile method for the preparation of various optically active α-hydroxy aldehydes (*12*) of desired configuration (Table 11) [33].

Thus, both enantiomers of frontalin, a pheromone of several species of beetles belonging to the genus *Dendroctonus*, were separately synthesized by applying this asymmetric reaction sequence. A reversal of order of the addition of the requisite Grignard reagents (Scheme 5) affords either enantiomer at will [34].

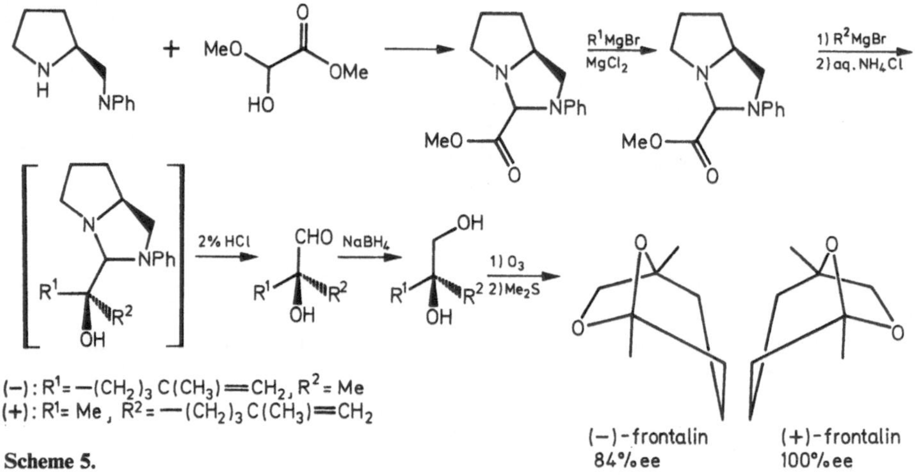

$(-): R^1 = -(CH_2)_3 C(CH_3) = CH_2, R^2 = Me$
$(+): R^1 = Me, R^2 = -(CH_2)_3 C(CH_3) = CH_2$

Scheme 5.

(−)-frontalin
84% ee

(+)-frontalin
100% ee

Further, a marine antibiotic, (−)-malyngolide, discovered recently in the marine blue alga *Lyngbya majuscula* Gomont [35], and the vitamin E precursor chromanmethanol [36], were synthesized in high e. e. by utilizing this asymmetric reaction (Scheme 6 and 7).

α-Benzyloxy aldehydes having a chiral tertiary center at the α-carbon atom were also synthesized in high e.e. by treating the aminal *11* with diisobutylaluminium hydride (DIBAL-H) in place of Grignard reagent, followed by a similar reaction sequence (Eq. 7) [37] as mentioned above. (+)-*exo*-Brevicomin, the principal

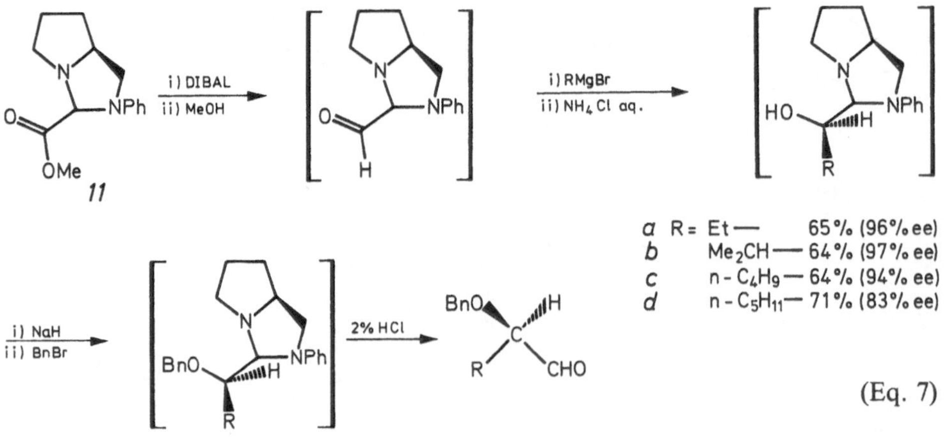

a	R = Et —	65% (96% ee)
b	Me₂CH —	64% (97% ee)
c	n-C₄H₉ —	64% (94% ee)
d	n-C₅H₁₁ —	71% (83% ee)

(Eq. 7)

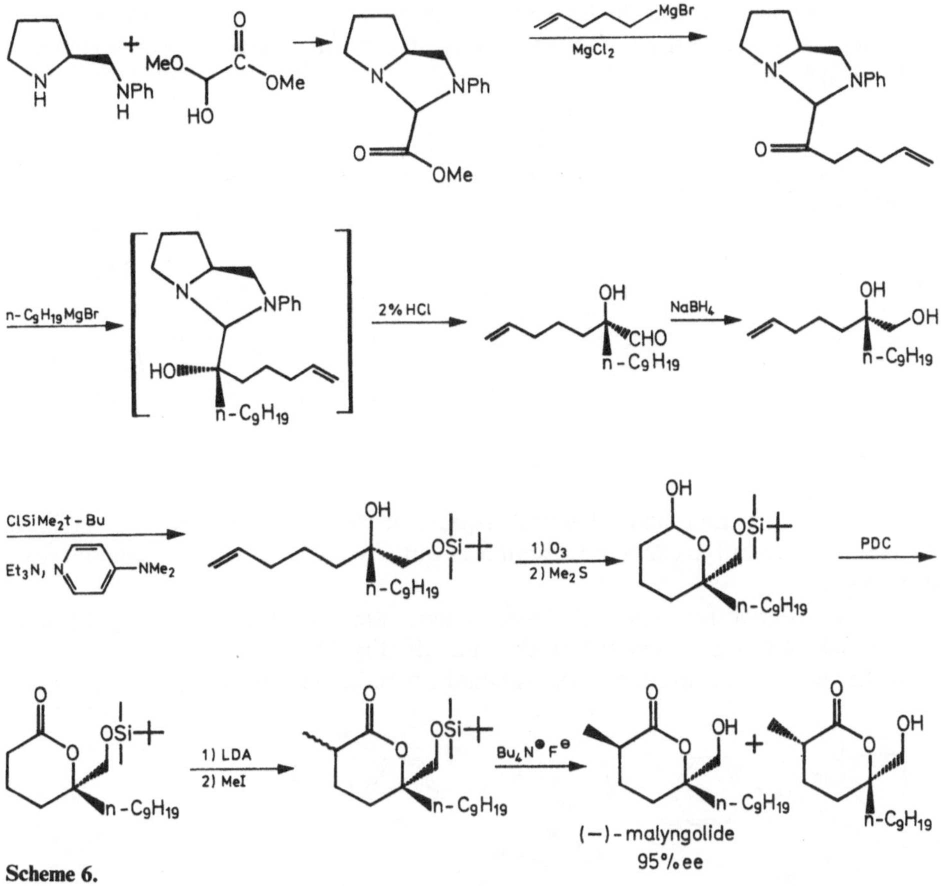

Scheme 6.

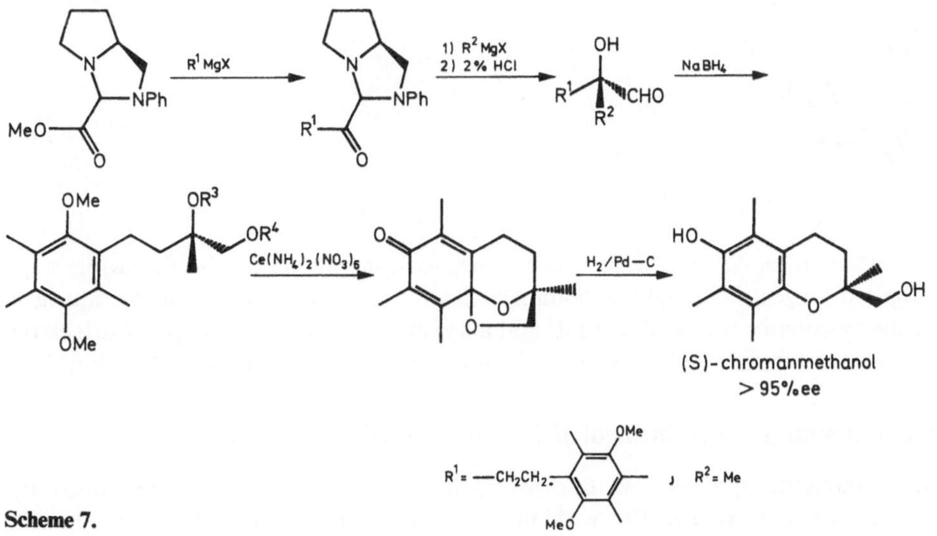

Scheme 7.

151

aggregation pheromone in the frass of the female western pine beetle, *Dendroctonus brevicomis*, was synthesized stereoselectively starting from (R)-2-benzyloxybutanal (Scheme 8) [37].

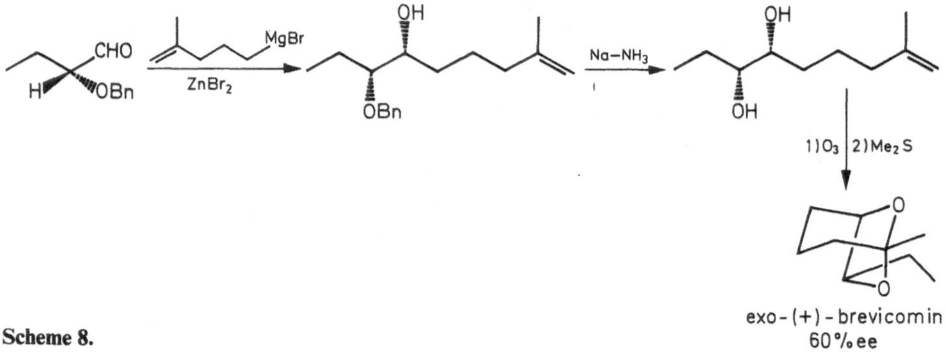

Scheme 8.

exo-(+)-brevicomin
60% ee

Although the actual stereochemical aspects in this reaction have not yet been clarified, the overall asymmetric reaction essentially consists of two stereoselective steps, namely,

    i) stereoselective formation of diastereomeric aminals *10* and *10'* (Fig. 6), and

    ii) attack of Grignard reagent on the aminals (Fig. 7).

Of the two possible diastereomeric aminals, predominant formation of *10* should

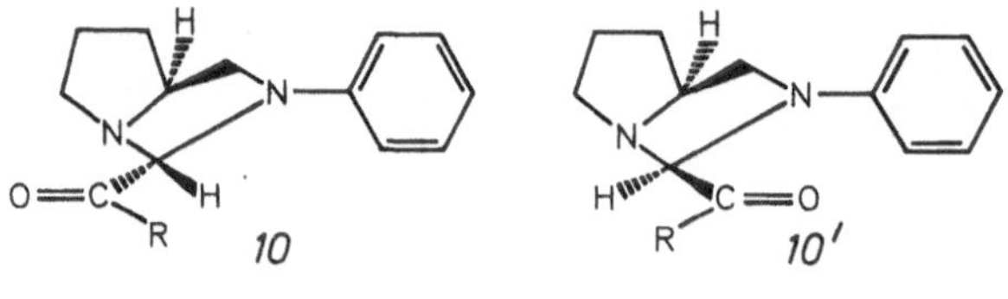

Fig. 6.

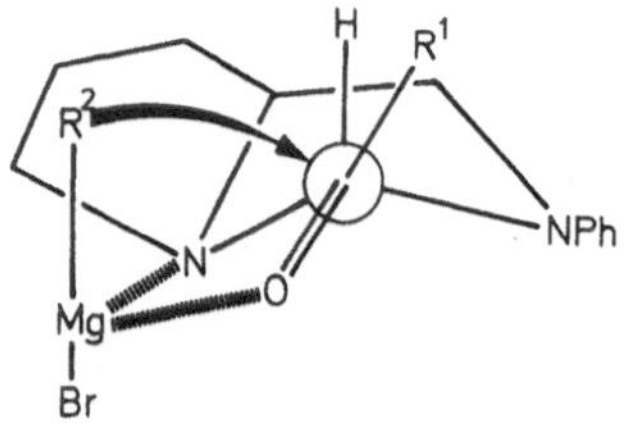

Fig. 7.

be favored over *10'*, based on steric considerations alone. In the following step, Grignard reagent is thought to complex with the carbonyl oxygen and the nitrogen on the pyrrolidine ring leading to a rigid alkyl metal complex, nucleophilic attack on carbonyl carbon being directed from the less hindered side, as illustrated in Fig. 7.

4.2 Asymmetric Synthesis of β-Formyl β-Hydroxy Esters

An asymmetric synthesis of β-formyl β-hydroxy esters was also examined by treating the keto aminal *10a* with metal enolates of ethyl acetate. Zinc enolates

Table 12. Asymmetric Synthesis of β-Formyl β-Hydroxy Esters

Entry	R^1	R^2	Reaction Conditions	Yield/%	Opt. Purity/%	Config.
1	C_6H_5	C_6H_5	a	58	88	S
2	C_6H_5	C_6H_5	b	85	62	R
3	C_6H_5	o-MeOC$_6$H$_4$	c	85	84	S
4	Me	o-MeOC$_6$H$_4$	c	70	92	R
5	Et	o-MeOC$_6$H$_4$	c	64	87	—
6	Me$_2$CH	o-MeOC$_6$H$_4$	c	50	92	—

[a] Zn/BrCH$_2$CO$_2$Et, benzene, reflux

[b] Lithium diisopropylamide/CH$_3$CO$_2$Et, THF, $-78\ °C$

[c] Lithium diisopropylamide/CH$_3$CO$_2$Et, toluene, $-78\ °C$

derived from zinc and the corresponding ester of haloacetic acids, the Reformatsky reaction, and ii) lithium enolates derived from lithium amide derivatives and esters of acetic acid were both investigated and the results are listed in Table 12.

The reaction of keto aminal *10a* with zinc and ethyl bromoacetate in refluxing benzene, followed by hydrolysis, afforded ethyl 3-formyl-3-hydroxy-3-phenylpropionate in 58% yield and 88% e.e. (entry 1). It is interesting that the corresponding β-formyl β-hydroxy ester is obtained in such high optical yield even under the seemingly drastic condition of refluxing benzene. The lithium enolate, generated from lithium diisopropylamide and ethyl acetate, reacts with keto aminal *10a* to give ethyl 3-formyl-3-hydroxy-3-phenylpropionate in high optical yield. Furthermore, both enantiomers are readily accessible by appropriate choice of the starting keto aminal (entries 2 and 3). By the introduction of a methoxy group at the ortho position of the phenyl group of the diamine (entry 3) improvement of the e.e. in this reaction was observed, possibly due to the stronger affinity of the aminal moiety for lithium cation. Various optically active β-formyl β-hydroxy esters are thus obtained in high e.e. by the reaction of the keto aminal *13*, prepared from diamine *1i*, with the lithium enolate of ethyl acetate [38].

4.3 Asymmetric Synthesis of 3-Substituted Succinaldehydic Acid Methyl Esters

In addition to the diastereoface-differentiating 1,2-addition of Grignard reagents to keto aminals, highly stereoselective 1,4-addition of Grignard reagents to α,β-unsaturated esters having an aminal moiety was also examined. Thus, the aminal *14*, easily prepared from the diamine *1a* and fumaraldehydic acid methyl ester, was treated with Grignard reagent in the presence of a catalytic amount of CuI to afford, after hydrolysis of the resulting aminal, 3-alkylsuccinaldehydic acid methyl ester in high e.e. (Eq. 8) [39].

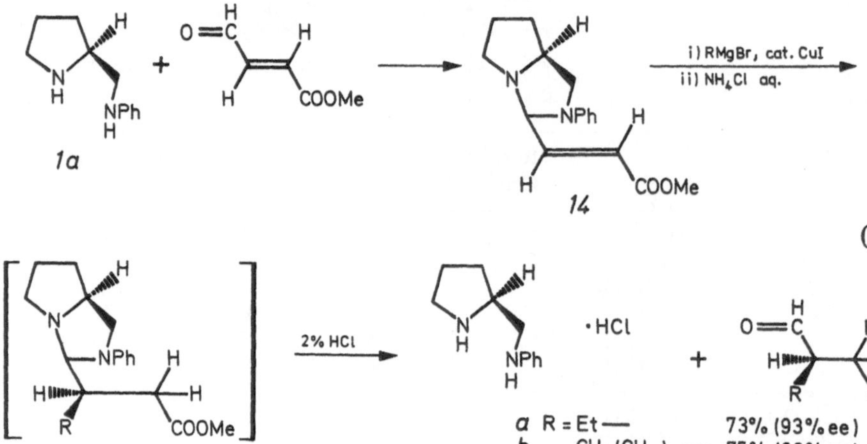

i) RMgBr, cat. CuI
ii) NH$_4$Cl aq.

14

1a

(Eq. 8)

2% HCl

$\cdot$ HCl

a R = Et —   73% (93% ee)
b   CH$_3$(CH$_2$)$_2$ —   75% (89% ee)
c   Me$_2$CH —   73% (85% ee)
d   CH$_3$(CH$_2$)$_3$ —   83% (93% ee)
e   CH$_3$(CH$_2$)$_4$ —   65% (92% ee)
f   PhCH$_2$ —   38% (35% ee)

4.4 Enantioselective Addition of a Chiral Aryllithium to Aldehydes

As demonstrated in Sect 4.1–4.3, it is clearly evident that the chiral aminal moiety is very effective for asymmetric induction. In contrast to the aforementioned

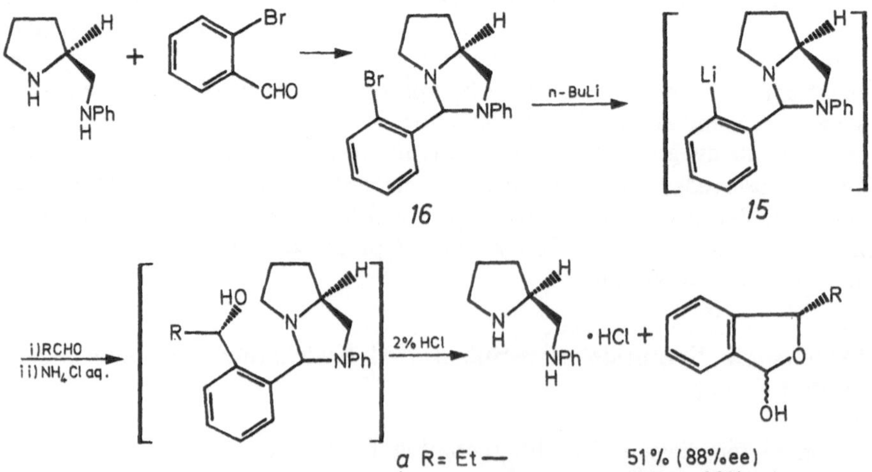

n-BuLi

16   15

i) RCHO
ii) NH$_4$Cl aq.

2% HCl

$\cdot$ HCl +

a R = Et —   51% (88% ee)
b   Me$_2$CH —   52% (>90% ee)
c   CH$_3$(CH$_2$)$_3$ —   62% (87% ee)
d   CH$_3$(CH$_2$)$_7$ —   58% (90% ee)
e   CH$_3$CH=CH —   70% (20% ee)

(Eq. 9)

diastereoface-differentiating reactions, examination of an enantioface-differentiating reaction using a chiral aryllithium aminal seemed justified. Thus, the chiral aryl-lithium *15*, generated by treating the aminal *16* with butyllithium, was treated with various aldehydes to afford the corresponding 3-alkyl-1-hydroxy-2-oxa-indanes in high e.e. (Eq. 9) after hydrolysis. (S)-3-Butylphthalide, an essential oil of celery, was obtained by the oxidation of (S)-3-butyl-1-hydroxy-2-oxaindane with silver oxide (Eq. 10) [40].

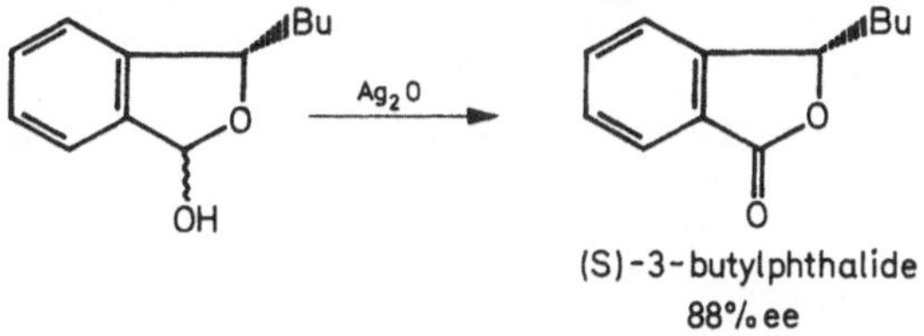

(S)-3-butylphthalide
88% ee

(Eq. 10)

The stereochemical integrity of this reactions is assumed as follows; the aminal *16* is formed stereoselectively by an analogous mechanism as dicussed in Sect. 4.1. In the addition of the aryllithium to aldehyde, a rigid tricyclic 5-membered ring structure is formed by intramolecular coordination of the nitrogen atom of the pyrrolidine ring to the lithium ion. The aldehyde is then directed to approach from the less hindered front-side in the manner depicted in Fig. 8. An approach as shown in Fig. 9 would be less favorable due to the steric repulsion between the alkyl group of the aldehyde and the pyrrolidine ring. Thus, the resulting 3-alkyl-1-hydroxy-2-oxaindanes are obtained in high e.e. and with S configuration.

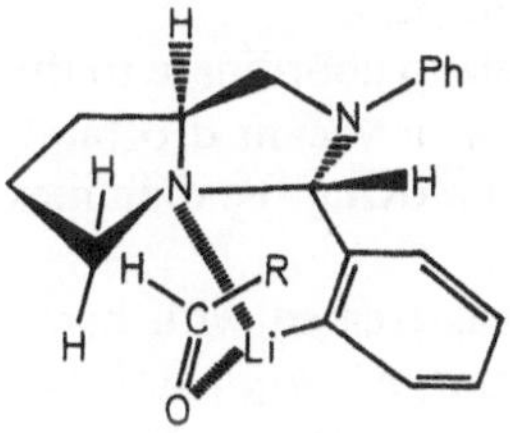

Fig. 8.

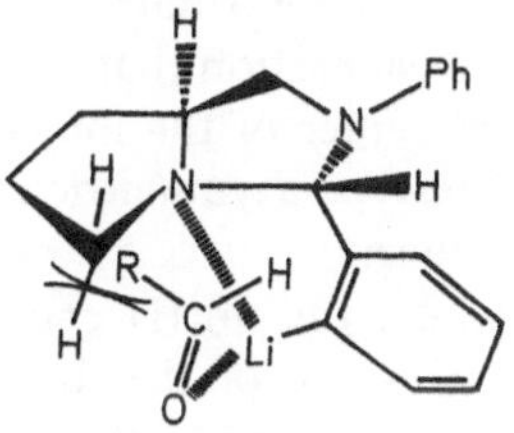

Fig. 9.

5 Asymmetric Cross Aldol Reactions via Divalent Tin Enolates

During the course of our investigation of synthetic reactions promoted by divalent tin species, divalent tin enolates, generated from stannous trifluoromethanesulfonate (stannous triflate) and ketones, were found to react with aldehydes to afford aldol products with high *erythro*-selectivity (Eq. 11) [40]. This enolate also gives cross aldols

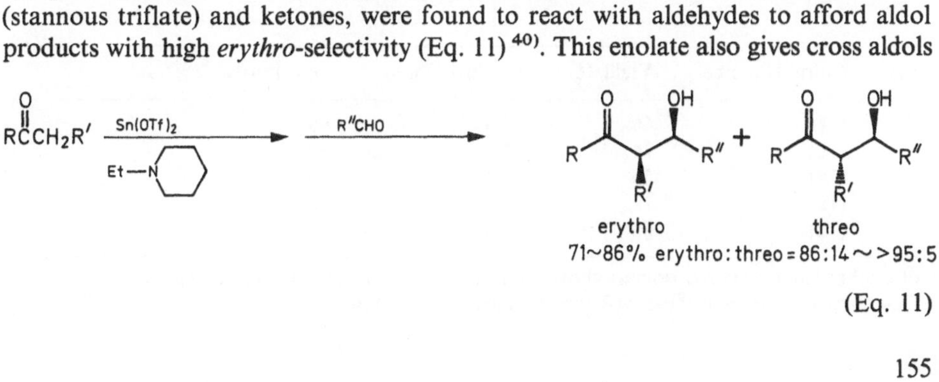

erythro   threo
71~86% erythro:threo = 86:14 ~ >95:5

(Eq. 11)

155

between two different ketones [42]. Further, divalent tin enolate can be formed from stannous triflate and 3-acylthiazolidine-2-thione, which on treatment with carbonyl compounds affords the corresponding aldol-type adducts in excellent yields and with high *erythro*-selectivity (Eq. 12). The adduct is easily transformed into β-hydroxy aldehyde and β-hydroxy carboxylic acid derivatives [43].

erythro
88~95%   erythro:threo = ≧ 97:3

$$(Eq.\ 12)$$

Recent rapid developments in stereoselective aldol reactions has resulted in the exploitation of the asymmetric version of this reaction. Although dramatic progress has been made in the area of asymmetric aldol reactions, only those reactions of achiral enolates with chiral aldehyde or of achiral aldehyde (ketone) with chiral enolates are of importance [44,45]. However, the efficiency of these reactions is greatly diminished by the necessity of tedious procedures for the attachment and removal of the chiral auxiliaries. Thus the development of a highly enantioselective aldol reaction employing chiral auxiliaries that are not covalently bonded to either reactant is strongly desirable, though the influence of such chiral addends in the aldol reaction has met with only modest success. An example employing spartein as a chiral auxiliary in the Reformatsky reaction has been reported [46].

Based on the consideration that suitable ligand should be able to coordinate to the divalent tin metal center of the intermediate enolate, which has a vacant d orbital, the achiral, chelate-type asymmetric reaction was examined employing the diamines *1 s–z* as chiral auxiliaries.

As a model study, the tin(II) enolate of propiophenone was treated with benzaldehyde in the presence of the diamine *1 t*. The corresponding *erythro*-aldol was obtained in 66% yield and with an optical purity of 60% when the reaction is

Table 13. The Effect of Employed Chiral Diamine on Optical Purity

Entry	Chiral Diamine	Yield/%	Erythro:Threo	Opt. Purity[a,b]/%
1	(1t)	66	6:1	65
2	(1u)	74	6:1	80
3	(1v)	56	6:1	50
4	(1s)	72	6:1	75
5	(1x)	66	20:1	20

[a] of erythro isomer; Threo isomers shows almost the same degree of enantioselection.
[b] Determined by ^{1}H- and ^{19}F-NMR measurement of its MTPA ester.

carried out in CH_2Cl_2 at $-95\,°C$ using an equimolar amount of diamine $1t$ to the tin(II) enolate. After examination of various chiral diamines, the diamine $1u$ was found to be most efficient, and the enantioselectivity of the cross aldol reaction of various aromatic ketones with aromatic aldehydes was found to range from 75 to 80% e.e. in the *erythro*-aldol product (Table 13). In the case of aliphatic aldehydes, suitable choice of chiral ligand afforded the corresponding cross aldols in high e.e. (Table 14) [5].

Table 14. Enantioselective Cross-Aldol Reaction between Aromatic Ketones and Various Aldehydes

Entry	Ketone	Aldehyde	Chiral Diamine	Yield/%	Erythro:Threo	Opt. Purity[a]/%
1	$PhCOCH_2CH_3$	PhCHO	(1u)	74	6:1	80[b]
2		*p*-Me-PhCHO	(1u)	72	8:1	80[b]
3		*p*-Cl-PhCHO	(1u)	72	6:1	85[b]
4		*p*-MeO-PhCHO	(1u)	78	8:1	80[b]
5	$PhCOCH_2CH_2CH_3$	PhCHO	(1u)	72	5:1	75[b]
6	$PhCOCH_3$	PhCHO	(1u)	35[e]	—	75[c]
7	$PhCOCH_2CH_3$	$(CH_3)_2CHCHO$	(1x)	69	>20:1	75[b]
8		$(CH_3)_3CCHO$	(1u)	57	*erythro* only	90[d]
9		cyclo-$C_6H_{11}CHO$	(1u)	67	4:1	80[b]

[a] of erythro isomer; threo isomer shows almost the same degree of enantioselection.
[b] Determined by ^1H- and ^{19}F-NMR measurement of its MTPA ester.
[c] Determined by the optical rotation of the acetate of cross-aldol product.
[d] Determined by using chiral shift reagent Eu(hfc)$_3$.
[e] Self-coupled product of acetophenone was obtained in 64% yield.

This asymmetric aldol reaction was further extended to the aldol-type reaction of 3-acetylthiazolidine-2-thione with various achiral aldehydes via divalent tin enolates using chiral diamine $1u$ as ligand. The corresponding aldol adduct 17 is obtained in high optical purity (Table 15) and, as previously mentioned, the adduct 17 is easily converted to the β-hydroxy aldehyde or β-hydroxy carboxylic acid derivative. Thus, this method constitutes a simple, versatile, and useful method for the preparation of a variety of optically active compounds [47].

Further, a highly enantioselective synthesis of 2-substituted malates is achieved by application of this reaction. The divalent tin enolate of 3-acetylthiazolidine-2-thione reacts with various α-ketoesters to afford the corresponding aldol-type products, generally in greater than 95% e.e. (Table 16) [48].

Moreover, we observed that in the reaction of 3-(2-benzyloxyacetyl)thiazolidine-2-thione (18) with aldehyde, reversal of relative stereochemistry was attainable commencing from the same reactants. That is, in the absence of diamine, the α-benzyloxy adduct was predominantly of *syn*-stereochemistry while in the presence of diamine (TMEDA),

Table 15. Enantioselective Aldol-type Reaction of 3-Acetylthiazolidone-2-thione

Entry	Aldehyde	Yield/%	$[\alpha]_{\hat{a}}^{21}$ (c,C_6H_6)	Opt. Purity[a]/%
1	PhCHO	79	−48.4° (1.1)	65[b]
2	$(CH_3)_2$CHCHO	63	−68.8° (0.8)	>90
3	$PhCH_2CH_2$CHO	76	−40.2° (1.4)	>90
4	cyclo-C_6H_{11}CHO	81	−61.5° (1.6)	88
5	C_2H_5CHO	70	−64.8° (1.0)	90
6	n-C_5H_{11}CHO	65	−51.6° (0.9)	90

[a] Determined by the measurement of the ^{1}H-NMR spectrum of the corresponding methyl ester using Eu(hfc)$_3$ as a chiral shift reagent. The -OCH$_3$ signal was completely separated.

[b] In this case, the absolute configuration of the adduct was determined to be S by the optical rotation of the β-hydroxy carboxylic acid. In other cases, the absolute configurations were not rigorously established; however, judging from the similarity in the chemical shifts of -OCH$_3$ signals using the chiral shift reagent, other aldol products are thought to have the same absolute configuration.

Table 16. Enantioselective Aldol-type Reaction with Various α-Ketoesters

Entry	R	Chiral Diamine	Yield/%	Opt. Purity[a]/%
1	CH_3	(1u)	53	18
2	CH_3	(1w)	44	≃ 2
3	CH_3	(1x)	61	54
4	CH_3	(1z)	66	61
5	CH_3	(1y)	74	85
6	C_6H_5-	(1y)	78	>95[b]
7	$(CH_3)_2$CH-	(1y)	75	>95[b]
8	$(CH_3)_2$CHCH$_2$-	(1y)	65	>95[b]
9	$CH_3O_2C(CH_2)_2$-	(1y)	80	>95[b]

[a] Determined by ^{1}H-NMR measurement of the corresponding dimethyl ester using chiral shift reagent, Eu(hfc)$_3$. The methoxy peak of methyl ester adjacent to the tertiary carbon center cleanly separated in all cases for *racemic* forms.

[b] No separation of the methoxy peak could be detected at all by ^{1}H-NMR analysis. Absolute configuration was assigned R based on determined sign of optical rotation. The stereochemical course of this reaction is thought to be the same in all examples.

Table 17. Aldol-type Reactions of 2-(Benzyloxyacetyl)-thiazolidine-2-thione

Aldehyde	Diamine	Yield/%	Syn : Anti	Opt. Purity[a]/%
Ph⌒CHO	—	62	74:26	—
	TMEDA	70	15:85	—
	Chiral Diamine[b]	81	13:87	94
n-C_5H_{11}CHO	—	62	74:26	—
	TMEDA	68	12:88	—
	Chiral Diamine[b]	75	17:83	90
(cyclohexyl)—CHO	—	74	56:44	—
	TMEDA	74	7:93	—
	Chiral Diamine[b]	68	7:93	87
(isopropyl)—CHO	—	60	61:39	—
	TMEDA	65	9:91	—
	Chiral Diamine[b]	70	9:91	87
PhCHO	—	95	23:77	—
	TMEDA	82	17:83	—
	Chiral Diamine[b]	93	19:81	90

[a] of *anti* isomer; determined by the measurement of the ^{1}H-NMR spectrum of the corresponding methyl ester using Eu(hfc)$_3$ as a chiral-shift reagent. The -OCH$_3$ signal is completely separated.
[b] (1 u) was employed.

anti-product prevailed. Furthermore, by employing the chiral diamine *1u*, up to 94% e.e. was observed in the *anti*-adduct (Table 17) [49].

6 Catalytic Asymmetric 1,4-Addition of Thiols to Cycloalkenone in the Presence of Chiral Diamino Alcohol

The asymmetric reactions described in Sect. 3 to 5 are all classified as "stoichiometric asymmetric reactions", that is, a stoichiometric amount of chiral auxiliary is essential. In contrast, "catalytic asymmetric reactions" are very attractive and highly desirable, since a chiral catalyst can greatly multiply chirality and are thus very economic.

With regard to the "catalytic asymmetric reaction", only a few successful examples, except those reactions using chiral transition metal complexes, have been reported. For example, the cinchona-alkaloid-catalyzed asymmetric 1,4-addition of thiols or β-keto esters to Michael acceptors [50], quinidine catalyzed the asymmetric addition of ketene to chloral [51], and the highly enantioselective 1,4-addition of β-keto esters in the presence of chiral crown ethers to Michael acceptors [52] have been most earnestly studied.

Table 18. Effects of the Structure of the Catalysts in the Asymmetric Addition of Benzenethiol to 2-Cyclohexen-1-one

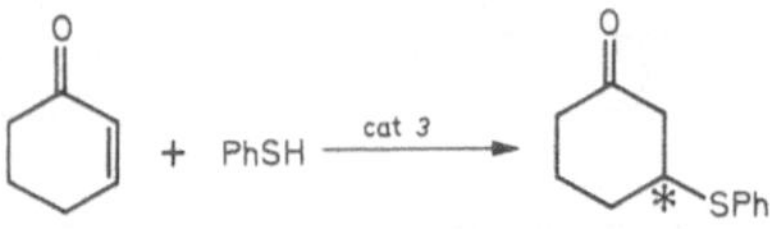

Entry	Catalyst	Yield/%	Opt. Yield/%
1	(3a)	86	67
2	(3b)	90	34
3	(3c)	67	69
4	(3d)	90	72
5	(3e)	81	57
6	(3f)	24	26
7	(3g)[a]	84	10
8	(3h)[b]	68	3
9	(3i)[c]	89	21

[a] (3g) is

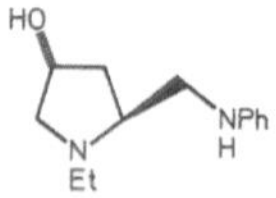

[b] (3h) is

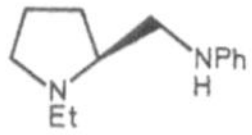

[c] (3i) is

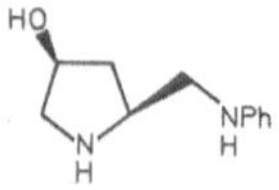

We examined optically active pyrrolidine derivatives as efficient chiral catalysts in the asymmetric 1,4-addition of thiols to α,β-unsaturated ketones. Thus, benzenethiol in the presence of 1 mol% of a chiral diamino alcohol, *3a–f*, as catalyst was added to 2-cyclohexen-1-one in toluene at −20 °C. The results are listed in Table 18 [7].

The most noteworthy point in this reaction is the pivotal role of the hydroxy group in the enantioselection (entries 1, 7, and 8). Obviously, the β-oriented hydroxy group of the catalyst *3a* directs the stereochemical course of this reaction, and either the absence or the difference in orientation of the hydroxy group of the pyrrolidine ring causes a dramatic decrease in the enantioselectivity of this reaction. While the optical yields are fairly good using catalysts having an aromatic ring (*3a*, *3c*, and *3d*), a substantial decrease in e.e. is observed with catalysts (*3b*) without aromatic moiety. When the catalysts *3e* and *3i* in which both of the amino groups are tertiary and secondary, respectively, are employed, the optical yields decrease as compared with the case of catalyst *3a* (entries 1, 5, and 9). The poor chemical yield found for entry 6 is attributable to the decrease of the basicity of the amine *3f* by virtue of the bulky substituent. Thus, it is noted that slight structural change in the catalyst causes a dramatic change in the optical yield in this reaction.

Table 19. Asymmetric Michael Addition of Various Thiols to 2-Cycloalkenones

Entry	R	Enone	Yield/%	Opt. Yield/%
1			83	77
2	Me—		75	73
3	Cl—		84	47
4	MeO—		75	83
5	t—Bu—		74	88
6	—CH₂—		22	1
7	t—Bu—		85	11
8			64	38
9			74	42

The results obtained by the reaction of various thiols and cycloalkenones using *3a* as catalyst are summarized in Table 19 [7b)].

The addition of aromatic thiols to 2-cyclohexen-1-one gave satisfactory results. Among them, the addition of *p-tert*-butylbenzenethiol furnished the adduct in 88 % e.e. with the R enantiomer being predominantly formed. On the other hand, the alkanethiol (entry 6) gave a poor result, and this may be attributable to the difference of the pK_a values of the thiols. The addition of *p-tert*-butylbenzenethiol to other cycloalkenones afforded adducts in moderate optical yields [7b)].

Concerning the stereochemical course of this reaction, Wynberg has proposed a mechanism based mainly on kinetic measurements and structures of substrates in pioneering work on the asymmetric thiol addition catalyzed by cinchona alkaloids. In our case, various sterically modified analogs of the base catalyst are readily achievable, and elucidation of the mechanism appears mainly dependent on the effects of the structure of the catalyst on the optical yields. From the experimental results, three notable points must be taken into consideration, namely,

I) the role of the hydroxy group of the catalysts,

II) the role of the aromatic ring,

III) the absolute configuration of the predominantly formed enantiomer, (R).

With these three points in mind, and from molecular model inspection, the following mechanistic pathway is proposed (Fig. 10).

In the initial stage of the reaction, an ammonium thiolate complex is formed, where the ammonium counterpart has a *cis*-fused 5-membered bicyclic ring structure by virtue of hydrogen bonding as illustrated in *A*. This rigid structure is proposed to be a crucial factor in the enantioselection, because the enantioselectivity decreased when the substituent of the amine became more bulky, where such a structure is no

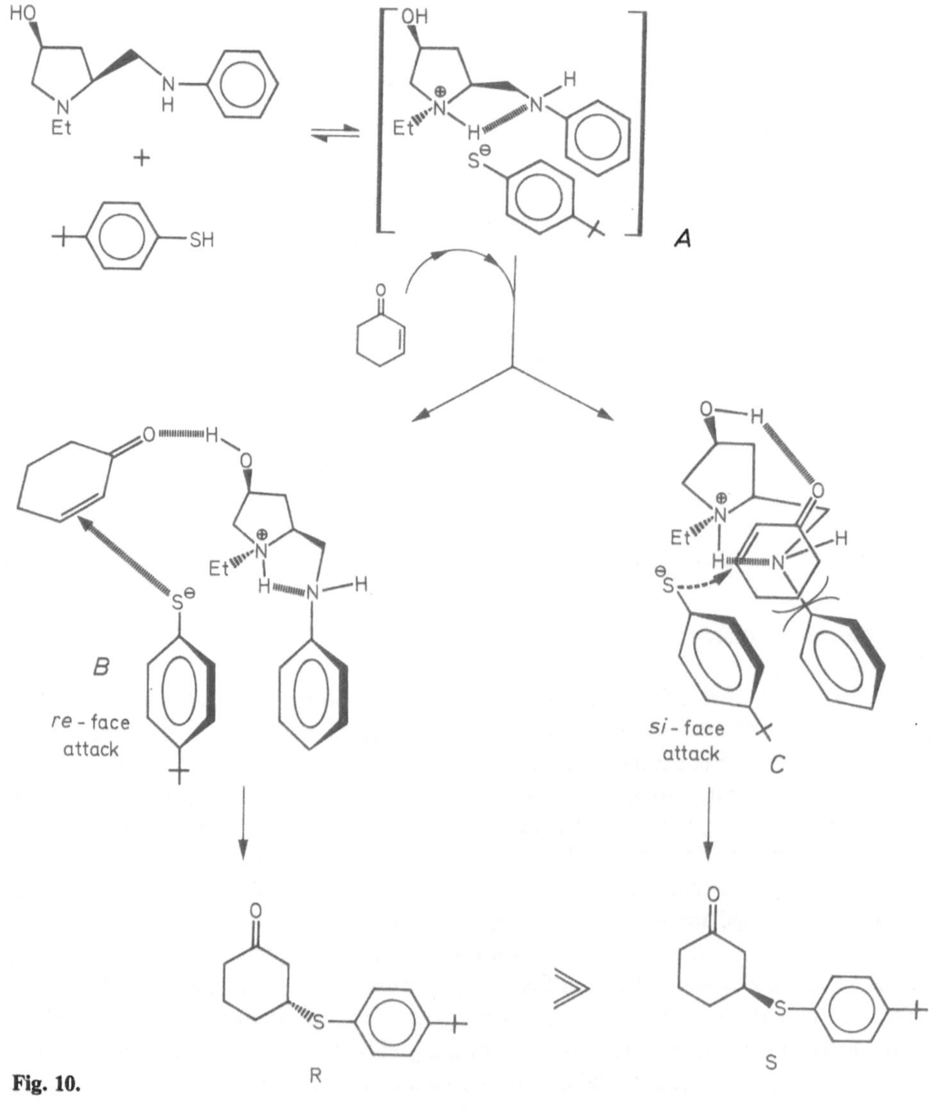

Fig. 10.

longer favorable. Moreover, in a solvent like toluene, the thiolate anion is stabilized by $\pi-\pi$ interaction, and freedom of orientation is further limited by this interaction which the aromatic ring of the catalyst. At the next stage, the approach of the cyclohexenone to the complex A takes place with a hydrogen-bond interaction between the carbonyl group and the hydroxy group of the catalyst. In a protic solvent like ethanol, this interaction is effectively cancelled with no enantioselection being observed. Finally, the attack of the thiolate anion to the enone can occur, via two possible pathways, B and C. In case of pathway C, the steric congestion of the anilino group and the cyclohexenone ring may prevent the complex A and the enone from reaching the desired transition state. Thus, the preferential attack on the re-face of 2-cyclohexen-1-one, through pathway B, takes place resulting in the predominant formation of the R enantiomer.

Furthermore, the stereospecific transformation of (R)-3-(p-tert-butylphenylthio)-cyclohexan-1-one, obtained in optically pure form by recrystallization, into cyclohexanol derivatives was achieved by employing several common reducing reagents. The results are summarized in Table 20 [53].

Numerous data have been recorded concerning the stereochemistry of the reduction of the substituted cyclohexanone derivatives, and in general, preferential axial attack occurs with LiAlH$_4$ derivatives, whereas equatorial attack takes place for trialkylborohydrides. This tendency is valid for this case and it is noteworthy that (1S, 3R)-3-(p-tert-butylphenylthio)cyclohexanol is obtained almost exclusively by employing LiAlH(OBu^t)$_3$ (entry 9). Thus, the 3-arylthio asymmetric center is effectively transferred to C-1, producing both configurational isomers (R and S) by suitable choice of reducing reagent.

Table 20. Stereoselective Reduction of 3-(p-tert-butylphenylthio)cyclohexanone

Entry	Reductant	Temp./°C	Yield[a]/%	Cis/Trans[b]
1	NaBH$_4$[c]	0	95	84/16
2	LiAlH$_4$		87	91/ 9
3		− 78	100	94/ 6
4	LiAlH(OBu^t)$_3$	− 78	85	98/ 2
5	L-Selectride	− 78	96	30/70
6		−100	84	17/83
7	N-Selectride	− 78	80	12/88
8	K-Selectride	− 78	90	9/91
9		−100	95	4/96

[a] Combined yield of both diastereomers after purification with salica-gel TLC.
[b] Determined by HPLC (Merck LiChrosorb SI60: AcOEt-hexane).
[c] Ethanol was used as a solvent.

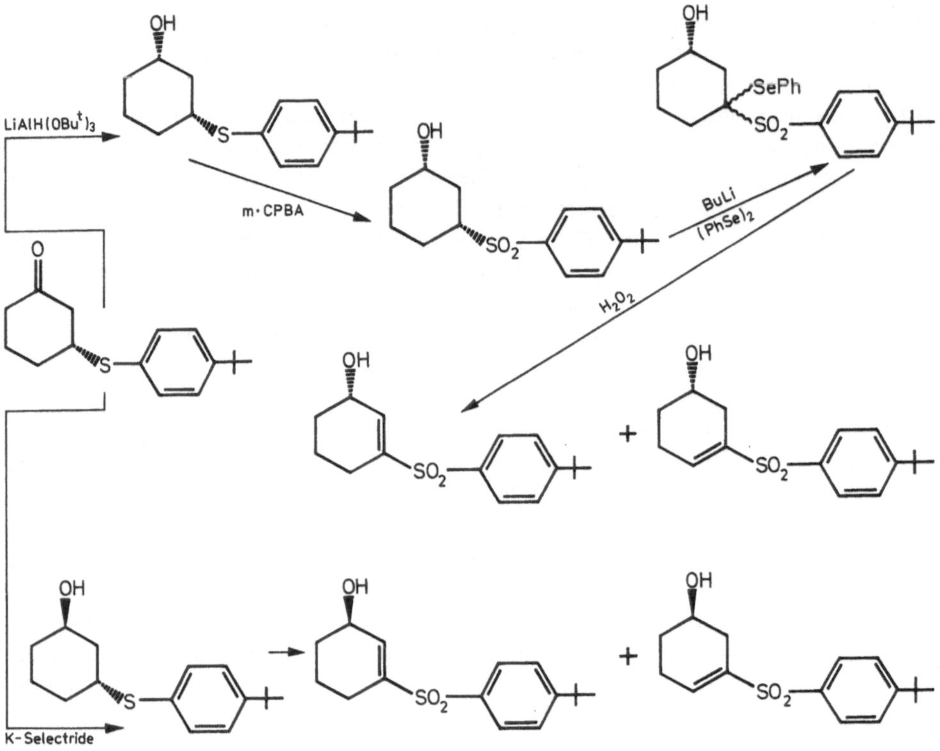

Scheme 9.

Optically pure 2-cyclohexen-1-ol derivatives were obtained from each 3-(*p-tert*-butylphenylthio)cyclohexanol enantiomer, as shown in Scheme 9.

7 Summary

As has been illustrated throughout this chapter the concept of non-covalent bonded interactions between organometallic reagents and chiral auxiliary has led to highly successful and versatile diastereoface and enantioface-differentiating reactions, which have in turn found fruitful application to the asymmetric synthesis of several natural products. In conclusion, the use of economically feasible, readily available, and easily recoverable chiral auxiliaries derived from (S)-proline (and L-4-hydroxy-proline) has elegantly demonstrated the following significant features for successful asymmetric synthesis based on this fundamental approach.

(i) *The generation of bidentate (or tridentate) "chelate complexes" by non-covalent bonded interaction with metal center;*

(ii) *"chelate complexes" are necessarily of conformationally restricted cis-fused 5-membered ring structure;*

(iii) *a rigid ring structure of this type efficiently controls orientation of reacting substrates leading to superior asymmetric induction.*

The overall simplicity of the asymmetric reactions discussed in this review offers a conceptually new dimension to the field of asymmetric synthesis.

164

8 List of Symbols and Abbreviations

Z = benzyloxycarbonyl
DCC = dicyclohexylcarbodiimide
$LiAlH_4$ = lithium aluminium hydride
BOC = *tert*-butoxycarbonyl
HBr = hydrogen bromide
AcOH = acetic acid
e.e. = enantiomeric excess
DIBAL-H = diisobutylaluminium hydride
CuI = cuprous iodide
CH_2Cl_2 = dichloromethane
TMEDA = N,N,N',N'-tetramethylethylenediamine
$LiAlH(OBu^t)_3$ = lithium tri-*tert*-butoxyaluminium hydride

9 References

1. (a) Morrison, J. D., Mosher, H. S.: Asymmetric Organic Reactions, Prentice-Hall. Englewood Cliffs, N.J. 1971
   (b) Asymmetric Reactions and Processes (Amer. Chem. Soc. Symp.) (eds. Eliel, E. L., Otsuka, S.): Amer. Chem. Soc., Wahington, D.C. 1982
   (c) Asymmetric Synthesis Vol. 2, (ed. Morrison, J. D.), Academic Press, New York 1983
2. (a) Valentine, D., Scott, J. D.: Synthesis 329 (1978)
   (b) Kagan, H. B., Fiaud, J. C.: New approach in asymmetric synthesis, in: Top. Stereochem., Vol 10, p. 175, (eds., Eliel, E. L., Allinger, N. L.) John Wiley, New York 1978
   (c) ApSimon, J. W., Seguin, R. P.: Tetrahedron *35*, 2797 (1979)
   (d) Mosher, H. S., Morrison, J. D.: Science *221*, 1013 (1983)
3. (a) Mukaiyama, T., Asami, M., Hanna, J., Kobayashi, S.: Chem. Lett. 783 (1977)
   (b) Asami, M., Ohno, H., Kobayashi, S., Mukaiyama, T.: Bull. Chem. Soc. Japan *51*, 1869 (1978)
   (c) Asami, M., Mukaiyama, T.: Heterocycles *12*, 499 (1979)
4. Asami, M.: Chem. Lett. 829 (1984)
5. Iwasawa, N., Mukaiyama, T.: ibid. 1441 (1982)
6. (a) Mukaiyama, T., Soai, K., Kobayashi, S.: ibid. 219 (1978)
   (b) Soai, K., Mukaiyama, T.: ibid. 491 (1978)
   (c) Mukaiyama, T., Soai, K., Sato, T., Shimizu, H., Suzuki, K.: J. Am. Chem. Soc. *101*, 1455 (1979)
   (d) Sato, T., Soai, K., Suzuki, K., Mukaiyama, T.: Chem. Lett. 601 (1978)
7. (a) Mukaiyama, T., Ikegawa, A., Suzuki, K.: ibid. 165 (1981)
   (b) Suzuki, K., Ikegawa, A., Mukaiyama, T.: Bull. Chem. Soc. Japan *55*, 3277 (1982)
8. (a) Hojo, K., Mukaiyama, T.: Chem. Lett. 619 (1976)
   (b) Hojo, K., Mukaiyama, T.: ibid. 893 (1976)
   (c) Hojo, K., Yoshino, H., Mukaiyama, T.: ibid. 133, 437 (1977)
   (d) Hojo, K., Kobayashi, S., Soai, K., Ikeda, S., Mukaiyama, T.: ibid. 635 (1977)
9. Cervinka, O., Malon, P., Prochazkova, H.: Collect. Czech. Chem. Commun. *39*, 1869 (1974) and references cited therein
10. (a) Landor, S. R., Miller, B. J., Tachell, A. R.: J. Chem. Soc. C. 197 (1967)
    (b) Baggett, N., Stribblehill, P.: J. Chem. Soc. Perkin Trans. *1*, 1123 (1977)
11. (a) Yamaguchi, S., Mosher, H. S.: J. Org. Chem. *38*, 1870 (1973)
    (b) Vigneron, J. P., Jaquet, I.: Tetrahedron 939 (1976)
12. Meyers, A. I., Kendall, P. M.: Tetrahedron Lett. 1337 (1974)
13. Seebach, D., Daume, H.: Chem. Ber. *107*, 1748 (1974)

14. Yamaguchi, S., Yasuhara, F., Kabuto, K.: J. Org. Chem. *42*, 1578 (1977)
15. Haller, R., Schneider, H. J.: Chem. Ber. *106*, 1312 (1973)
16. (a) Noyori, R., Tamino, I., Tanimoto, Y.: J. Am. Chem. Soc. *101*, 3129 (1979)
    (b) Terashima, S., Tanno, N., Koga, K.: Chem. Lett. 981 (1980)
17. (a) Nozaki, H., Aratani, T., Toraya, T.: Tetrahedron Lett. 4097 (1968)
    (b) Nozaki, H., Aratani, T., Toraya, T., Noyori, R.: Tetrahedron *27*, 905 (1971)
    (c) Seebach, D., Dörr, H., Bastani, B., Ehrig, V.: Angew. Chem. *81*, 1002 (1969)
    (d) Seebach, D., Kalinowski, H., Bastani, B., Crass, G., Daum, H., Dörr, H., DuPreez, N. P., Ehrig, V., Langer, W., Nussler, C., Oei, H.-A., Schmidt, M.: Helv. Chim. Acta *60*, 301 (1977)
18. Zweig, J. S., Luche, J. L., Barreiro, E., Crabbé, P.: Tetrahedron Lett. 2355 (1975)
19. (a) Blomberg, C., Coops, J.: Rec. Trav. Chim. Pays-Bas *83*, 1083 (1964)
    (b) French, W., Wright, G. F.: Can. J. Chem. *42*, 2474 (1964)
    (c) Inch, T. D., Lewis, G. J., Sainsbury, G. L., Sellers, D. J.: Tetrahedron Lett. 3657 (1969)
    (d) Meyers, A. I., Ford, M. E.: ibid. 1341 (1974)
    (e) Iffland, D. C., Davis, J. E.: J. Org. Chem. *42*, 4150 (1977)
20. Bruer, H.-J., Haller, R.: Tetrahedron Lett., 5227 (1972)
21. Boireau, G., Abenheim, D., Bourdais, J., Henry-Basch, E.: ibid. 4781 (1976)
22. Mazaleyrat, J. P., Cram, D. J.: J. Am. Chem. Soc. *103*, 4585 (1981)
23. Soai, K., Mukaiyama, T.: Bull. Chem. Soc. Japan *52*, 3371 (1979)
24. Mukaiyama, T., Suzuki, K., Soai, K., Sato, T.: Chem. Lett. 447 (1979)
25. Mukaiyama, T., Suzuki, K.: ibid. 255 (1980)
26. Thummel, R. P., Rickborn, B.: J. Am. Chem. Soc. *92*, 2064 (1970)
27. Whitesell, J. K., Felman, S. W.: J. Org. Chem. *45*, 755 (1980)
28. (a) Sone, T., Terashima, S., Yamada, S.: Chem. Pharm. Bull. *24*, 1273 (1976)
    (b) Sone, T., Terashima, S., Yamada, S.: ibid. *24*, 1288 (1976)
29. (a) Meyers, A. I., Poindexter, G. S., Brich, Z.: J. Org. Chem. *43*, 892 (1978)
    (b) Hashimoto, S., Yamada, S., Koga, K.: J. Am. Chem. Soc. *98*, 7450 (1976)
    (c) Hashimoto, S., Komeshima, N., Yamada, S., Koga, K.: Tetrahedron Lett. 2907 (1977)
30. Enders, D., Eichennauer, H.: ibid. 191 (1977)
31. (a) Eliel, E. L., Koskimies, J. K., Lohri, B.: J. Am. Chem. Soc. *100*, 1614 (1978)
    (b) Eliel, E. L., Frazee, W. J.: J. Org. Chem. *44*, 3598 (1979)
    (c) Eliel, E. L., Soai, K.: Tetrahedron Lett. *22*, 2859 (1981)
32. Mukaiyama, T., Sakito, Y., Asami, M.: Chem. Lett. 1253 (1978)
33. Mukaiyama, T., Sakito, Y., Asami, M.: ibid. 705 (1979)
34. Sakito, Y., Mukaiyama, T.: ibid. 1027 (1979)
35. Sakito, Y., Tanaka, S., Asami, M., Mukaiyama, T.: ibid. 1223 (1980)
36. Sakito, Y., Suzukamo, G.: Tetrahedron Lett. 4953 (1982)
37. Asami, M., Mukaiyama, T.: Chem. Lett., 93 (1983)
38. Sakito, Y., Asami, M., Mukaiyama, T.: ibid. 455 (1980)
39. Asami, M., Mukaiyama, T.: ibid. 569 (1979)
40. Asami, M., Mukaiyama, T.: ibid. 17 (1980)
41. Mukaiyama, T., Stevens, R. W., Iwasawa, N.: ibid. 353 (1982)
42. Stevens, R. W., Iwasawa, N., Mukaiyama, T.: ibid. 1459 (1982)
43. Mukaiyama, T., Iwasawa, N.: ibid. 1903 (1982)
44. (a) Eichenauer, H., Friedrich, E., Lutz, W., Enders, D.: Angew. Chem. (Int. Ed. Engl.), *17*, 206 (1978)
    (b) Sugasawa, T., Toyoda, T.: Tetrahedron Lett. 1423 (1979)
    (c) Heathcock, C. H., Pirrung, M. C., Buse, C. T., Hagen, J. P., Young, S. D., Sohn, J. E.: J. Am. Chem. Soc. *101*, 7077 (1979)
    (d) Heatcock, C. H., White, C. T., Morrison, J. J., VanDerveer, D.: J. Org. Chem. *46*, 1296 (1981)
    (e) Heathcock, C. H., Pirrung, M. C., Lampe, J., Buse, C. T., Young, S. D.: ibid. *46*, 2290 (1981)
    (f) Evans, D. A., Bartroli, J., Shih, T. L.: J. Am. Chem. Soc. *103*, 2127 (1981)
    (g) Evans, D. A., McGee, L. R.: ibid, *103*, 2876 (1981)
    (h) Evans, D. A., Nelson, J. V., Vogel, E., Taber, T. R.: ibid. *103*, 3099 (1981)

(i) Masamune, S., Ali, Sk. A., Snitman, D. L., Garvey, D. S.: Angew. Chem. (Int. Ed. Engl.), *19*, 557 (1980)

(j) Masamune, S., Choy, W., Kerdeskey, F. A. J., Imperiali, B.: J. Am. Chem. Soc. *103*, 1566 (1981)

(k) Masamune, S., Hirama, M., Mori, S., Ali, Sk. A., Garvey, D. S.: ibid. *103*, 1568 (1981)

45. Meyers, A. I., Yamamoto, Y.: ibid. *103*, 4278 (1981)
46. Guetté, M., Guetté, J. P., Capillon, J.: Tetrahedron Lett. 2863 (1971)
47. Iwasawa, N., Mukaiyama, T.: Chem. Lett. 297 (1983)
48. Stevens, R. W., Mukaiyama, T.: ibid. 1799 (1983)
49. Mukaiyama, T., Iwasawa, N.: ibid. 753 (1984)
50. (a) Hiemstra, H., Wynberg, H.: J. Am. Chem. Soc. *103*, 417 (1981)
    (b) Wynberg, H., Greijdanus, B.: J. Chem. Soc. Chem. Commun. 427 (1978)
51. Wynberg, H., Staring, E. G. J.: J. Am. Chem. Soc. *104*, 166 (1982)
52. Cram, D. J., Sogah, G. D. Y.: J. Chem. Soc. Chem. Commun. 625 (1981)
53. Suzuki, K., Ikegawa, A., Mukaiyama, T.: Chem. Lett. 899 (1982)

Solid Charge-Transfer Complexes of Phenazines

Heimo J. Keller[1] and Zoltán G. Soos[2]

1 Anorganisch-Chemisches Institut der Universität Heidelberg, Im Neuenheimer Feld 270, 6900 Heidelberg 1, FRG
2 Department of Chemistry, Princeton University, Princeton, N.J. 08544, USA

Table of Contents

169

Topics in Current Chemistry, Vol. 127
Managing Editor Dr. F. L. Boschke
© Springer-Verlag Berlin Heidelberg 1985

1 Introduction

1.1 Phenazine Derivatives as π-Donors

This review summarizes the preparation, structural chemistry, and selected physical properties of organic molecular solids in which substituted phenazines act as π-electron donors (D). The parent phenazine (P) is a weak π-acceptor (A).

Alkyl- or aryl-addition at the 5 position yields donor radicals of the form $R\dot{P}$, which produce diamagnetic RP^+ cations in solid-state complexes with strong π-acceptors. 5.10-Disubstituted 5.10-dihydrophenazines, $RR'P$, are among the most powerful π-donors known and readily form charge-transfer (CT) solids containing $RR'\dot{P}^+$ cation radicals.

Substituted phenazines yield free-radical solids, ion-radical salts, CT complexes,

Table 1. Molecular structure and abbreviation of neutral phenazine donors discussed in this review

DONOR	R	DONOR	R	R'
$H\dot{P}$	H	H_2P	H	H
$M\dot{P}$	CH_3	HMP	H	CH_3
		M_2P	CH_3	CH_3
		MEP	CH_3	C_2H_5
$E\dot{P}$	C_2H_5	HEP	H	C_2H_5
		E_2P	C_2H_5	C_2H_5
$P\dot{P}$	$n\text{-}C_3H_7$	P_2P	$n\text{-}C_3H_7$	$n\text{-}C_3H_7$
$B\dot{P}$	$n\text{-}C_4H_9$	B_2P	$n\text{-}C_4H_9$	$n\text{-}C_4H_9$
		Ph_2P	C_6H_5	C_6H_5
$Be\dot{P}$	$CH_2\text{-}C_6H_5$	Be_2P	$CH_2\text{-}C_6H_5$	$CH_2\text{-}C_6H_5$

and organic alloys which show enormous structural variety, different degrees of ionicity, and novel paramagnetic properties. We list in Table 1 monosubstituted phenazine radicals $R\dot{P}$ and disubstituted $RR'P$ donors.

Several individual solid phenazines have previously been of interest.

Solid 5-ethylphenazinyl [1] (EṖ) is a stable free-radical solid [2] with antiferro-magnetic spin exchange and a low-lying CT excitation corresponding

$$2E\dot{P} \xrightarrow{h\nu_{CT}} EP^+ + EP^-$$ (1)

to a Mulliken self complex [3]. Solid EṖ closely resembles [2] the more extensively studied [4] Wurster's blue perchlorate, based on the cation radical TMPD$^{+\cdot}$ of NNN'N'-tetramethyl-1.4-diaminobenzene. The structure of phenazine [5] and its largely neutral complexes [6] have been pursued within the larger context of π-molecular organic solids [7]. Optically excited triplet states, for example in solid phenazine pyromellithic acid, have also received attention [8]. The paramagnetism of RR P$^+$ salts, by contrast, has been noted [9] in passing but not previously investigated as a separate topic.

Interest in organic conductors focused initially on the powerful π-acceptor TCNQ, 2.2'-(cyclohexadiene-1.4-diylidene)bispropanedinitrile, prepared by the Du Pont group [10,11]. Their best conductor [12] was MP-TCNQ, the first organic "metal". MP-TCNQ was subsequently studied in detail by Heeger and coworkers [13-15], whose application of solid-state techniques to organic systems has been central to the characterization of even better molecular conductors [16,17], polymeric conductors [18,19], and organic superconductors [20]. The MP-TCNQ analysis in terms of complete charge transfer, or MP$^+$TCNQ$^{\pm}$, need reinterpretation, however, since later work [21] demonstrates partial ionicity q < 1 in MP^{+q}TCNQ^{-q}. The degree of ionicity, q, in π-molecular conductors and CT complexes has emerged [22,23] as the crucial variable for high conductivity. We will frequently attempt to assign q in phenazine-related solids from structural or spectroscopic data.

The enormous literature [24] on the biological role of phenazines, which deals mainly with CT complexes in solution, will not be discussed. Their facile one-electron redox reactions [24,25] in conjunction with proton transfer and H-atom abstraction, are also important in the solid state [26], as shown in Fig. 1 for HMP. Such electron, proton, or H-atom transfers rationalize in retrospect their structural variety, although several specific aspects elude such generalizations. Substituted phenazine donors

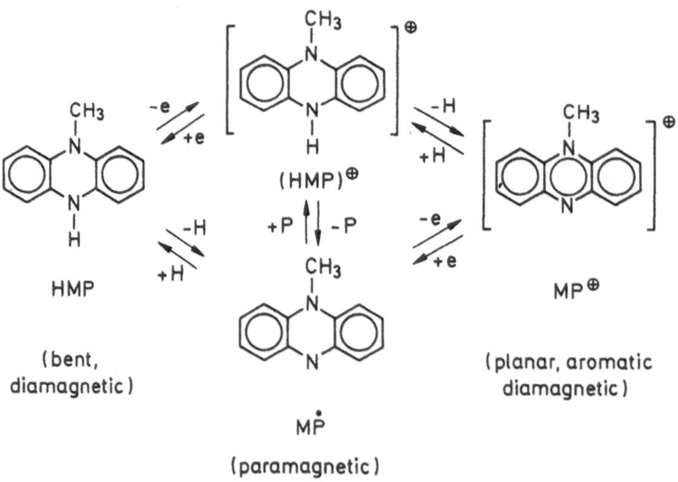

Fig. 1. Diamagnetic and paramagnetic species in the series HMP, HMP$^+$, MP and MP$^+$. (With permission from Ref. [26])

in the solid state thus occur as neutral, partly-ionic, or fully ionic species that, in addition, may also be diamagnetic or paramagnetic.

1.2 Scope and Goals

Both chemists and physicists have become interested in the remarkable properties of ion-radical organic solids and related inorganic and polymeric systems. Recent reviews and conference proceedings document [17-19, 27-35] the sometimes parallel, sometimes divergent viewpoints of current workers. The field is very active, interdisciplinary, recent, and often difficult. There are questions about the preparation and crystallization of chemically pure materials, with the interpretation or relevance of subsequent measurements, and with applications of theoretical models whose implications are far from clear. The description of ion-radical solids, solid CT complexes, mixed-valent conductors, and conjugated polymers share some qualitative features mentioned in Section 2.2. Their incorporation then reveal crucial differences in detailed transport, structural, chemical, and spectroscopic analyses. Some prototypical systems are consequently in the limelight at any time, as illustrated by the early focus on MP-TCNQ.

Subsequent work on phenazine falls into a broad second category of relatively well-characterized systems. The conducting alloys [35] $P_x(MP)_{1-x}TCNQ$, with $0 \leq x \leq 0.5$, have been studied by a variety of physical methods. The intrinsic complications of alloys, as well as the specific chemical and structural questions about MP-TCNQ in Section 3.2., led to a smaller effort and less definitive results than for the better conductor TTF-TCNQ. The transport and physical properties of these materials are nevertheless the most extensively studied among the phenazines.

The present work deals with solid CT complexes of phenazines which, in the most favorable cases, include the crystal structure, single-crystal electron spin resonance (epr) data, the temperature dependence of the magnetic susceptibility or of the powder conductivity, and a CT absorption spectrum. We have not pursued complexes that could not be crystallized, nor have we investigated completely similar materials.

We note in Section 3, 4 and 5 many $R\dot{P}$ and $RR'\dot{P}^+$ systems for which the data are scanty, even though crystals could probably be grown. The chemistry of less favored donors is just as interesting. Many new systems besides phenazines illustrate the infinite variety of synthetic organic chemistry. Such molecular engineering is however more discussed than pursued, judging by the concentration of physical measurements to a few systems. As there are so few phenazine complexes with high conductivity or other exploitable properties, we chose complexes with novel magnetic or structural features.

2 Classification of Phenazine Complexes: Structure and Ionicity

2.1 Chemistry of Phenazines

The parent phenazine (P) is [5] a typical van der Waals molecular crystal. Its CT complexes [36] have long been known, especially with its 5.10-dihydrogenated form, H_2P in Table 1, the so-called phenazhydrines [37-39]. P acts as a weak π-acceptor here and in combination with potent π-donors like phenylenediamines [40] it behaves as a weak donor with strong acceptors like iodine [41], TCNQ [6], or Lewis acids [42,43].

The rich redox chemistry of phenazine was already recognized in early work [42-45], and the reduced forms RR'P were identified to be potent donors [36-39]. Thus P can be reduced to H_2P with [42,43] H_2S or [44] NaHS in ethanol solutions containing ammonia or photolytically [46]. H_2P can also be obtained by direct synthesis from 1.2-diaminobenzene and 1.2-dihydroxybenzene [43]. It readily forms CT complexes, some of which can be crystallized, with a variety of acceptors [37-39,47]. With strong acceptors (oxidation agents), 5.10-dihydrophenaziniumyl ($H_2\overset{+}{P}$) salts are obtained [45-47].

The chemistry of 5-alkyl- or 5-arylphenazines (R$\overset{.}{P}$) is more complicated. They occur as stable, diamagnetic RP$^+$ salts with counterions like ClO_4^-, PF_6^-, or monoalkylsulfates. The most famous example is 5-methylphenazinium (MP$^+$) methosulfate ($CH_3SO_3^-$), first prepared by Kehrmann and Havas [48] in 1913, which has become a well-known redox catalyst in biological systems [24]. RP$^+$ salts can be reduced chemically [24,49] or photolytically [50-52] to the "semiquinoidal" 5.10-dihydro-5-alkylphenaziniumyl radical cations (HR$\overset{.}{P}$+). RP$^+$ salts can be oxidized chemically [53] or photolytically[1] to the biologically important pyocyanine dyes (1-hydroxy-5-methylphenazine) or reduced to 5-alkyl- or 5-aryl-5.10-dihydrophenazines (HRP) [39, 49a]. Hydrogen abstraction from the latter with PbO$_2$ yields R$\overset{.}{P}$ radicals. The stable free-radical solid [1] E$\overset{.}{P}$ has already been mentioned in connection with magnetic and optical studies [2]. Replacing methyl in Fig. 1 with alkylgroups (R) yields a general scheme for preparing phenazine donors.

Finally, 5.10-dialkyl-5.10-dihydrophenazine (RR'P) compounds can be prepared by dialkylating phenazine dianions [54], by alkylating RP$^+$ salts[1], by reducing 5.10-dimethoxymethyl derivatives with Grignards [55], by direct synthesis from N,N'-dialkyl-1.2-diaminobenzenes and cyclohexane-1.2-dione [38], or by reacting the substituents as *free radicals* with neutral phenazine [56]. Neutral R$_2$P compounds are strong π-donors and form fully ionic solids with the best π-acceptors (like TCNQF$_4$). They are oxidized by weaker acceptors to form complexes whose partial ionicity is discussed more completely in Section 4. We consider principally phenazine complexes with the strong π-acceptors like TCNQ and TCNQF$_4$.

TCNQ   TCNQF$_4$

CT complexes are prepared either through the usual metathesis reaction [10, 12]

$$LiTCNQ + RPClO_4 \rightarrow RP\text{-}TCNQ + LiClO_4$$

or directly by mixing solutions of neutral donors and acceptors and precipitating the complex. The formation of single crystals can be enhanced by using slow diffusion for the mixing and by careful control of the temperature.

Herbstein [7] has discussed extensively the packing and energetics of π-molecular organic solids, in which many weak secondary interactions compete on nearly equal footing. Thus dispersion forces, dipole-dipole interactions, hydrogen bonds, and CT stabilization all contribute in largely neutral solids [57, 58]. Partial ionicity and open electronic shells require including electrostatic (Madelung) and band contributions [22]. The cohesive energy of partly-ionic systems [58], including MP-TCNQ and other phenazines, still poses unsolved problems. There is no question, at the present, of controlling or predicting the interplay among these forces. The structural chemistry of substituted phenazines is limited instead to classifying the major observed packing motifs.

2.2 Open-Shell Organic Molecular Solids

Molecular solids containing free radicals like $R\dot{P}$, cation radicals D^+, or anion radicals A^- constitute a distinct new class of elementary solids in which modest electronic overlap coexists with open electronic shells [22]. The resulting open-shell organic solids have spectacularly extended [27-35] the phenomena associated with typical van der Waals organic solids. The past several decades has seen the characterization of several families of good organic conductors, a few superconductors, and a host of paramagnetic semiconductors, CT complexes and photoconductors, and low-dimensional magnets.

The modest π-π overlap in open-shell molecular solids suggests an approach based on separated molecular fragments [59]. Structural and spectroscopic evidence supports the occurrence of essentially unperturbed molecules or molecular ions in the solid state. Since valence-bond (VB) treatments of molecules become exact in the dissociated-atom limit, a diagrammatic VB approach has been developed [60-65] for open-shell molecular solids. The resulting correlated crystal states are simply weighted linear combinations of VB structures for the entire solid.

The most telling evidence for correlated states, rather than the simple band theory of metals, is the semiconduction of half-filled regular D^+ or A^- stacks. As first realized by Mott [66], sufficiently small overlap then stabilizes a covalent VB state with one valence electron per site. Electrical conduction requires a finite activation energy that is related to the ionization potential, I, and electron affinity, A. Partial occupancy, with q either greater or smaller than unity, results in many degenerate VB structures that permit nonactivated conduction without ever requiring a net ionization. Several closely related [67, 68] isostructural pairs of organic solids confirm high conduction for fractional q and semiconduction for q = 1. The first demonstration [21] of partial ionicity in MP-TCNQ was motivated by such considerations.

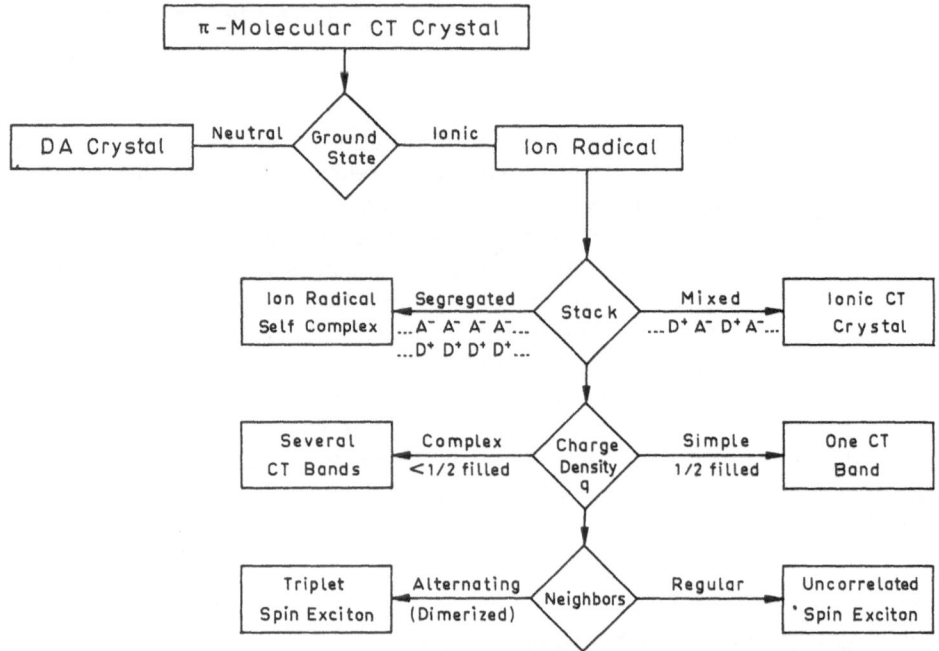

Fig. 2. Schematic classification of π-molecular CT crystals, with emphasis on the structural variations in ion-radical crystals. (With permission from Ref. [22])

A number of different structure/ionicity possibilities arise for packing open-shell molecular subunits. Typical closed-shell π-molecular solids have a lowest-energy intramolecular excitation involving a $\pi \rightarrow \pi^*$ or $n \rightarrow \pi^*$ transition. CT and ion-radical solids, by contrast, have a lowest excitation involving an electron transfer, as illustrated in Eq. (1) for EṖ. The classification [22] of π-molecular organic solids in Fig. 2 then follows naturally. We distinguish first between *segregated* stacks of donors (...DDDD....) or acceptors (...AAAA...) and mixed stacks (...DADA...) in CT complexes. The degree of ionicity, q, of D^{+q} and A^{-q} sites describes the filling of the valence band. *Simple* systems with integral q have a single CT excitation and are semiconductors; *complex* systems have nonintegral q. Equally strong overlap with both neighbors yields a *regular* stack, while preferential interaction with one neighbor results in *alternating* arrays of dimers, trimers, tetramers, etc. whose triplet (Frenkel) spin excitons were the focus of early work [69].

The structural diversity of substituted phenazines is illustrated by their CT complexes with TCNQ and its fluorinated analogue, TCNQF₄. The latter is an even stronger acceptor [70] with similar steric requirements. In contrast to segregated stacks in MP-TCNQ, EP-TCNQ crystallizes [71] in mixed ionic stacks with a long (1.631 Å) σ-bond between TCNQ⁻ ions in adjacent stacks. This unusual and unexpected structure is discussed in Section 3.2. The 1-butyl complexes BP-TCNQ and BP-TCNQF₄, on the other hand, form novel ...D⁺D⁺A⁻A⁻D⁺D⁺A⁻A⁻... stacks [72, 73] in which D⁺ is diamagnetic and A⁻ is paramagnetic. Such structures also require going beyond the possibilities in Fig. 2.

Several HRP-TCNQ and RR'P-TCNQ complexes [74] illustrate mixed dimerized stacking, as confirmed by their triplet spin excitons (TSE) spectra [74, 75]. Such dimerized CT complexes are apparently common, but previous examples had essentially neutral ground states and required optically excited triplets. The partial ionicity of phenazine-TCNQ complexes results in a lower, thermally accessible triplet state.

The classification in Fig. 2 contains other oversimplifications. For example, the asymmetry of MP$^+$ cations results in a disordered MP-TCNQ structure [76]. Such disorder leads to characteristic power laws in thermodynamical properties [77, 78, 63] as illustrated most completely for quinolinium (TCNQ)$_2$. Chemical disorder provides another possibility. Miller and Epstein have exploited phenazine doping [35, 79-83] in segregated-stack alloys of the type P$_x$(MP)$_{1-x}$TCNQ for $0 \leqq x \leqq 0.5$. Phenazine doping [84] of M$_2$P-TCNQ, which has a mixed dimerized stack, restores segregated stacking and high conductivity in P(M$_2$P)(TCNQ)$_2$, a system that closely resembles MP-TCNQ. The M$_6$P-TCNE complex [85] in Section 4.4. probably shows yet another kind of disorder involving preferential TCNE-allyl interactions.

Additional structural possibilities in Sections 3–5 include complex salts, with 2:3 or 1:2 stoichiometry for phenazine and TCNQ, and many polyiodides. Phenazines thus hardly fit into convenient isostructural families that facilitate systematic investigations. The organic donors related to tetrathiafulvalene (TTF) and macrocyclic organometallic donors [34] involving porphyrins and phthalocyanines have been notably successful in producing isostructural series. The more numerous TCNQ salts and substituted phenazines display instead structural variety, although isostructural pairs undoubtedly exist. Now differences rather than similarities become interesting.

2.3 Partial Ionicity and Magnetic Gap

The sharp separation [86] of CT solids in Fig. 2 into neutral (...DADA...) and ionic (...D$^+$A$^-$D$^+$A$^-$...) is not satisfied by several phenazine-TCNQ complexes whose ground states are best characterized [87] as partly-ionic, with fractional q in mixed stacks. Partial ionicity raises both theoretical and experimental questions. The main theoretical problem for weakly-overlapping sites is that any one-electron treatment, even in the Hartree-Fock limit, yields [58, 65] a minimum energy for integral, rather than fractional q. Perturbing the individual molecular sites is not an attractive solution for small overlap.

The separated-molecule limit, with vanishing transfer integrals t, results in Wigner lattices for the energetically best distribution of Nq charges and N(1 − q) neutral sites in an N-site solid. Such perfect correlations have been discussed [58, 88-93] in connection with both mixed and segregated organic stacks. Wigner lattices, while suggestive, cannot model the antiferromagnetism, the optical properties, or any other feature that requires finite t and π-π overlaps. Correlated crystal states [65], on the other hand, involve linear combinations of VB structures, of which the Wigner lattice may have the largest coefficient. Now fractional q can be stabilized in narrow bands whenever the first coordination sphere contain DD or AA contacts. In contrast to inorganic salts, where anions are surrounded by cations and vice versa, partly-ionic molecular conductors and CT complexes always have such repulsive nearest-neighbor contacts. The correlated states lower the electrostatic energy by increasing the weight of VB structures with D^+A^- or DD, AA contributions and decreasing the weight of DA, D^+D^+, or A^-A^- structures. Fractional ionicity in the solid state resembles polar covalent bonding in heteronuclear diatomics. The exact degree of ionicity cannot easily be assigned there either.

The experimental determination of q in the solid state can take many forms. Structural differences [94, 95] between D and D^+ or A and A^- are a useful and convincing guide; but the small changes in bond lengths and angles and crystal packing forces hamper the accuracy. Interpolating vibrational differences [96–99] between D and D^+ or A and A^- provide potentially more precise information that nevertheless suffers from similar limitations. Electronic differences probed by photoelectron and optical spectroscopies tend to be less precise but perhaps more straightforward. The observation [100, 101] of superlattices by neutrons or diffuse x-rays yields q very precisely, provided that partial ionicity is known to drive the low-temperature formation of the superlattice. The paramagnetism of RP, D^+, or A^- radicals provides further approaches [33, 102] which are direct but not very precise. In some cases, the stoichiometry is sufficient. Agreement among several methods is obviously desirable.

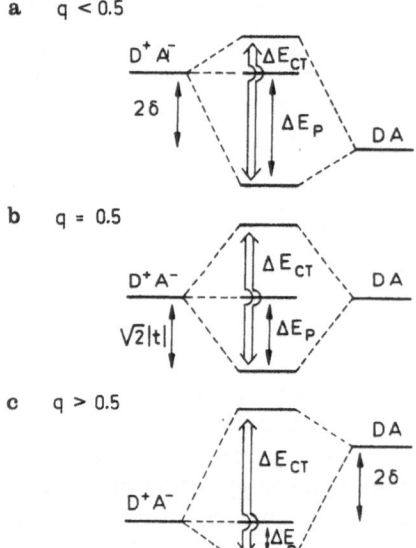

Fig. 3. Optical and magnetic transitions with their excitation energies in DA dimers with different amounts of charge transfer. (With permission from Ref. [87])

We assign q primarily from the paramagnetism, although other techniques are cited when possible. The Mulliken dimer in Fig. 3 shows the standard configuration interaction (CI) between the singlet VB structures $|DA\rangle$ and $^1|D^+A^-\rangle$. The triplet $^3|D^+A^-\rangle$ is not shifted in this case and is necessarily ionic. The admixture depends on the CT integral

$$t = \langle D^+A^- | \mathcal{H} | DA \rangle \tag{2}$$

and the energy difference 2δ between D^+A^- and DA. The ground state is

$$|\zeta\rangle = \cos \zeta |DA\rangle - \sin \zeta \, ^1|D^+A^-\rangle \tag{3}$$

where $0 \leq \zeta < \pi/2$ and $\tan 2\zeta = |t|/\sqrt{2}\,\delta$ determines the mixing. The dimer ionicity is simply $q = \sin^2\zeta$ and the electrostatic energy of $|\zeta\rangle$ is $-q\,e^2/R$ for point charges at R. We note that this differs from the molecular-orbital, or one-electron, approach of fractional charges $D^{+q}A^{-q}$, where the electrostatic energy is $-q^2\,e^2/R$. The allowed CT excitation, ΔE_{CT}, and paramagnetic (singlet-triplet) gap, ΔE_p, are given by

$$\Delta E_p = \sqrt{2}\,|t|\cot\zeta$$

$$\Delta E_{CT} = \frac{\sqrt{2}\,|t|}{\sin\zeta\cos\zeta} = 2(\delta^2 + 2t^2)^{1/2} \tag{4}$$

$$\Delta E_p\,\Delta E_{CT} = 2t^2/q$$

Largely neutral, diamagnetic ground states with small ζ have $\Delta E_p \sim \Delta E_{CT}$, as shown in Fig. 3 and found from Eq. (4). Conversely, ion-radical ground states with large ζ have $\Delta E_p \ll \Delta E_{CT}$. It takes far less energy to change the relative orientation of neighboring spins than to transfer the electron back form D^+A^- to DA.

The dimer analysis cannot readily be extended to correlated states in extended systems like solid CT complexes. The observation of small ΔE_p via static susceptibility or electron paramagnetic resonance (epr) nevertheless suggest a paramagnetic ground state. The observed CT transition at ΔE_{CT} then supports correlated states when it greatly exceeds ΔE_p and affords, through the last Eq. (4), an estimate for q when $|t|$ is known. Face-to-face overlap of π complexes yields [103] $|t| \sim 0.1$–0.3 ev, with solid-state data favoring the lower value [22, 61]. The bandwidth, $4|t|$, is typically estimated [27−32] between 0.5 and 1 ev.

CI treatments for mixed regular DA stacks rely on diagrammatic VB methods [60]. Accurate results are possible on neglecting Coulomb interactions between ion radicals, on including them self-consistently, or on treating nearest-neighbors exactly [33]. The principal result is that mixed regular stacks have a critical ionicity, $q_c \sim 0.7$ depending on the precise model, above which ΔE_p vanishes. Thus the susceptibility $\chi(T)$ is activated for $q < q_c$ and the complex becomes diamagnetic at temperatures $kT \ll \Delta E_p$. But $\chi(T)$ is not activated for $q > q_c$ and the limit $q \to 1$ is known from the Bonner-Fisher analysis [104] of the linear Heisenberg antiferromagnetic chain, with $J = t^2/\Delta E_{CT}$. The M_2P-$TCNQF_4$ complex [87] discussed in Sect. 4.2. is the first $q \sim 1$ complex that confirms early predictions [86] about magnetic excitations in such ion-radical solids.

The dimer energies from Eq. 4 are plotted in Fig. 4 as a function of ionicity, together with ΔE_p for mixed regular stacks. Explicit results for ΔE_{CT} remain approximate for extended systems [62]. The relative values of ΔE_p and ΔE_{CT} still yield q approximately, within the dimer picture, for solid CT complexes. In addition, the minimum CT energy probably also occurs for $q = {}^1/_2$ in Fig. 4, or degenerate D^+A^- and DA, in the solid. Torrance and coworkers [105−108] have effectively exploited the idea of minimum ΔE_{CT} in borderline complexes. They have identified several CT solids which show rapid changes in q as a function of pressure. The TTF-chloranil complex also changes from a largely neutral to a largely ionic state as a function of temperature [105, 106].

The elucidation of neutral-ionic transitions is an important current topic. Several phenazine complexes are likely candidates. The diamagnetic-paramagnetic interface

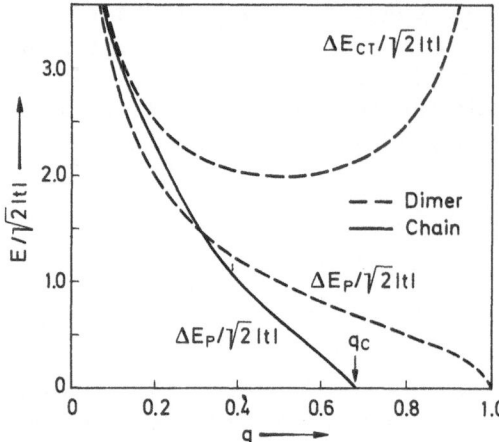

Fig. 4. Dependence between the excitation energies and amounts of ionicity, q. (With permission from Ref. [87])

at the critical ionicity q_c is a related topic for mixed regular stacks. CT complexes with partly-ionic ground states have now been identified. Their detailed investigation, together with improved theoretical models, should yield more accurate and consistent determinations of q.

3 Structural Diversity of RṖ and RP⁺ Complexes

Before discussing particular RṖ and RP⁺ complexes, we survey in Table 2 the available materials. The reference for the preparation includes at least a chemical analysis. We further note in Table 2 references to powder or single-crystal structural work and to epr or magnetic susceptibility studies. The more extensively studied entries such as MP-TCNQ are discussed separately.

Table 2. RP Complexes references for the preparation, structural data and results of other physical investigations on RP-TCNQ compounds

Complex	Preparation	Crystal structure	epr	susceptibility	Other physical measurements
MP-TCNQ	12	76, 109, 110		21	13–15, 21, 26, 35
(MP)₂(TCNQ)₃		111			
EP-TCNQ	12	71	112	74	
PP-TCNQ	74	74			
(PP)₂(TCNQ)₃	74	113	113		
BP-TCNQ	74	72	72		
MP-TCNQF₄	74	74	116		
EP-TCNQF₄	74	74	116		
PP-TCNQF₄	74	74	116		
BP-TCNQF₄	74	73	73		
MP-TCNDQ	114		114		114
MBP-TCNQ	115		115		115

3.1 MP-TCNQ: Segregated-Stack Conductor

The 1:1 5-methylphenazinium (MP) complex with TCNQ was for a time the best organic "metal" and was the first organic complex whose transport, optical, and magnetic properties were systematically subjected to the full panoply of modern solid-state methods [13-15]. MP-TCNQ nevertheless remains elusive and ambiguous.

Its chemical composition, for example, is still uncertain. The powerful H-atom abstracting ability of MP$^+$ was overlooked in the otherwise careful early work and has yet to be controlled convincingly. The so-called second phase [117] of MP-TCNQ turned out to be [47,118] HMP-TCNQ, which as summarized in Section 4.1 has typical mixed stacks of D$^+$ and A$^-$ ion radicals. Photochemical generation of HMP$^+$ and the concomitant degradation of MP$^+$ in solution are apparently general phenomena [50, 119,120]. Slow crystallization to enhance crystal perfection then becomes counter-productive. Sandman [119, 120] has addressed such questions about the kinetic control of crystallization, and very much more remains to be done in this difficult area.

The MP-TCNQ structure also remains unresolved. Fritchie's early finding [76] of segregated regular TCNQ and MP stacks, as shown in Fig. 5, has been confirmed in all subsequent studies. His repeat unit of 3.8682(4) Å along the MP stack is based on disordering the methyl between the 5 and 10 positions of phenazine. Kobayashi [109] interpreted diffuse streaks along the c* direction in other MP-TCNQ samples as one-dimensionally ordered methyls. This doubles the repeat unit and restricts disorder to different stacks. Morosin [121] associates diffuse streaks with rapid crystallization and interpenetrating phases. Detailed analysis [122] reveals two types of diffuse scattering whose wavevectors are related by a factor of two. Different one- and three-dimensional superlattices are invoked, with possibly different modulation of the MP

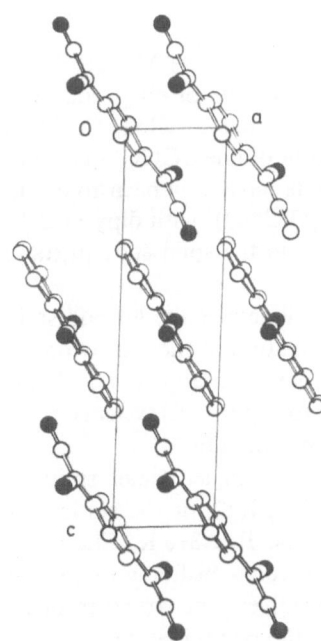

Fig. 5. Projection of the MP-TCNQ structure along the crystallographic b-axis (With permission from Ref. [76])

and TCNQ stacks. The unresolved degree of structural disorder was crucial initially because disorder and high conductivity [123–126] were linked. Now that the best organic conductors have no obvious disorder, the MP-TCNQ disorder is less urgent.

The low-temperature (10 mK < T < 20 K) thermodynamics [78] of quinolinium (TCNQ)$_2$ are now rather well represented [63] by random-exchange Heisenberg anti-ferromagnetic chains (REHACs). While MP-TCNQ has similar power laws [77], its magnetic specific heat is quite different [127], possibly due to facile methyl rotation about the 6-fold MP$^+$ barrier [128]. Disorder is surely important, but MP-TCNQ is complicated in this respect also.

It should consequently be of no surprise that the degree of ionicity is also open. Indeed, the surprise may be that all recent investigations support partial ionicity, with $0.5 < q < 0.9$ in MP^{+q}TCNQ^{-q}. Proton NMR on MP-TCNQ and its analogue with deuterated methyls clearly showed [21] paramagnetism associated with the MP moiety. The value of $q = 0.94$ suggests some 6% paramagnetic MṖ sites. Analysis of the TCNQ bond lengths and angles yields [94] the far smaller value of $q = 0.42$. The most precise measurements, again on different samples than either of the above, involve diffuse x-ray scattering and yield $q = 0.67$ on associating the superlattice [83] with the degree of band filling. Still further results are, respectively, $q \sim 0.9$ from the Seebeck coefficient [129], 0.9 from resonance Raman [97], 0.63 from infrared [99]. Thermopower [130] and CT spectra [131] do not determine q, but do rule out $q = 1$ and a half-filled TCNQ$^-$ band. Given the chemical and structural problems with MP-TCNQ, it is perhaps gratifying that widely different techniques converge to $q = 0.7 \pm 0.2$. Our subsequent assignments of the ionicity in less extensively studied phenazine complexes are hardly more accurate.

3.2 EP-TCNQ: Triplet Spin Excitons in σ-Bonded Dimers

The slight change of adding a CH$_2$ group to the side chain converts MP-TCNQ to EP-TCNQ. The 5-ethylphenazinium complex is some ten orders of magnitude less conducting [12] and has mixed stacks which, as shown in Fig. 6, feature a long (1.631 Å) σ-bond and distorted TCNQ$^-$ ions [71]. The atomic position of one TCNQ$^-$-TCNQ$^-$ dimer are shaded in Fig. 7 for clarity. Such long σ-bonds have now been found in two other complexes [132, 133], the first [132] being Pt(dipy)$_2$(TCNQ)$_2$ with dipy = 2.2'-dipyridyl. The planar TCNQ$^-$ ions are clearly perturbed in the solid state in these exceptional structures.

The diamagnetic ground state of (TCNQ)$_2^{-2}$ formally resembles the ground state in Fig. 3, since the electrons are spin paired. Thermal activation of the long σ-bonds produces an excited triplet, again as indicated in Fig. 3, based on unpaired TCNQ$^-$ spins on adjacent stacks. As discussed for other triplet spin exciton (TSE) systems [33, 69], singel-crystal epr provides detailed confirmation for such a picture.

The magnetic gap in Fig. 3, for example, becomes the singlet-triplet splitting associated with breaking the σ-bond. The epr intensity [112], I(T), or the static susceptibility yield $\Delta E_p = 0.27 \pm 0.01$ ev in the range $110 < T < 370$ K. The usual occurrence of $g \sim 2$ doublets in Fig. 7, which probably involve misfits in dimerized chains, does not complicate the TSE intensity in the resolved fine-structure lines. The reason is that magnetic dipole-dipole interactions of the two unpaired electrons

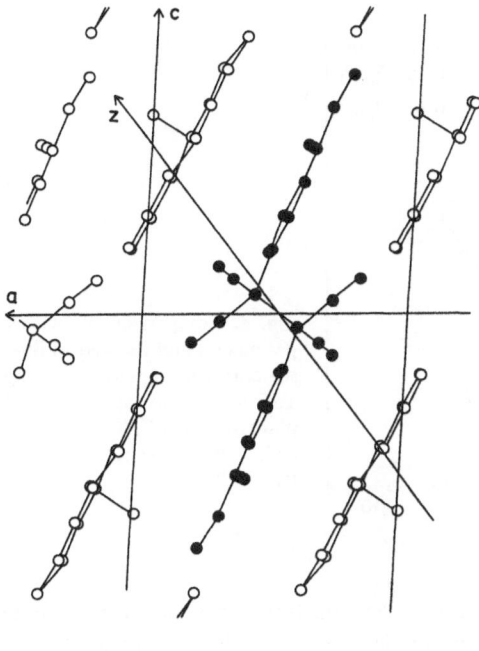

Fig. 6. Projection of the EP-TCNQ structure along the crystallographic b-axis. (With permission from Ref. [71])

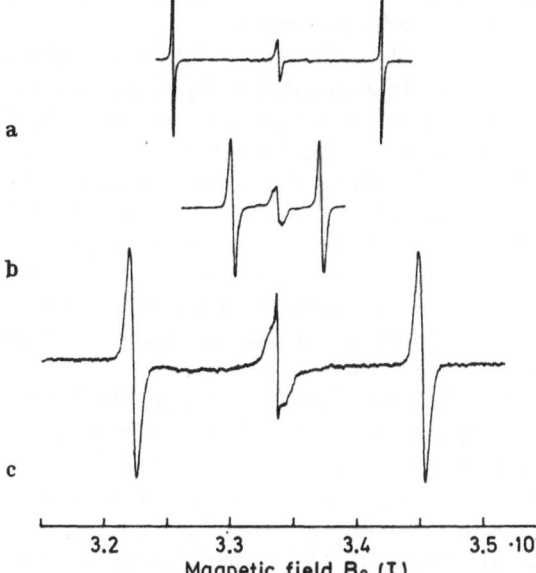

Fig. 7. Fine-structure splittings in single crystals of EP-TCNQ along the principal axis directions x, y, z. In figures 7b, 7c, the "narrow" S = 1/2 signal can be seen in addition to the impurity signal. (With permission from Ref. [112])

add and subtract from the external field, thus producing the characteristic TSE splitting in Fig. 7. The fine structure Hamiltonian for any triplet (S = 1) can be written in the principal axes (X, Y, Z) as [134]

$$\mathscr{H}_T = D\left(S_Z^2 - \frac{2}{3}\right) + E(S_X^2 - S_Y^2) \tag{5}$$

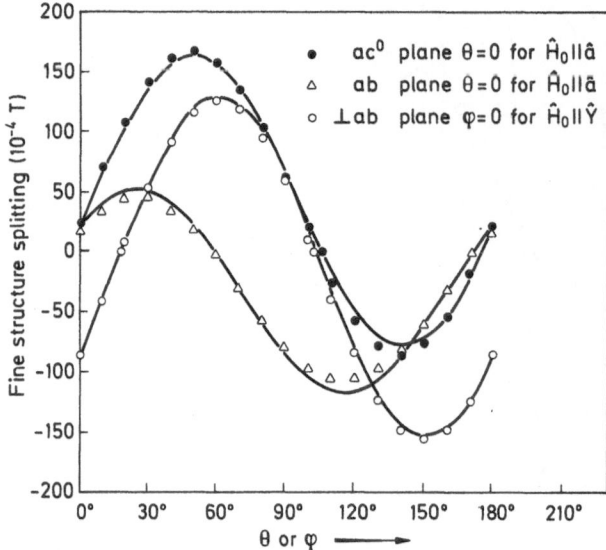

Fig. 8. Comparison of the experimental and theoretical fine-structure splittings for the strong TSE lines in orthogonal crystal planes of EP-TCNQ single crystals. (With permission from Ref. [112])

As shown in Fig. 8, the angular variations of the TSE splittings fixes the parameters D and E and the axes (X, Y, Z). The Z principal axis is shown in Fig. 6; the X, Y axes closely follow the dimer's geometry. The unpaired electrons in the thermally-activated triplet consequently reside on $TCNQ^-$ ions in adjacent stacks.

The light atoms in π-radicals results in negligible spin-orbit coupling, in contrast to transition metal complexes. Thus D and E are dominated by dipole-dipole interactions, which can be [135, 136] computed from the crystal structure and delocalized π-electron spin densities. The successful fit of D, E and principal axes in many TSE crystals[22] provides strong support for essentially unperturbed molecular ions in the solid state, since the theoretical or solution hyperfine spin densities are for isolated radicals. Activation of the π-bond in EP-TCNQ, on the other hand, should produce more planar and distant ion-radicals in the triplet state. The calculated D and E for the ground-state geometry are indeed too high, although the calculated principal axes agree with experiment [112]. The r^{-3} distance dependence of D and E would reduce them at the larger separation in the triplet state.

Large geometrical changes on activating the σ-bond produce unfavorable Franck-Condon factors for moving the triplet along the a axis in Fig. 6. The triplet is quasi-immobilized, in contrast to mobile TSE in many other organic ion-radical solids, and the fine structure lines in Fig. 7 are relatively broad. However the absence of hyperfine structure demonstrates some motion. As reviewed elsewhere [33], there remain open questions about the central doublets in TSE systems, about the temperature and angular variation of the linewidth, and about the broadening and eventual destruction of the fine structure with increasing exciton concentration on heating. The central conclusions about the triplet state are, on the other hand, very well documented.

The mixed stacks in pure EP-TCNQ are energetically not very different from a lattice with segregated stacks. The possibility of nearly equivalent minima for widely different structures is suggested naturally by the structural diversity of RP systems.

Doping EP-TCNQ with phenazine does in fact yield a new phase [119]. The conductivity [119] is increased by nine orders of magnitude, presumably because segregated stacks are formed as in phenazine-doped MP-TCNQ. The fact such slight steric changes can produce drastic structural and physical consequences, is one of the attractions of phenazine CT complexes.

3.3 PP and BP Complexes: π-Dimers in Novel Mixed Stacks

In order to obtain systematic information about related molecular complexes, 5-(1-propyl)phenazine (PP) and 5-(1-butyl)phenazine (BP) were prepared [74]. Their 1:1 complexes with TCNQ and TCNQF$_4$ again show TSE spectra [116]. Their singlet-triplet splitting, ΔE_p, and fine-structure constants, D and E, are seen in Table 3 to resemble EP-TCNQ. The BP complexes nevertheless have entirely different structures based on mixed $...D^+D^+A^-A^-D^+D^+A^-A^-...$ stacks of diamagnetic D^+ cations and face-to-face π-dimers of A^- anion radicals. Such packing goes beyond the classification in Fig. 2 and again suggests that electrostatic interactions are important but not dominant, in view of the unfavorable D^+D^+ and A^-A^- contacts.

Table 3. Singlet-triplet splittings, ΔE_p, and fine structure constants, D and E, for different RP-TCNQ TSE systems

Sample	T-range (K)	E_p (eV)	D/hc (cm^{-1})	E/hc (cm^{-1})
EP-TCNQ[a]	190–370	0.27	0.0104	0.0014
PP-TCNQ	180–320	0.12	0.0130[b]	0.0016[b]
BP-TCNQ	160–340	0.14	0.0124	0.0020
MP-TCNQF$_4$	280–380	0.27	0.0142[b]	0.0016[b]
EP-TCNQF$_4$	—	—	0.0115[b]	0.0020[b]
PP-TCNQF$_4$	220–410	0.33	0.0138[b]	0.0016[b]
BP-TCNQF$_4$	150–350	0.20	0.0120	0.0021

[a] Data from Ref. [112]
[b] Obtained by analysis of a powder spectrum.

Table 4. Structural data for RP-TCNQ complexes with π-π dimers

Complex	Lattice Parameters								Ref.
	a*	b*	'c*	α^+	β^+	γ^+	A*	B*	
PP TCNQ	9	40	19.9	—	?	—			74)
PP TCNQF$_4$	10.487(3)	10.723(6)	13.35(1)	101.63(6)	103.62(5)	116.41(3)			74)
BP TCNQ	8.315	9.657	16.333	70.54	71.51	76.78	3.16	3.44	72) 74) 116)
BP TCNQF$_4$	10.981	17.544	13.544	—	111.90°	—	3.15(4)	3.40(2)	73, 74, 116)

A = distance between dimerized acceptor planes,
B = distance between dimerized donor planes,
* in Å, $^+$ in °

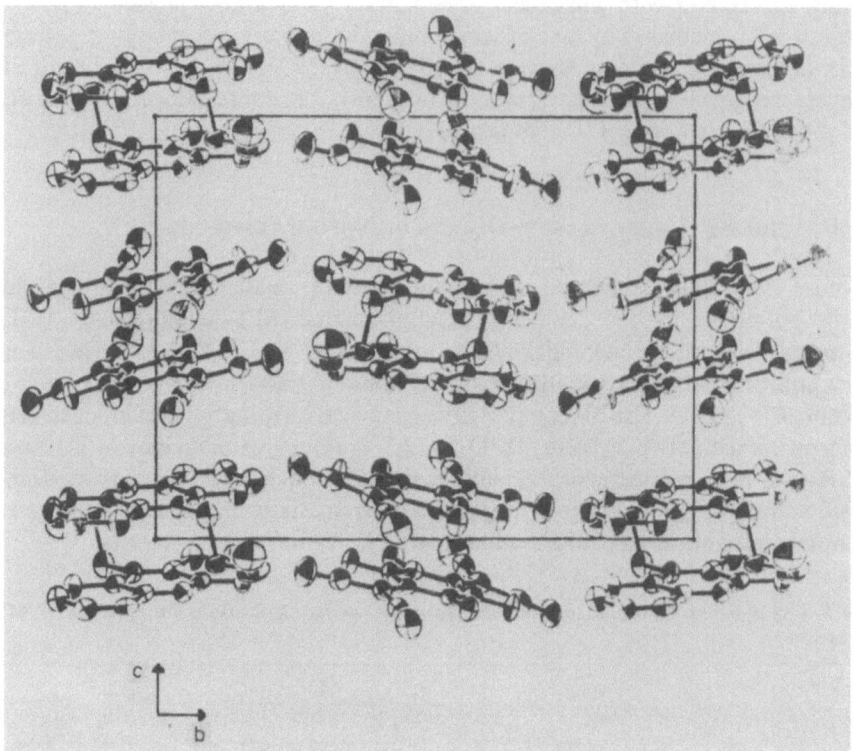

Fig. 9. A segment of three stacks in the structure of BP-TCNQ projected onto the bc plane. (With permission from Ref. [73])

PP-TCNQ could not be obtained as crystals suitable for a full structure, but unit cell parameters are listed in Table 4. The structure of BP-TCNQ was determined and related to the TSE spectra [72]. With the exception of the "folded" butyl group and a slight torsion (2.3° and 4.3° respectively) of the C = (CN)$_2$ groups, all the molecules are planar and there is effective π-π overlap of the TCNQ$^-$ ions at 3.15(3) Å. The complete structure and single crystal epr was also obtained [73] for BP-TCNQF$_4$, whose novel mixed stack in Fig. 9 clearly shows the face-to-face TCNQ$^-$ dimers.

The ground state is again diamagnetic, while the thermally accessible triplets are paramegnetic TCNQ$^-$ radicals. The magnetic data in Table 3 and the structural data in Table 4 show PP-TCNQ, PP-TCNQF$_4$, BP-TCNQ, and BP-TCNQF$_4$ to be closely related. The expected "normal" behavior after making small changes in the side chain finally occurs.

The reaction of RP$^+$ salts with LiTCNQF$_4$ solutions yields analytically pure 1:1 complexes. In addition to PP and BP complexes, solid MP-TCNQF$_4$ and EP-TCNQF$_4$ were prepared [74], though not in a form suitable for structure determination. Their magnetic properties in Table 3 resemble the other TCNQF$_4$ complexes. This confirms dimers with a thermally accessible triplet state, but not the dimer's constituents.

TSE spectra have been observed in many ion-radical organic systems with alternating stacks [22, 33]. The fine-structure constants of D$^+$D$^+$, D$^+$A$^-$, and either π or σ-bonded

A^-A^- dimers tend to be similar, since the molecular sizes and intermolecular separations are comparable. Dimers involving supermolecular A_2^- radicals have reduced spin densities and reduced D and E. The absence of hyperfine structure always implies exciton motion or relaxation faster than the width of the hyperfine spectrum in solution. Many subtle features of TSE systems have been noted in passing, without detailed analysis.

3.4 Miscellaneous Complexes

7-methylbenzo(a)phenazinium-TCNQ (MBP-TCNQ). Matsunaga prepared this benzophenazinium complex [115] in connection with investigations of the neutral-ionis interface. MBP-TCNQ is obtained in two modifications, violet crystals wIlich were

MBP⁺

considered to be ionic (MBP⁺TCNQ⁻) and green crystals which were considered non-ionic (MBP-TCNQ). The TSE spectra of the violet form are quite similar to the dimerized alkylphenazine-TCNQ complexes in Table 3, which are also violet. The single, g ~ 2 epr resonance of the green modification, on the other hand, resemble the single resonance of mixed regular stacks such as HMP-TCNQ or TMPD-TCNQ. Structural data will be required for a convincing assignment of the ionicity.

MP-TCNDQ (role of polarizability). [4-[4-(dicyanomethylene)-2,5-cyclohexadien-1-ylidene]-2,5-cyclohexadien-1-ylidene]-propanedinitrile (TCNDQ)

TCNDQ

has an expanded central region, and is more polarizable than TCNQ. Early speculations [137] about high conductivity focused on polarizability to reduce electron-electron correlations. This idea has not been supported by subsequent experiments. Rather, the stabilization of partly-ionic segregated regular stacks [22, 23] has produced many good conductors, with counterions of widely different polarizabilities. The MP-TCNDQ complex does not really test the polarizability argument. Since its pressed-pellet conductivity [114] is some nine orders of magnitude lower than that of MP-TCNQ, its structure is probably quite different.

"Complex" stoichiometry: $(MP)_2(TCNQ)_3$ and $(PP)_2(TCNQ)_3$. A few complex salts of 5-alkyl- or 5-arylphenazines and TCNQ have been reported. The $(MP)_2(TCNQ)_3$ and $(PP)_2(TCNQ)_3$ structures have been [111, 113] determined. Struc-

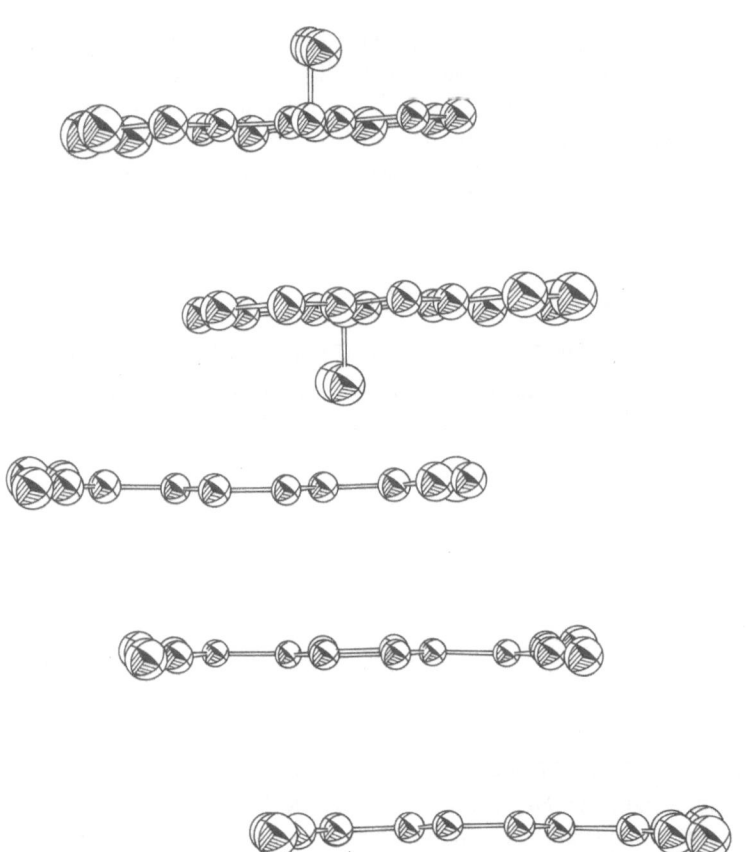

Fig. 10. Repeat unit in the mixed stacks of $(PP)_2(TCNQ)_3$ projected parallel to the TCNQ ring plane. The stacks run along [0 1 1]. (With permission from Ref. [113])

tural work is needed to establish complex stoichiometries, since unambiguous analytical proofs are difficult. The following remarks are restricted to the well-characterized, similar $(MP)_2(TCNQ)_3$ and $(PP)_2(TCNQ)_3$ complexes. As shown in Fig. 10, planar MP (or PP) dimers and planar TCNQ trimers run along a unit cell diagonal. There is probably complete charge transfer, so that the paramagnetism is associated with $TCNQ^{-2/3}$ ion radicals. Trimerized stacks spoil high conductivity. Complex alternating TCNQ stacks are nevertheless moderately good semiconductors with small energy gaps, as expected for correlated crystal states.

4 Partial Ionicity in RR′P-TCNQ Complexes

We proceed as before and list in Table 5 the available RR′P complexes, again noting structural and magnetic references. The most extensively characterized systems are then discussed separately.

Table 5. References for the preparation, structural data and results of other physical investigations on RR'P-TCNQ compounds

Complex	Preparation	Crystal data		epr	suscept.	Other
		Lattice data	Molecular structure			
HMP-TCNQ	138		118	139	—	47, 118, 119
M_2P-TCNQ	12		140	75	—	—
M_2P-$(TCNQ)_2$	12, 142a		—	—	—	142a
MEP-TCNQ	142a		—	—	—	142a
E_2P-TCNQ	138, 142a		142	142	—	142a
E_2P-$(TCNQ)_2$	138		143	—	—	—
P_2P-TCNQ	138, 142a		144	—	—	142a
B_2P-TCNQ	142a		—	—	—	—
MPhP-TCNQ	142a		—	—	—	142a
Ph_2P-TCNQ	142a		—	—	—	142a
Be_2P-TCNQ	142a		—	—	—	—
M_2P-$TCNQF_4$	145		87	—	87	—
E_2P-$TCNQF_4$	142a	144	—	—	—	—
M_6P-$TCNQF_4$	145		—	—	—	—
M_2P-TCNE	146	144, 146	—	146	—	—
M_6P-TCNE	85		85	85	—	—

4.1 M_2P-TCNQ: Triplet Spin Excitons in Mixed Alternating Stacks

Alkylating both nitrogens of phenazine dramatically increases the donor strength and eliminates the asymmetry. Fully reduced 5.10-dihydrophenazine (H_2P) reduces TCNQ to H_2TCNQ rather than forming a complex [138]. The next entry in Table 1 is HMP, whose 1:1 complex with TCNQ was initially thought [117] to be a second phase of MP-TCNQ. The missing H site was found in difference Fourier maps [118] after HMP-TCNQ was synthesized separately [138]. HMP$^+$ is the primary photo-product [25, 51] on irradiating MP$^+$ solutions in protic solvents and occurs in small concentrations in all MP-TCNQ samples. Slow crystallization of MP-TCNQ under light in protic solvents increases the HMP content and is one approach to high yields of HMP-TCNQ.

The HMP-TCNQ structure [118] consists of mixed alternating stacks, although the HMP contacts with either TCNQ neighbor are similar. The strong paramagnetism [139] and low activation energy ($\Delta E_p = \sim 0.1$ ev) suggest a predominantly ionic ground state. The electrical conductivity is also activated [117] ($\Delta E_c = 0.405$ ev). The epr spectrum has a single $g \sim 2$ doublet, whose analysis is complicated by crystal twinning [139]. These fragmentary reports on HMP-TCNQ will have to be integrated and extended to assign the ionicity more precisely.

M_2P-TCNQ has probably received more attention than any phenazine complex except MP-TCNQ. First prepared by Melby [12], its structure was determined by Goldberg and Shmueli [140]. The mixed alternating M_2P-TCNQ stack is shown in Fig. 11. The bent M_2P was interpreted as favoring a neutral ground state, although the interesting remark [140] that "epr absorbtion is obtained only when the crystals are powdered" suggests facile formation of $M_2\dot{P}^+$ and TCNQ$^{\pm}$ ion radicals.

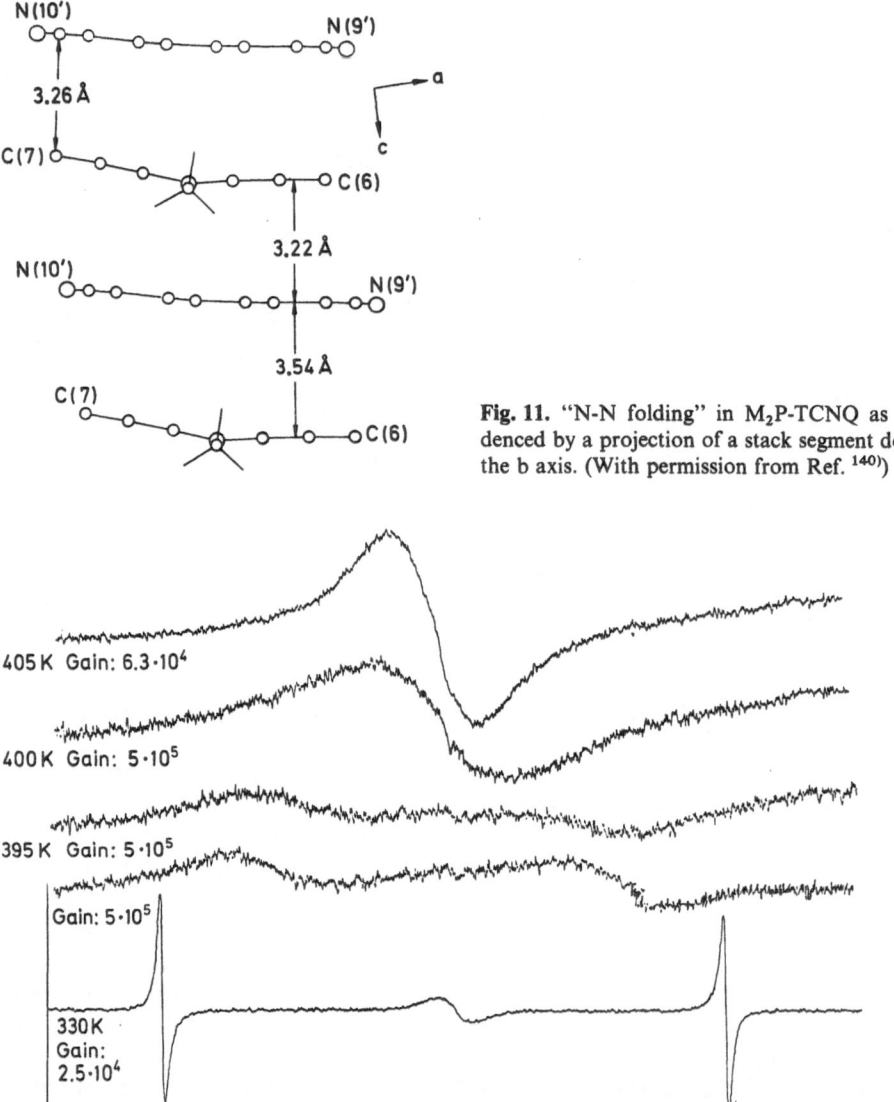

Fig. 11. "N-N folding" in M_2P-TCNQ as evidenced by a projection of a stack segment down the b axis. (With permission from Ref. [140])

Fig. 12. Temperature dependence of TSE spectra of M_2P-TCNQ between 300 K and 420 K

Our subsequent investigations [75] of M_2P-TCNQ identified thermally activated triplet spin excitons (TSE) by single-crystal epr, as shown in Fig. 12. The activation energy is so high, $\Delta E_p = 0.60 \pm 0.06$ ev, that heating above 300 K is required for observing TSE at X-band (3300 G); the TSE are seen even at 300 K at Q-band (11 000 G). The fine structure constants in Eq. (5) are D = 144 G and E = 20 G, in good agreement with computations based on adjacent M_2P^+ and $TCNQ^-$ ion-radical dimers along the mixed alternating stack. Previous TSE spectra were restricted to segregated alternating stacks.

The CT absorption of M_2P-TCNQ has been reported [147] at 0.84 ev or [107] 0.89 ev. Comparison with Figs. 3 and 4 then suggests [87] an intermediate ionicity of q ~ 0.4 ± 0.1, since ΔE_{CT} exceeds ΔE_p by a small amount. The ionic triplet state is probably formed from a partly-ionic singlet. Detailed ir and Raman vibrational studies [141] have recently led to q = 0.49 ± 0.10, thereby confirming the intermediate ionicity of M_2P-TCNQ.

The epr signals on powdering the samples suggests interesting pressure dependences, but these have not been pursued [140]. The collapse [75] of the fine structure around 390–410 K, as shown in Fig. 12, does not follow usual TSE behaviour. A more ionic ground state is possibly formed at elevated temperatures, but this remains a conjecture. The TCNQ bond angles and bond length in M_2P-TCNQ yield [94] q = 0.30. But its infrared has been assigned [147] as "ionic", without giving q or following individual vibrational shifts. The more complete vibration analysis [141] yields q = 0.49. Partial ionicity evidently requires several independent estimates, to minimize the inevitable extraneous contributions due to the crystalline environment.

4.2 M_2P-TCNQF$_4$: Ionic Mixed Regular Stacks

The most ionic CT complex identified in early work on TCNQ salts involved the π-donor TMPD. TMPD-TCNQ apparently has a mixed regular structure [148] (albeit with large thermal parameters), a sharp g ~ 2 epr signal, and an activated susceptibility [149] above 200 K,

$$\chi(T) \, T \propto \exp - \Delta E_p/kT \tag{6}$$

with $\Delta E_p \simeq 0.07$ ev. This differs from the $\chi(T)$ for a linear Heisenberg antiferromagnetic chain, the predicted [86] behavior for q ~ 1, mixed regular stacks. Soos and Mazumdar [60] have modeled the q-dependence of ΔE_p, as shown in Fig. 4, and associated the small but finite ΔE_p in TMPD-TCNQ with partial ionicity close to to the critical value, $q_c \sim 0.7$, where ΔE_p vanishes. Girlando et al. [150] have recently found q = 0.92 ± 0.10 on the basis of combined ir and Raman shift. They also find evidence for dimerization or disorder in TMPD-TCNQ at 300 K, thus removing the difficulty with finite ΔE_p. The larger ΔE_p of TMPD-Chloranil is consistent with q = 0.63 as found [151] from vibrational analysis.

The more powerful π-acceptor TCNQF$_4$ is expected to produce more ionic CT complexes. TMPD-TCNQF$_4$ is indeed highly ionic, with q ~ 0.7 indicated simply from the magnitude of $\chi(300)$ for two mols of s = 1/2 ion radicals [87]. The temperature dependence of $\chi(T)$ is not activated, but follows roughly the power law $T^{-0.75}$ down to 4 K. Such behavior is indicative of disordered antiferromagnetic exchanges that reduce $\chi(T)$. Thus q > 0.9 can safely be assigned to TMPD-TCNQF$_4$ and the complex is fully ionic. Unfortunately, crystals suitable for structural work could not be grown.

The M_2P-TCNQF$_4$ complex could be crystallized, however, and its structure resembles [87] that of TMPD-TCNQ rather than the dimerized M_2P-TCNQ stacks in Fig. 11. As shown in Fig. 13, M_2P-TCNQF$_4$ forms mixed regular stacks containing planar M_2P^+ ions. Inversion centers at each D^+ and A^- site ensure equal interaction with both neighbors. The susceptibility, $\chi(T)$, in Fig. 14 follows the Bonner-Fisher [104]

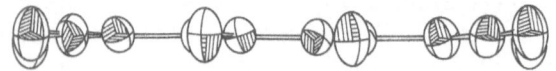

Fig. 13. Projection of a stack segment with two donor and one "sandwiched" acceptor molecule of M_2P-TCNQF$_4$ parallel to the molecular planes. (With permission from Ref. [87])

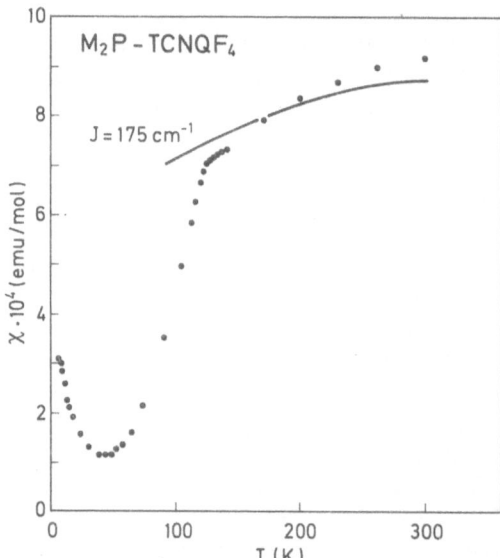

Fig. 14. $\chi_p \times 10^4$ vs T (K) plot for M_2P TCNQF$_4$. Full line: Bonner-Fisher fit with J = 175 cm^{-1}. (With permission from Ref. [87])

result for a Heisenberg chain with antiferromagnetic exchange J = 175 cm^{-1} (0.02 ev). The transition around 122 K is to an uncharacterized, and probably dimerized, phase. M_2P-TCNQF$_4$ is consequently ionic, with q $\gtrsim$ 0.9, a conclusion confirmed by vibrational analysis [141].

In mixed regular stacks, where CT to D^-A^+ involves far higher energy, the exchange J is approximately given by $t^2/\Delta E_{CT}$. The transfer integral $|t| \sim 0.1$ ev is defined in Eq. (2), while $\Delta E_{CT} \sim 1.0$ ev is typical of the strongest complexes. The observed J value for M_2P-TCNQF$_4$ is thus reasonable. The interplanar separation of 3.36 $\pm$ 0.07 Å in Fig. 13 indicates less π-π overlap than for the dimers in Table 3, where the separation is 3.15 A and there is more nearly face-to-face overlap. In dimers, the exchange is simply $J = \Delta E_p/2 \approx 1000\text{--}2000 \text{ cm}^{-1}$, an estimate that remains useful for alternating stacks.

4.3 MEP, E_2P, P_2P Complexes and P Doping

The 5.10-dihydro-5-methyl-10-ethylphenazine (MEP) complex of TCNQ has been prepared [142] but not characterized in detail. We anticipate the MEP-TCNQ structure and physical properties to resemble M_2P-TCNQ and E_2P-TCNQ. The 5.10-dihydro-5.10-diethylphenazine (E_2P)-TCNQ complex's structure has been determined [142]. The E_2P moeities are folded along the N-N axis, as shown in Fig. 11 for M_2P-TCNQ; the two benzene halves have a dihedral angle of 12.7°. The mixed stack is again dimerized. Both ethyls are bent to one side of the average phenazine plane, a configuration that reinforces the dimerization or is reinforced by the dimerization.

The 5.10-dihydro-5.10-bis(1propyl)phenazine (P_2P) complex with TCNQ has similar structural and physical properties to other R_2P-TNCQ complexes, in contrast to the structural changes discussed in Section 3 for RP-TCNQ systems. There are two crystallographically inequivalent, mixed alternating stacks in P_2P-TCNQ [144]. The dihedral angles for bending P_2P units along N-N are 14.0° and 4.5°, respectively. The dimerization pattern is different in the two different stacks. Again both propyl groups of a distinct molecule are bent to one side of the average phenazine plane, but in opposite directions on the different stacks.

E_2P also forms a 1:2 complex TCNQ salt, $E_2P(TCNQ)_2$, whose structure projected onto the ac plane is shown in Fig. 15. Segregated stacks of planar molecules are

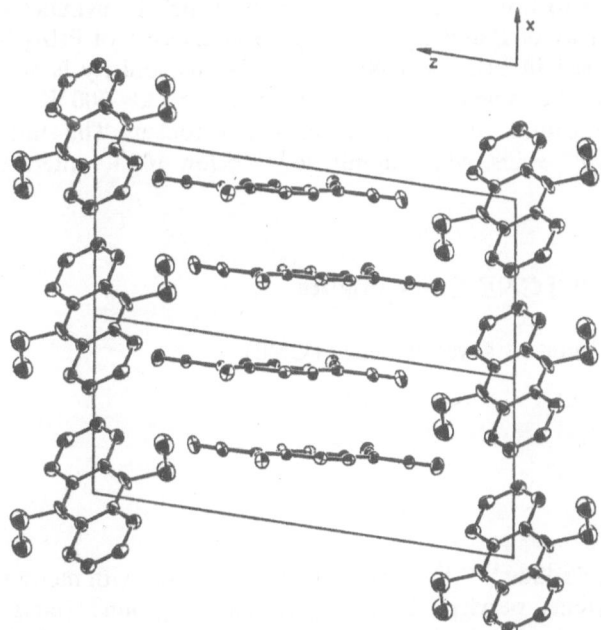

Fig. 15. Projection of the E_2P (TCNQ)$_2$ structure onto the ac plane. (With permission from Ref. [143])

193

found [143]. The inclination of the molecular planes to the stacking direction is different. The TCNQ planes are normal to the stack, while the E_2P planes are at 52.4°. There are two TCNQs along the stack for every E_2P. Such packing has previously been described [152] in TMPD(TCNQ)$_2$.

Doping M_2P-TCNQ with pure phenazine yields a good organic conductor [84]. Rather than the variable composition expected in alloys, however, analytical and static susceptibility data indicate [153] a single homogeneous solid, P(M_2P)(TCNQ)$_2$, with a 1:1 ratio of M_2P:P. Its crystal structure [84] is virtually indistinguishable from the MP-TCNQ structure in Fig. 5. The interplanar TCNQ separation is 3.26 Å in both, while the interplanar separations are 3.44 Å in P(M_2P)(TCNQ)$_2$ and 3.36 Å in MP-TCNQ. The chemically disordered P/M_2P stack thus replaces a structurally disordered MP$^+$ stack. Both types of disorder involve some (unknown) short-range ordering, rather than naive random packing. Since neither P or M_2P has a dipole moment, the disorder potential in P(M_2P)(TCNQ)$_2$ must be quite different.

Other properties of P(M_2P)(TCNQ)$_2$ and MP-TCNQ besides their structure are also similar. The four-probe single-crystal conductivity [153] of the former is about 50–100 Ω^{-1} cm^{-1} at 300 K, about 2–3 times lower than the latter, and also increases slightly on cooling to 200 K. The magnetic susceptibility of P(M_2P)(TCNQ)$_2$ is high and can be described between 10 and 100 K by a Curies-Weiss model [153]. Such fits in either system are preliminary, pending applications of more realistic disorder models. The strong donor M_2P is presumably fully ionized, while the weak acceptor P is neutral in P(M_2P)(TCNQ)$_2$. The average ionicity of the TCNQ stack is then $q = 0.50$ and is modulated by the chemical disorder of the phenazine stack. Comparison with $\chi(T)$ data for MP-TCNQ between 10 and 100 K yield [153] $q = 0.7 \pm 0.1$ and constitute an unexpected determination of the MP-TCNQ ionicity.

Phenazine doping of E_2P-TCNQ also yields segregated stacks [142]. The bulkier ethyl groups apparently suffice to order the phenazine stack, which by symmetry now consists [142] of a mixed array of P and E_2P. The chemical disorder of P(M_2P)(TCNQ)$_2$ is therefore suppressed in P(E_2P)(TCNQ)$_2$. The TCNQ ionicity is still around $q = 0.5$. Its four-probe single-crystal conductivity [153] at 300 K is 1–10 Ω^{-1} cm^{-1}, an order of magnitude lower than its disordered cousins. The static susceptibility is higher. The $E_2\dot{P}^+$ spins, which dominate [153] below 100 K, interact less strongly.

4.4 TCNE Complexes: M_6P-TCNE Susceptibility

Before TCNQ, the smaller acceptor tetracyanoethene (TCNE),

TCNE

was a popular π-acceptor for CT solids [154–159]. Most TCNE complexes with modest π-donors like naphthalene, pyrene, perylene have largely neutral ground states. Even stronger π-donors [158] are only slightly ionic. But alkali salts such as Na$_n$TCNE demonstrate [159] the occurrence of $TCNE^{-n}$ for n = 1, 2, 3. The smaller molecular size of TCNE indicates smaller polarizability and stronger electronic correlation in excited $TCNE^{-2}$ than in excited $TCNQ^{-2}$. Several phenazine-TCNE complexes were prepared [141] in connection with the systematic study of phenazine donors.

TCNE reacts with M_2P to give three different products [146]. One is a violet CT solid whose persistent twinning precluded a full structure determination; the repeat distance of 7.63 Å along needle supports mixed M_2P and TCNE stacking. The second product, a deep blue solid, is a condensation product (TMPP).

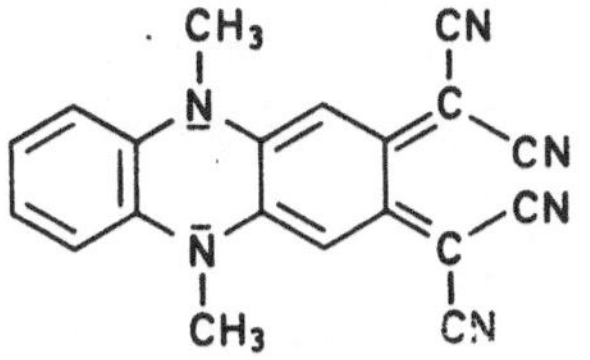

TMPP

Its structure and optical properties [146] indicate both inter- and intramolecular CT bands. The third product is a dark green solid whose paramagnetism points to $M_2\overset{.}{P}{}^+$ cation radicals and whose stoichiometry is consistent with partlyhydrolized TCNE counterions.

To suppress the TMPP condensation at C_2 and C_3, the corresponding 2,3,7,8-tetramethyl analogue of M_2P was prepared [85, 145],

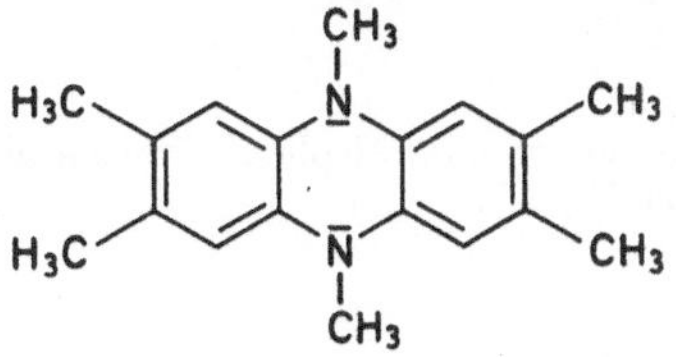

M_6P

M_6P and TCNE form a 1:1 CT complex in high yield [145]. The mixed regular M_6P-TCNE stacks [144] are shown in Fig. 16. There is an inversion center at each D and A. The static susceptibility [35] in Fig. 17 indicates that M_6P-TCNE is strongly paramagnetic at 300 K, with some 25% of the possible spins contributing. We plot $\log \chi(T)/\chi_c$ vs. $\log T$ in Fig. 17, where χ_c is the Curie paramagnetism for two mols of noninteracting s = 1/2 spins,

$$\chi_c = 2N_A \left(\frac{g^2 u_B^2}{4kT} \right) \tag{7}$$

At any temperature, $\chi(T)/\chi_c$ gives the effective number of free spins and is related [63] by the fluctuation-dissipation theorem to the sum of all static spin correlations. The complicated $\chi(T)/\chi_c$ behavior in Fig. 17 thus points to disordered exchanges in the power-law regime below 40 °K; the flat region 40 < T < 140 K suggests that some 8% are essentially free and follow the Curie law, Eq. (7); the T > 140 K regime

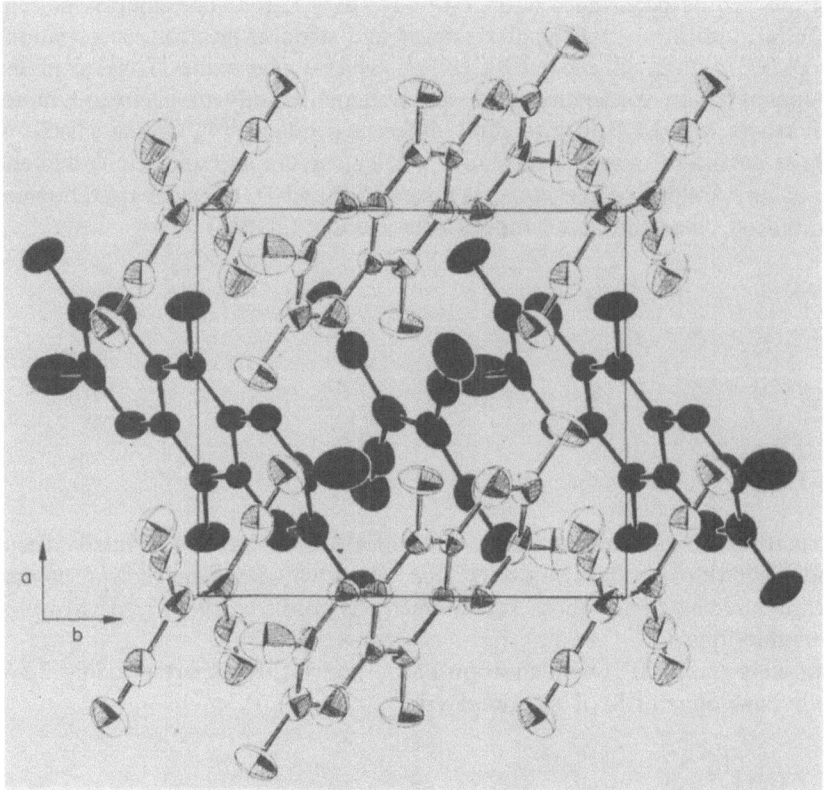

Fig. 16. Projection of three stack segments of the M_6P-TCNE structure onto the ab plane. The central stack is drawn with filled elipsoids. (With permission from Ref. [85])

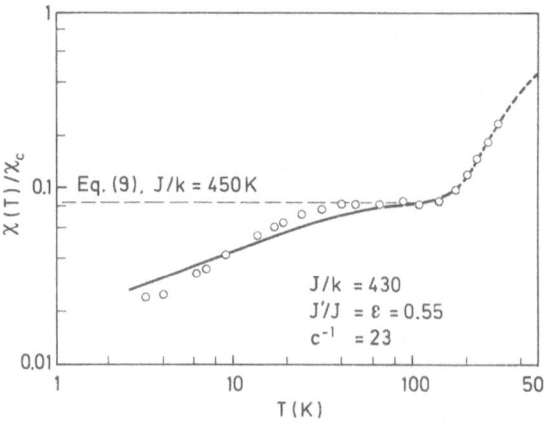

J/k = 430
J'/J = ε = 0.55
c^{-1} = 23

Eq. (9), J/k = 450 K

Fig. 17. log χ/χ_c vs log T plot of the susceptibility, χ, for M_6P-TCNE

then describes thermal activation of antiferromagnetic exchanges of the order of $J \sim 300$ cm^{-1}. The paramagnetism suggests that M_6P-TCNQ is ionic, with $q = 1$, and has significant antiferromagnetic exchange interactions.

196

To unravel the paramagnetism, we postulate an exchange Hamiltonian,

$$\mathscr{H}_s = \Sigma_n \, 2J_n(\vec{S}_n \cdot \vec{S}_{n+1} - 1/4) \tag{8}$$

for the M_6P^+ and $TCNE^-$ ion radicals along the chain, each with $S_n = 1/2$. The antiferromagnetic exchanges $J_n > 0$ are treated as adjustable parameters. While any successful model must reproduce the $\chi(T)$ data, a fit hardly establishes the validity of the model. The following remarks are consequently preliminary and illustrative.

The M_6P-TCNE structure in Fig. 16 has mixed regular stacks. But a single J along the chain leads to a regular Heisenberg antiferromagnetic chain whose susceptibility is entirely different, as shown in Fig. 14 for M_2P-$TCNQF_4$. A rather different approach is consequently needed. The observed power law for $T < 40$ K in Fig. 17 strongly suggests disordered exchanges in M_6P-TCNE. Examination of the DA overlap in Fig. 18 reveals a possible $TCNE^-$ interaction with an allyl fragment whose central carbon has a noticeably larger thermal ellipsoid. Neutral TCNE complexes with naphthalene [155], pyrene [156], or perylene [157] have different overlaps, with TCNE essentially centered on a ring.

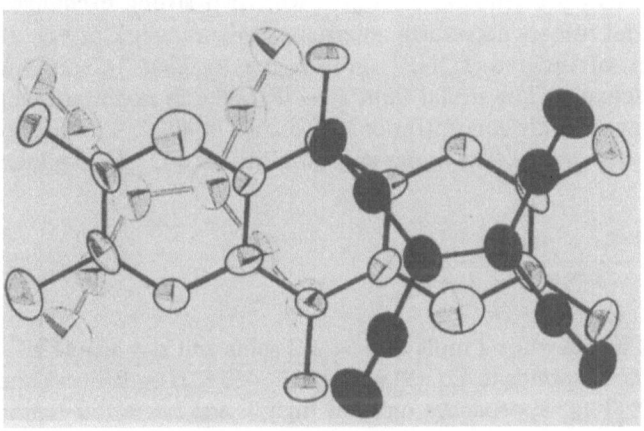

Fig. 18. A M_6P unit "sandwiched" by TCNE ions.

To pursue the idea of specific TCNE-allyl interactions, we suppose [85] in Fig. 19 that each $TCNE^-$ has a strong contact (J) with one MP^+ and a weaker one $(J' < J)$ with the other. This results in a dimerized stack if J and J' alternate precisely when the unit cell is doubled and there is no reason for an enlarged thermal ellipsoid. A more promising hypothesis is that J and J' do not alternate precisely, so that sometimes two $TCNE^-$ form strong contacts with the same M_6P^+, to give trimers in Fig. 19. M_6P^+ sites with two weak (J') contacts are monomers. There are equal numbers of J and J', but now the distribution is random, subject to the constraint that no more than two successive J' or J can occur.

A systematic analysis of such disordered J, J' exchanges offers many possibilities. It is more instructive to consider the $\chi(T)/\chi_c$ plateau in Fig. 17, which indicates a Curie-law behavior above 40 K for some 8% of the spins. In zeroth order, we take $J' \ll J$ and describe the susceptibility in terms of a concentration, 2c, of disordered

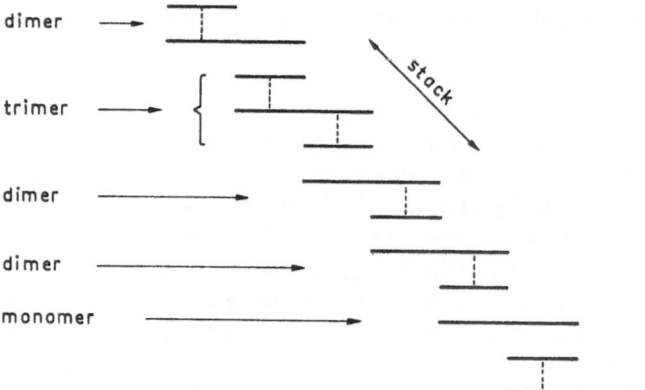

Fig. 19. Schematics of possible exchanges in M_6P-TCNE leading to monomer or trimer $S^. = 1/2$ and dimer $S = 0$ systems

TCNE⁻. It follows immediately that half of them are monomers, or $M_6\dot{P}^+$ with two weak exchanges, and half are trimers, or $M_6\dot{P}^+$ with two strong exchanges. Furthermore, monomers and trimers necessarily alternate along the stack in Fig. 19 with an arbitrary number of ordered TCNE⁻ in between forming TCNE-M_6P dimers with one strong exchange. The trivial limit $J' \to 0$ results in noninteracting monomer, trimers, and dimers with concentrations of $c:c:(1-4c)/2$. The energy levels depend on J and are elementary. The molar susceptibility for a D^+A^- complex is

$$\frac{\chi^{(0)}(T)}{\chi_c} = 2c + \frac{4(1-4c)}{3+z^2} + \frac{8c}{2+z+z^3} \tag{9}$$

where χ_c in Eq. (6) is the Curie-law for 2 mols of $S = 1/2$ spins and $z = \exp(J/kT)$. The dashed line in Fig. 17 corresponds to Eq. (9) with $J/k = 450$ K ($J \sim 0.04$ ev) and $c^{-1} = 24$. As expected, $\chi^{(0)}(T)/\chi_c$ approaches unity at high T and has a flat region below $kT \gtrsim 2J$.

More realistically, J' does not vanish and results in reduced exchange among the paramagnetic ground states of monomers and trimers. Such reduced effective exchanges [63] account for the power-law thermodynamics of $Q(TCNQ)_2$. We see that the largest effective exchanges must be less than about 20 K in order to decouple the spins for $T > 40$ K. We can certainly exclude adjacent monomers and trimers on physical grounds, since shifting a single TCNE⁻ leads to a more stable pair of dimers. Quite generally the collective motion of $p+1$ TCNE⁻ generate $p+2$ dimers from a monomer-trimer sequence with p intervening dimers. For comparable values of J' and J, the energy gain is small and the collective motion becomes improbable, thus freezing in the disorder. From the known reduction for intervening ground-state dimers, we estimate that there are always at least two intervening dimers and that J'/k is of the order of 300 K. This produces small effective exchanges needed for the power-law region below 40 K in Fig. 17, as shown by the solid curve. The actual analysis [85], which requires solving various finite segments, is essentially the implementation of the preceding discussion.

5 Phenazine Salts with Inorganic Ions

As before, we list in Table 6 some representative RP^+ $RR'\overset{.}{P}{}^+$ salts with inorganic counterions, as well as polyiodide $RR'P$ complexes whose partial ionicity results in interesting structural and transport properties. There are of course many other possible entries to Table 6.

5.1 Diamagnetism of RP^+ and Paramagnetism of $RR'\overset{.}{P}{}^+$ Salts

Powerful inorganic acceptors like the halogens, certain transition metal complexes, and negatively charged counterions with negligible nucleophilic tendency like BF_4^-, ClO_4^-, PF_6^-, AsF_6^-, NO_3^-, $CH_3^-SO_3^-$, etc. result in fully ionic phenazine compounds. The physical properties of phenazine salts with inorganic anions have not been extensively or systematically studied. We mention below materials whose structures have been determined.

The 5-alkyl- or 5-arylphenazinium salts with the above mentioned acceptors are used as starting materials for most of the compounds described in this review. Since these RP^+ salts are apparently insulating, diamagnetic, ionic solids expected from closed shell ions, they need no further discussion here. One exceptional case is worth mentioning: $(MP)_2[Ni(mnt)_2]$; (mnt^- = maleonitrildithiolato anion). In this case the anion is planar, making a kind of mixed stacking possible. The compound crystallizes [160] in DAD triads which are stacked along the crystallographic a-axis in a DAD-DAD pattern [160].

In searching for new conducting solids based on segregated or $RR'P$ stacks, the neutral donors were oxidized, in most cases chemically or electrolitically, and combined with non-nucleophilic anions like ClO_4^-, PF_6^-, BF_4^-, etc. The resulting 1:1 salts with q = 1 were found to have rather low powder conductivities. Our usual doping procedures were not successful for the inorganic $RR'\overset{.}{P}{}^+$ salts. Some of the analytically and structurally identified species [142, 145] are summarized in Table 6. The partly ionic, complex iodide salts discussed in the next section did show far higher conductivity, as expected from the elementary considerations in Section 2.2.

Tabelle 6. Some analytically and structurally identified $RR'P^+X^-$ species with references to their preparation and physical properties

Complex	Preparation	Crystal structure	Paramagnetism	Other physical properties
HMP-ClO$_4$	138	118	138	
$M_2P \cdot I_3$	165	165		
$E_2P \cdot I_3$	165	165		165
MEP $\cdot I_{1.5}$	142a	168, 170		
$E_2P \cdot I_{1.6}$	142a	166, 168, 169		167
$B_2P \cdot I_{1.6}$	142a	166, 168, 169, 170		
M_2P-PF$_6$	161			161
E_2P-Br(H$_2$O)$_2$	161	161		142a

The basic structural data [144] for these compounds are summarized in Table 7. It is apparent ·that $M_2P\text{-}PF_6$ and its AsF_6^- and BF_4^- homologues are isostructural, differing appreciably from $M_2P\text{-}NO_3$, $E_2P\text{-}Br(H_2O)_2$ and $HMP\text{-}ClO_4$. The structure of $M_2P\text{-}PF_6$ and its

Table 7. Structural data for single crystals of $RR'P^+X^-$ salts

Compound	Space group	a (Å)	b (Å)	c (Å)	β °	Ref.
HMP-ClO$_4$	Pcab	17.286(4)	17.964(3)	8.0852(9)	—	[118]
M$_2$P-PF$_6$	Cc	14.555(3)	6.323(1)	18.542(4)	121.32(2)	[161]
M$_2$P-AsF$_6$	Cc	14.90	6.41	18.18	121	[144]
M$_2$P-BF$_4$	Cc	13.27	6.33	19.90	118	[144]
M$_2$P-NO$_3$	P2,Pm	9.91	7.02	10.26	114	[144]
E$_2$P-Br(H$_2$O)$_2$	C2/c	21.732(5)	5.3464(9)	14.436(3)	105.01	[161]

BF_4^-, ClO_4 homologues can be described as "rock-salt-like" [161], with dominant electrostatic binding. They do not appear to have strong intermolecular π-π interactions, in marked contrast to the structures, for example, of carbocyclic radical cations of perylene [162] or TMPD [163]. There is no explanation for the absence of strong π-π overlap in these $RR'\overset{+}{P}$ cation-radical salts. $HMP\text{-}ClO_4$ on the other hand crystallizes [118] in segregated stacks resulting in physical properties which are strongly influenced by the intermolecular exchange along the stacking axis.

5.2 Linear Polyiodides of RR'P Donors

Reacting solutions of neutral RR'P donors with elemental iodine results in the precipitation of iodides with complex stoichiometries, which depend furthermore on the mode of preparation. The complexity of the iodine-RR'P reaction was already recognized in the early phenazine work [164] and gave rise to many discussions about the character of these materials [49]. Principally two different types can be discerned:
(i) Iodides or triiodides with rock-salt-like structures and minor intermolecular interactions.
(ii) "Linear" segregated stacked structures with polyiodide chains filling the channels built by the organic radical cation stacks.
Type (i) materials — which could be considered as $RR'\overset{+}{P}$ "*salts*" also (Sect. 5.1) — are of interest mainly because of the different types of iodides and polyiodides which can act as counterion and because of the different configurations of the phenazine system (bent or planar) found in these ionic solids [165], but not because of any collective physical properties.

Extensive X-ray [166−171], Raman Resonance (RR) [142a] conductivity [167, 172, 34], epr [167], and optical [172] investigations on type (ii) polyiodides were motivated by quite different questions concerning these solids. Their special *crystal structure* poses difficult and interesting problems. They consist of linear regular stacks of the planar $RR'\overset{+}{P}$ cations and more or less linear chains of iodine atoms forming polyiodide ions of different lengths, as shown in Fig. 20 for $Be_2P\text{-}I_{1.62}$ in the plane perpendicular to the stack. The stability of linear chains of I_3^-, I_5^- ions in the solid state has been amply documented by Resonance Raman spectroscopy [142a, 173–175] and X-ray investigations [176–179]. There are many channel compounds, quite aside of phenazines, based on stacks of donors whose interstitial channels are filled with I^-, I_3^-, etc., the most famous example being the blue "Starch-Iodine" complex [173, 176] in which the matrix is made up of glucoses. Other matrices consist of planar metal complexes or organic molecules of quite different geometry and chemical character. Both fields were reviewed recently [177, 178].

Most cases investigated so far are complicated by the fact that the iodines and the $RR'P^+$ units form "separate" incommensurate lattices whose lattice constants a, a' along the stack are unrelated. The $RR'P$ and iodine sites consequently have small inequivalences that should be reflected in their ionicities. Further more, in most cases the polyiodide ions are disordered within the channels and resemble [166] more a one-dimensional liquid rather than an ordered crystal [166]. Therefore, only a projection of the structure in a plane perpendicular to the stacks can readily be found. Such a "simplified" structure of a typical compound — 5.10-dibenzyl-5.10-dihydro-phenaziniumiodide ($Be_2P \cdot I_{1.62}$) — is shown in Fig. 20. The phenazine: iodine lattice constants along the stack are $17.781 : 9.6$ Å. So far, four different compounds have been identified [166–170], the structural data of which are summarized in Table 8. The elucidation of the full three-dimensional structure is difficult and is further complicated by the fact that the degree of disorder is strongly temperature dependent [166, 171]. Incommensurate structures driven by partial filling are also found in many molecular conductors [27, 34, 35, 102].

b = 12.24

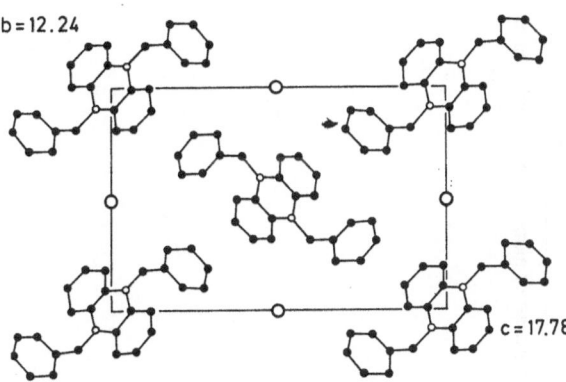

c = 17.78

Fig. 20. Projection of a polyiodide structure along the stacks. (With permission from Ref. [168])

The collective physical properties of linear polyiodides are, as expected, very anisotropic. Several molecular conductors have been identified [34, 167]. Since these physical properties clearly depend on the degree of order along the chains, they are again a complicated function of the temperature. A detailed interpretation of the epr, optical and conductivity data must await progress with the structure determinations. Both disorder within channels and incommensurate stacks are important topics for current investigations.

Table 8. Structural data of linear chain polyiodides with stacks of RR'P$^+$ cations as counterions

Stoichiometry	RR'P lattice				Iodide lattice			electrical conductivity	Ref.
	a	b	c	β in °	iodine distances		iodine		
					intra I$_3^-$	inter I$_3^-$	c	σ (Ω^{-1}·cm^{-1})	
	(all in Å)								
MEP · I$_{1.62}$	19.072(2)	15.143(1)	5.133(3)		3.211	—	25.691(2)*	0.2	
MEP · I$_{1.52}$	15.16(1)	12.14(2)	9.76(1)	125.9	3.11	3.54	9.76	3–6·10^{-4}	[165]
DEP · I$_{1.62}$	12.321(2)		5.330(2)				9.7	0.1	[165]
Be$_2$P · I$_{1.62}$	5.192(4)	12.237(3)	17.781(1)	103.5(3)	2.93	3.74	9.6	1–4	[165]
	5.216(4)						9.70(6)		

c* is the repeating distance of the superlattice along the polyiodide chains.

6 Discussion

6.1 Molecular Structure and Ionicity of RP and RR'P Donors

The structural assignment of ionicity was discussed by Herbstein [7] for the then available TCNQ structures. Flandrois and Chasseau [44] consider a more restricted TCNQ series and follow subtle structural changes. Shibaeva [95] has recently summarized structure-ionicity relations for π-donors like TTF and TTT (tetrathiatetracene). The underlying idea is simple. The quinoidal TCNQ structure (page 174) becomes more benzenoid in TCNQ⁻, thereby lengthening and shortening specific bonds on adding an electron. Linear interpolation among reference structures with definite ionicity, such as neutral TCNQ or an alkali-TCNQ, then fixes q in any compound. The assumption is that ionicity rather than crystal packing dominates the bond lengths or angles sensitive to q. It is also important to retain the planar D_{2h} geometry of TCNQ on adding an electron. The same is true for acceptors like chloranil and donors like TMPD, TTF or TTT.

Substituted phenazines are more complicated. Simple bonding arguments indicate that fully reduced, neutral RR'P donors are bent along the N-N axis, with some sp^3 hybridization at N. Such bending for M_2P-TCNQ is shown in Fig. 11 and the resulting dihedral angles for other RR'P-TCNQ complexes are mentioned in Section 4.1 and 4.3. Ionic RP⁺ cations, on the other hand, have π-systems that are isoelectronic with anthracene and are planar, as found for several systems in Section 3. Neutral RṖ and ionic RR'Ṗ⁺ radicals are intermediate. Their possible bending requires changes in molecular packing and is consequently sensitive to weak crystal forces in addition to the ionicity. Most molecular changes are expected in the central phenazine ring, without clear trends for the adjacent phenyl rings.

These caveats are borne out in Table 9, where selected bond lengths and angles, including the dihedral angle for N-N folding, are listed for phenazine-TCNQ complexes whose structures and estimated ionicities were mentioned individually in Sections 3–5. The selected features are the most sensitive to ionicity variations. The change from planar to bent geometry, and possibly the reduced symmetry of RP donors, are probably responsible for the limited structure-ionicity relations in Table 9.

The CT integral t, Eq. (2), reflects the detailed geometry of the π-π overlaps in either mixed or segregated stacks. Different overlap patterns in TCNQ salts, however, yield only qualitative information. For example, ring over external-bond overlap is favored [7] and presumably results in large |t|. Theoretical treatments [103, 180] are still more oversimplified and focus primarily on the plane-to-plane separation and tilt of simple π-donors and acceptors. As discussed in Section 3.3, the π-dimers of TCNQ⁻ in PP-TCNQ and BP-TCNQ have close face-to-face geometry. On changing the DA dimer in Fig. 3 to an A⁻A⁻ dimer, we find

$$2\Delta E_p = [\Delta E_{CT}^2 + 16|t|^2]^{1/2} - \Delta E_{CT} \tag{10}$$

The CT absorption in simple TCNQ salts are around 1.3 ev, which gives $|t| \sim 0.1$ ev for $\Delta E_p \sim 0.25$ ev in Table 3. Similar estimates are possible in other dimerized TSE systems. The analysis becomes more approximate for alternating stacks with com-

Table 9. Selected bond lengths, angles and "N-N" folding in solid phenazine TCNQ compounds

Complex	N-N-bending (angle in °)	A (Å)	B (Å)	C (Å)	Ref.
$(MP)_2$-Ni(mnt)$_2$	4	1.42	1.36	1.39	160)
HMP-ClO$_4$	1.5; 2.1 (twist)	1.389	1.370	1.395	118)
HMP-TCNQ	5.6	1.391; 1.395	1.357; 1.380	1.373; 1.289	118)
M$_2$P-PF$_6$	3.3	1.44; 1.48	1.35; 1.29	1.50; 1.27	161)
M$_2$P-I$_3$	14.5	1.412	1.364	1.437	165)
M$_2$P-TCNQ	15	1.398; 1.384	1.376; 1.377	1.393; 1.367	140)
M$_2$P-TCNQF$_4$	0	1.406	1.364	1.384	87)
M$_6$P-TCNE	0	1.410	1.378	1.395	85)
E$_2$P-TCNQ	12.7	1.407	1.380	1.356	142)
E$_2$P-Br(H$_2$O)$_2$	0	1.404	1.364	1.398	161)
P$_2$P-TCNQ	14.0; 4.5	1.36; 1.34	1.35; 1.33	1.30; 1.35	144)
BP-TCNQ	0	1.414	1.348	1.411	72)
BP-TCNQF$_4$	0	1.400	1.360	1.397	73)

This valence bond picture corresponds best to the found interatomic distances.

parable π-π overlaps with both neighbors. The different observed phenazine-TCNQ overlap patterns thus remain, for the time being, largely uninterpreted.

Typical infrared or Raman assignments of partial ionicity also depend on preserving the molecular point group symmetry [96]. The number and types of normal frequencies are then fixed and their q-dependence can be monitored, again assuming, crystal perturbations to be small. However, the point group symmetry of phenazines changes with ionicity. While no systematic spectroscopic data is available, we would not expect studies to be nearly as successful as for TCNQ, TTF and other D_{2h} donors or acceptors.

6.2 Multiple Minima and Crystal Cohesive Energy

The structural diversity of RP-TCNQ complexes in Section 3 suggests a complicated interplay among comparable crystal forces. The restoration of segregated stacking on phenazine doping of mixed-stack RR'P-TCNQ complexes, as discussed in Section 4.3, also points to nearly degenerate energy minima in widely separated regions of phase space. We indicate schematically in Fig. 21 several different packing arrangements with comparable crystal cohesive energies. The free energy differences ΔG_{12}, ΔG_{13}, etc. are far smaller than the "barriers" interconnecting the structures. We expect such valleys for any phenazine-TCNQ complex, even though the actual minimum will differ from one complex to the next. As expected for van der Waals molecular solids, the observed structures all fill space efficiently and imply similar densities for RP-TCNQ and RR'P-TCNQ complexes.

The crystal cohesive energy, E, in Fig. 21 could also be plotted along an ionicity axis, q, normal to the paper. The crystal electrostatic (Madelung) energy of mixed stacks clearly increase with q, so that the second valley may deepen with increasing q. The first valley for segregated stacks may initially deepen but becomes far less favorable

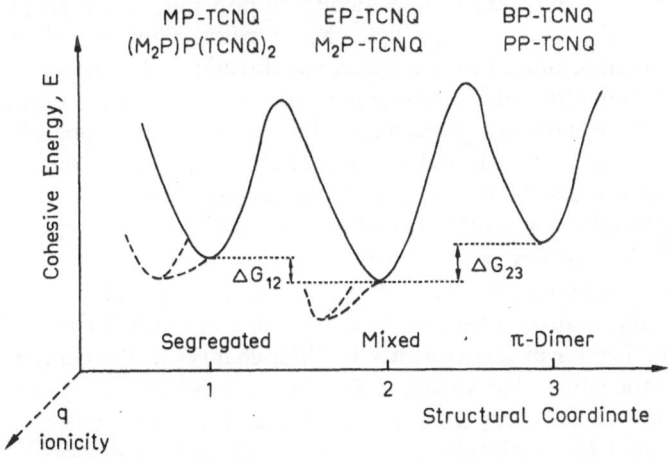

Fig. 21. Schematic dependence of lattice free energies on a structural coordinate or on the ionicity q. The drawing indicates the high activation energies for structural modifications and the smaller changes with q

at $q = 1$. This corresponds to a partly-ionic ground state. All three structures in Fig. 21 become typical neutral solids at $q = 0$. Now atom-atom potentials [58, 181, 182] could be used in principle to assess the most favorable, but hypothetical, neutral packing. Such analysis is far from simple even in closed-shell molecular solids. The additional complications of partly-filled bands ($|t| > 0$) and of long-range Coulomb interactions must be included for phenazine-TCNQ complexes. In our opinion, there is no dominant or controlling contribution to E.

Bandwidth contributions are unique to open-shell solids. Finite $|t|$ stabilizes the CT dimer by $\sim t^2/\Delta E_{CT}$ and was initially thought to control the geometry in solution. Subsequent work, as reviewed critically by Hanna and Lippert [183], demonstrates that CT stabilization does not determine the dimer geometry. It was also noted [184] in connection with largely neutral TCNE complexes, that the solid-state geometry does not maximize CT interactions. Since intermolecular forces usually depend on the total electron density, they are similar for singlet and triplet states of ionradicals. The measured magnetic gap, ΔE_p, then gives the CT contribution in the solid [185]. As discussed in Sections 3.3 and 4.1, typical ΔE_p values are around 0.1 ev and may range up to ~ 0.6 ev.

The Madelung constant, M per D^+A^- pair, is certainly the largest intermolecular interaction. Typical values [58] are $M \sim 2$–4 ev. Most of the energy goes into the I-A term for transfering an electron from D to A. The small net contribution is comparable to other terms and does not control the structure. A linear q-dependence is expected for I-A in the limit of separated sites, when ionicity q implies a fraction q of the donors and acceptors are fully ionized. The Hartree-Fock limit of fractionally charged sites, by contrast, yields a q^2 dependence for M(q) and yields [86] minima at $q = 0$ or 1, but not in between. Correlated crystal states [65], or VB structures for the entire solid, improve significantly on the Hartree-Fock electrostatic energy around $q \sim 1/2$. Favorable D^+A^- contacts are enhanced, while unfavorable D^+D^+ or A^-A^- contacts are suppressed through short-range charge correlations. The electrostatic energy is linear, or even sublinear, in q. Now M(q) and q(I — A) may almost cancel over a range of q and there is no difficulty, in principle, with a partly-ionic ground state [65]. Furthermore, short-range charge correlations are three-dimensional and do not require finite t in the transverse directions. The result for correlated states in partly-ionic ($q \sim 0.5$–0.8) organic solids is to minimize electrostatic differences between mixed and segregated stacks. This supports the comparable minima in Fig. 21 for segregated, mixed, and π-dimer stacking of various ionicity.

The crystal cohesive energy can be determined [58] thermochemically, no matter how complicated the intermolecular forces. The energy for partly-ionic complexes is only slightly lower [186], by <0.5 ev per DA pair, than that of the neutral solids of pure A and pure D. The ready formation of neutral CT complexes of weak D and A suggests that mixed stacking is already favored at $q \sim 0$; this accounts for part of the $\lesssim 0.5$ ev stabilization. There can consequently be little changes in the cohesive energy on partly-ionizing the lattice. The valleys in Fig. 21 are rather flat along the q axis, in contrast to the high barriers along the structural coordinate. This is consistent with significant pressure-induced ionicity changes in segregated-stack conductors [187] like TTF-TCNQ and mixed-stack insulators [105] like TTF-chloranil, both for P < 20 k bar. Such behavior under pressure is likely among phenazine-TCNQ complexes and offers promising extensions.

We have emphasized that minor chemical modifications stabilize different structural minima in Fig. 21. The MBP-TCNQ complexes [115] in Section 3.4 may illustrate, if supported by structure determinations, different structures for the same complex, presumably due to different preparative conditions. That one modification is meta-stable hardly matters in the solid state where there is no interconversion between different minima in Fig. 21. The precise control of crystallization conditions is obviously crucial. The impurity effects leading to MP-TCNQ or HMP-TCNQ, or to different M_2P-TCNE adducts [146] in Section 4.4, represent different structures for *different* complexes. We are not aware of definitive evidence for different structures of the *same* complex, but suspect such behavior to be possible among phenazine complexes.

6.3 Antiferromagnetic Exchange

The stabilization of the singlet $^1|D^+A^-\rangle$ state in Fig. 3 by configuration interaction with $|DA\rangle$ can also be represented by an effective Hamiltonian,

$$\mathcal{H}_e = 2J\left(S_1 \cdot S_2 - \frac{1}{4}\right) \tag{11}$$

for $S = 1/2$ sites. $\mathcal{H}_e$ describes only the singlet-triplet splitting, with $2J = \Delta E_p$. Exchange Hamiltonians are widely used in extended systems [33], including transition-metal complexes and inorganic solids in addition to organic ion-radicals. The antiferromagnetic ($J > 0$) exchange in Eq. (11) is then treated as a phenomenological parameter whose origins are a separate problem. CT stabilization of mixed ionic dimers, D^+A^-, goes at $J = t^2/\Delta E_{CT}$ in lowest order. The high-energy $|D^-A^+\rangle$ singlet is neglected. Virtual transfer in either direction is possible for A^-A^- or D^+D^+ dimers. The leading term of $J = 2t^2/\Delta E_{CT}$ then follows from expanding Eq. (10) for $|t|/\Delta E_{CT} \ll 1$.

The crystal structure usually dictates the appropriate exchanges, but not their magnitude. The π-dimer structures of PP- or BP-TCNQ in Section 3.3, for example, unambiguously point to Eq. (11). The mixed regular stacks of M_2P-TCNQF$_4$ in Section 4.2 imply a single J along the stack, in contrast to the alternating exchanges expected in dimerized stacks like M_2P-TCNQ. Additional exchange parameters are needed for trimerized or tetramerized stacks, whose susceptibility analysis is consequently less decisive. A normalized distribution, f(J), of exchanges is introduced [63] for TCNQ salts with disordered cations. As discussed for M_6P-TCNE in Section 4.4, the choice of f(J) may be arbitrary when there is limited data. In practice, weak exchanges in either one or both transverse directions can be included as additional parameters [188], thereby generating an anisotropic exchange network in two or three dimensions. Corrections [189] to isotropic exchange in Eq. (11) also require additional parameters and are important for transition-metal complexes with substantial spin-orbit coupling. The procedure [33, 190] is to choose exchanges consistent with structural data, solve the resulting $\mathcal{H}_e$ for the extended system and compare with susceptibility, magnetic specific heat, spin resonance, or other data.

Since $|t|$ depends sensitively on the overlap, molecular vibrations or thermal expansion modulate the transfer integral and, through Eq. (11), the exchange J. The usual [191)] Taylor expansion in the intermolecular separation, R, yields

$$J(R) = J(R_0) + \left(\frac{\partial J}{\partial R}\right)_{R0} \delta R + \dots \tag{12}$$

where $\delta R = R - R_0$ is the displacement. Only the linear δR term is typically retained. Such spin-phonon interactions [192)] have long been known for TSE motion in ion-radical organic solids. Similar expansion of $|t(R)|$ are important [27−29)] for the electronic structure of partly-filled organic conductors and polymers. As reviewed by Bray et al. [193)], one consequence of Eq. (12) is that the regular $s = 1/2$. Heisenberg antiferromagnetic chain is unstable at low temperature and becomes dimerized below the spin-Peierls temperature, T_e. Only a few well-characterized organic spin-Peierls systems were identified [193)], all with $T_c < 20$ K.

Transition-metal ions also form many linear chain magnets [194, 195)]. Instead of spin-Peierls distortions, however, three-dimensional magnetic ordering at low temperature typically results from small transverse interactions. These inorganic solids are evidently stiffer than organic molecular crystals with ~ 3.2–3.5 Å face-to-face separations. The gain in electronic (exchange) energy on dimerization is at most of order J and competes with the distortion energy. The larger ΔE_p for organic solids and their lower Debye temperatures thus both favor spin-Peierls distortions, although there is no quantitative discussion of the relative stiffness and exchange in organic and inorganic linear chains. The spin-Peierls systems identified by Bray et al. [193)] are in fact based on TTF^+ and $(TCNQ)_2^-$ molecular ion-radicals.

Torrance [196)] has suggested, on the contrary, that spin-Peierls transitions are common among organic ion-radicals and have substantially higher $T_c \sim 100$–400 K. The argument involves the frequent occurrence of $\chi(T)$ knees [196, 197,] for example at $T_0 = 122$ K for M_2P-$TCNQF_4$ in Fig. 14, and the correlation between higher T_0 and lower $\chi(T)$. Weak activation $\Delta E_p > 0$ above T_0, as indicated in Eq. (6), may reflect pseudogap effects of the type invoked [198)] for charge transport in inorganic linear chains. A deformable regular lattice may possibly support a finite ΔE_p, in contrast to the known, nonactivated Bonner-Fisher [104)] susceptibility for a rigid regular lattice. We interpreted the TMPD-TCNQ gap in Section 4.2 as evidence for partial ionity in a rigid mixed regular stack. The alternative suggestion for a deformable, ionic lattice also applies to segregated stacks [196, 197)].

The high value of T_0 is not understood. Indeed, organic molecular solids have so many possible transitions that we hesitate to assign them to any single mechanism. Whatever the mechanism, modeling a relatively soft, deformable linear chain with exchanges given by Eq. (12) is both possible and desirable. We note here that spin-phonon interactions open new possibilities for the relatively well-understood magnetic properties and that such interactions are particularly relevant to organic molecular solids. Antiferromagnetic exchanges in partly-ionic phenazine-TCNQ complexes or in M_6P-TCNE may then be assigned more confidently.

6.4 Summary and Extension

This initial survey of solid CT complexes based on phenazine donors has revealed a rich structural variety. Many of the $R\dot{P}$ and $RR'P$ complexes in Sections 3 and 4 were not characterized in detail. We nevertheless believe to have sampled the major structural motifs for phenazine complexes with π-acceptors. These include mixed and segregated stacks with either regular or alternating transfer integrals and π-dimers in ...$D^+D^+A^-A^-D^+D^+$... arrays with diamagnetic RP^+ cations. Inorganic counterions in Section 5 afford other structural possibilities, some standard. The interesting incommensurate structure of iodine complexes will attract further attention.

As suggested [22] a decade ago, modeling ion-radical organic solids must begin with the observed crystal structure, which then implies certain magnetic, optical, or electrical possibilities mentioned in Section 2.2. Phenazine-TCNQ structures clearly elude prediction by current theories of intermolecular forces, a situation that is not likely to change soon. The interplay among comparable intermolecular forces in phenazine complexes often complicates even the rationalization of observed physical properties. Partial ionicity is particularly important and difficult. It must be better understood for the cohesive energy of narrow-band solids, for magnetic and optical excitations, and for transitions from largely neutral to largely ionic ground states. Correlated crystal states [65] offer a promising description of partial ionicity, but their manipulation and properties still require extensive development.

In addition to their structures, we have focused on the magnetic properties of solid CT complexes of phenazines. Various triplet spin exciton (TSE) systems were identified, together with exchange-narrowed $g \sim 2$ resonances. Previous theory suffices for the principal features [33]. Numerous extensions are indicated, however, for any quantitative treatment of secondary features. In the context of modeling narrow-band organic solids, it is more logical to describe the optical and transport properties of selected phenazine-TCNQ complexes than to extend purely magnetic investigations. Promising complexes must be amenable to single-crystal studies. It is clearly desirable to study pressure as well as temperature dependences, as shown by work on neutral-ionic transitions [105].

We mention several extensions in closing. More work is needed on the interconversion in Section 4.3 of mixed to segregated stacks with phenazine doping. The contrast between structural disorder in MP-TCNQ and chemical disorder in $(M_2P)P(TCNQ)_2$ allows a novel study of disorder in virtually identical crystals. Another important extension is the preparation of the same complex in distinctly different structural modifications. As for doping studies, scrupulous care must be exercised in the preparation and characterization. We have often benefited from preparing several related complexes, rather than selecting a single candidate, and such a strategy is also indicated for future work. For example, TMPD-TCNQF$_4$ was a prime candidate for ionic mixed regular stacks, while the dimerized M$_2$P-TCNQ structure made the TCNQF$_4$ complex less attractive; nevertheless, M$_2$P-TCNQF$_4$ formed the desired ionic regular stacks. Promising extensions may well involve unexpected phenazines or acceptors, at least at the present level of understanding these extremely diverse solid CT complexes.

7 Acknowledgement

We gratefully acknowledge financial support by the Scientific Committee of NATO (Research Grant No. 137.81), to make this collaborative effort possible. The experimental work in Heidelberg was supported by Deutsche Forschungsgemeinschaft

(Ke 135/18 and 135/25) and Stiftung Volkswagenwerk (Az.: I/37244), the work in Princeton by NSF DMR-7727418. Z. G. Soos thanks the Guggenheim Foundation for a fellowship and the Physics Department, University of California, Los Angeles, for their warm hospitality in 1981–82. The bibliography for this review was completed in June 1982.

We would like to thank K. Dietz, D. Wehe, D. Nöthe, R. Harms, K. Ludolf, B. M. Metzger and J. B. Tortance for collaboration and/or many helpful discussions.

8 Notation

A	acceptor
Å	Angstrom
a, b, c	unit cell dimensions
α, β, γ	unit cell angles
a, b, c	crystallographic axis
cm^{-1}	wavenumber
CT	charge transfer
D	donor
ΔE_{CT}	charge transfer energy
ΔE_p	activation energy for paramagnetism
epr	electron paramagnetic resonance
eV	electron volt
I	Ionization energy
J	exchange integral
M	Madelung constant
q	ionicity
RR	resonance Raman
σ	specific conductivity
T_c	critical temperature for phase transition
t	charge transfer integral
TCNQ	7,7,8,8-tetracyano-p-quinodimethane, 2,2′-(2,5-cyclohexadiene-1,4-diylidene)-bispropanedinitrile
TCNE	Tetracyanoethane
TMPD	N,N,N′,N′-tetramethyl-1,4-diaminobenzene
TSE	triplet spin exciton
X	halide ion: Cl^-, Br^-, or I^- or anion like ClO_4^-, BF_4^-, PF_6^-

9 References

1. Mc Ilwain, H.: J. Chem. Soc. *1937*, 1704
2. a) Hausser, K. H., Murrell, J. N.: J. Chem. Phys. *27*, 500 (1957)
   b) Hausser, K. H.: Z. Naturforsch. *11a*, 20 (1956)
   c) Hausser, K. H., Birkofer, L.: Naturwiss. *42*, 97 (1955)
   d) Sakata, T., Nagakura, S.: Bull. Chem. Soc. Japan *42*, 1497 (1969)
   e) Serafimov, O., Zimmermann, H.: Ber. Bunsenges. Phys. Chem. *76*, 904 (1972)
3. Mulliken, R. S.: J. Amer. Chem. Soc. *74*, 811 (1952);
   J. Phys. Chem. *56*, 801 (1952); J. Chim. Phys. *61*, 20 (1964)
4. Thomas, D. D., Keller, H. J., McConnell, H. M.: J. Chem. Phys. *39*, 2321 (1963)
5. Herbstein, F. H., Schmidt, G. M. J.: Acta Cryst. *8*, 406 (1955)

6. a) Goldberg, I., Shmueli, U.: ibid. *B29*, 440 (1973)
   b) Goldberg, I., Shmueli, U.: Nature Phys. Sci. *234(45)*, 36 (1971)
7. Herbstein, F. H.: in Perspectives in Structural Chemistry, Vol. IV (eds. Dunitz, J. D., Ibers, J. A.), Wiley, New York, 1971, pp. 166
8. a) Nissler, H.: Diplomarbeit, Univ. of Stuttgart
   b) Karl, N., Ketterer, W., Stezowski, J. J.: Acta Cryst. 1982, in print
9. a) Clark-Lewis, J. W., Moody, K.: Austral. J. Chem. *24*, 2593 (1971)
   b) Takagi, J., Akasaka, K., Imoto, T., Kawai, H., Ishuzu, K.: Chemi. Lett. *1972*, 847
   c) Selbin, J., Durrett, D. G., Sherrill, H. J., Newkome, G. R., Collins, M.: J. Inorg. Nucl. Chem. *35*, 3467 (1973)
   d) Bandish, B. K., Shine, H. J.: J. Org. Chem. *42*, 561 (1977)
10. Melby, L. R., Harder, R. J., Hertler, W. R., Mahler, W., Benson, R. E., Mochel, W. E.: J. Amer. Chem. Soc. *84*, 3374 (1962)
11. Chesnut, D. B., Phillips, W. D.: J. Chem. Phys. *35*, 1002 (1961);
    Siemons, W. J., Bierstedt, P. E., Kepler, R. G.: ibid. *39*, 3523 (1963)
12. Melby, L. R.: Can. J. Chem. *43*, 1448 (1965)
13. a) Epstein, A. J., Etemad, S., Garito, A. F., Heeger, A. J.: Phys. Rev. *B5*, 952 (1972)
    b) Epstein, A. J., Etemad, S., Garito, A. F., Heeger, A. J.: Solid State Commun. *9*, 1803 (1971)
14. a) Coleman, L. B., Cohen, J. A., Garito, A. G., Heeger, A. J.: Phys. Rev. *B7*, 2122 (1973)
    b) Ehrenfreund, E., Etemad, S., Coleman, L. B., Rybaczewski, E. F., Garito, A. F., Heeger, A. J.: Phys. Rev. Lett. *29*, 269 (1972)
15. Garito, A. F., Heeger, A. J.: Accts. Chem. Res. *7*, 232 (1974)
16. a) Heeger, A. J.: in Chemistry and Physics of One-Dimensional Metals (ed. Keller, H. J.), NaTO-ASI Series Vol. *25B*, Plenum Press, New York 1977, 87
    b) Bloch, A. N., Carruthers, T. F., Poehler, T. O., Cowan, D. O.: ibid., p. 47
    c) Jérome, D., Weger, M.: ibid., p. 341
    d) Bechgaard, K., Andersen, J. R.: in The Physics and Chemistry of Low-Dimensional Solids (Alcácer, L., ed.), NATO-ASI Series *C56*, 1980, D. Reidel Publ. Corp. Dordrecht, 247
17. a) Bechgaard, K.: in Molecular Metals (Hatfield, W. E., ed.), NATO-Conf. Series VI, *1*, 1979, Plenum Press N. Y., p. 1
    b) Engler, E. M., Schumaker, R. R., Kaufmann, F. B.: ibid., p. 31
    c) Tomkiewicz, Y., Engler, E. M., Scott, B. A., La Placa, S. J., Brom, J.: ibid., p. 43
    d) Proceedings of the Int. Colloquium on "The Physics and Chemistry of Synthetic and Organic Metal", Journal de Physique *44*, C3 (1983)
18. a) Heeger, A. J., Mc Diarmid, A. G.: in The Physics and Chemistry of Low-Dimensional Solids (Alcácer, L., ed.), NATO-ASI Series, *C56*, 1980, D. Reidel Publ. Corp. Dordrecht, 343
    b) Proceedings of the Internat. Conf. on Low-Dimensional Conductors, Part A. Mol. Cryst. Liq. Cryst. *77*, pp. 1–356 (1981)
19. a) Proceedings of the Internat. Conf. on "Physics and Chemistry of Conducting Polymers", Journal de Physique *44*, (1983)
    b) Proceedings of the Int. Conf. on "The Physics and Chemistry of Low-dimensional Synthetic Metals", Mol. Cryst. Liq. Cryst. (1985), in print
20. a) Bechgaard, K., Carneiro, K., Olsen, M., Rasmussen, F. B., Jacobsen, C. S.: Phys. Rev. Letters *46*, 852 (1981)
    b) Crabtree, G. W., Carlson, K. D., Hall, L. N., Copps, P. T., Wang, H. H., Enge, I. J., Beno, M. A., Williams, J. M.: Phys. Rev. *B30*, p. 2958 (1984)
21. Butler, M. A., Wudl, F., Soos, Z. G.: ibid. *B12*, 470 (1975)
22. Soos, Z. G., Klein, D. J.: in Molecular Association (Foster, R., ed.), Vol. 1, 1975, Academic N.Y., 1–109;
    Soos, Z. G.: Ann. Rev. Phys. Chem. *25*, 121 (1974)
23. Torrance, J. B.: Acct. Chem. Res. *12*, 79 (1979)
24. a) Swan, G. A., Felton, D. G. I.: in The Phenazines in the Series Chem. Heterocycl. Comp. 1957
    b) Evans, C. A., Bolton, J. R.: J. Amer. Chem. Soc. *99*, 4502 (1977)
    c) Morrison, M. M., Seo, T. E., Howie, J. K., Sawyer, D. T.: ibid. *100*, 207 (1978)
    d) Nosaka, Y., Akasaka, K., Hatano, H.: ibid. *100*, 706 (1978)

211

25. a) Rubaszewska, W., Grabowski, Z. R.: J. Chem. Soc. Perkin Trans. II *1975*, 417
    b) Dobkowski, J., Rubaszewska, W.: Roczniki Chem. *50*, 1435 (1976)
26. Soos, Z. G., Keller, H. J., Moroni, W., Nöthe, D.: Ann. N.Y. Acad. Sci. *313*, 442 (1978)
27. The Physics and Chemistry of Low-Dimensional Solids (Alcácer, L., ed.), NATO-ASI Series 1980, *C56*; D. Reidel Publ. Corp. Dordrecht
28. Internat. Conf. of Low-Dimensional Synthetic Metals, Chem. Scripta 1981, *17*
29. Proceedings of the Internat. Conf. on Low-Dimensional Materials, J. Mol. Cryst. Liq. Cryst. 1982, *77*
30. Highly Conducting One-Dimensional Solids (Devreese, J. T., Evrard, R. P., van Doren, V. E., eds.), Plenum Press N.Y. 1979
31. Molecular Metals (Hatfield, W. E., ed.), NATO-Conf. Series VI, Vol. 1, 1979, Plenum Press N.Y.
32. Proceedings of the Symp. on Solid State Chemistry, Mol. Cryst. Liq. Cryst. *107*, p. 1–53 (1984)
33. Soos, Z. G., Bondeson, S. R.: in Extended Linear Chain Compounds (Miller, J. S., ed.), Vol. 3, 1983, Plenum Press N.Y. 193–261
34. Hoffman, B. M., Martinsen, J., Pace, L. J., Ibers, J. A.: in Extended Linear Chain Compounds (Miller J. S., ed.), Vol. 3, 1983, Plenum Press N.Y. 459–549
35. Epstein, A. J.: in Extended Linear Chain Compounds (Miller, J. S., ed.), Vol. 4, 1983, Plenum Press N.Y., to be published
36. Inoue, H., Hayashi, S., Imoto, E.: Bull. Chem. Soc. Japan *37*, 336 (1964)
37. Schlenk, W., Bermann, E.: Ann. *463*, 306 (1928)
38. Clemo, G. R., Mc Ilwain, H.: Chem. Soc. *1934*, 1991
39. Morley, J. S.: J. Chem. Soc. (London) *1952*, 4008
40. Clemo, G. R., Mc Ilwain, H.: J. Chem. Soc. *1935*, 738
41. Uchida, T.: Bull. Chem. Soc. Japan *40*, 2244 (1967)
42. Claus, A.: Ber. *8*, 601 (1875)
43. Ris, C.: ibid. *19*, 2206 (1886)
44. Scholl, R.: Monatsh. *39*, 238 (1918)
45. Inoue, H., Tamaki, K., Kida, Y., Imoto, E.: Bull. Chem. Soc. Japan *39*, 555 (1966), and literature cited therein
46. a) Japar, S. M., Abrahamson, E. W.: J. Amer. Chem. Soc. *93*, 4140 (1971)
    b) Baily, D. N., Roe, D. K., Hercules, D. M.: ibid. *90*, 6291 (1968)
47. Soos, Z. G., Keller, H. J., Moroni, W., Nöthe, D.: ibid. *99*, 5040 (1977)
48. Kehrmann, F., Havas, E.: Ber. *46*, 341 (1913)
49. a) Hantzsch, A.: ibid. *49*, 511 (1916)
    b) Keller, H. J., Maier, K., Nöthe, D.: Z. Naturforsch. *25b*, 1058 (1970)
50. Takagi, Y., Akasaka, K., Imoto, T., Kawai, H., Ishizu, K.: Chem. Lett. *1972*, 847
51. Keller, H. J., Nöthe, D., Moroni, W., Soos, Z. G.: J.C.S. Chem. Comm. *1978*, 331
52. a) Chew, V. S. F., Bolton, J. R.: J. Phys. Chem. *84*, 1903 (1980)
    b) Chew, V. S. F., Bolton, J. R., Brown, R. G., Porter, G.: ibid. *84*, 1909 (1980)
53. Kuhn, R., Schön, K.: Ber. *68*, 1537 (1935)
54. Gilman, H., Dietrich, J.: J. Amer. Chem. Soc. *79*, 6178 (1957)
55. Bettinetti, G. F., Maffei, S., Pietra, S.: Org. Synthesis *11*, 748 (1976)
56. Waters, W. A., Watson, D. H.: J. Chem. Soc. (London) *1959*, 2085
57. Kitaigorodsky, A. I.: in Molecular Crystals and Molecules, 1973, Academic Press, New York
58. Metzger, R. M.: in Crystal Cohesion and Conformational Energies, Topics in Current Physics, Vol. 26, 1981 (Metzger, R. M. ed.), Springer, Berlin, 80–107 and references therein
59. Klein, D. J., Soos, Z. G.: Mol. Phys. *20*, 1013 (1971)
60. Soos, Z. G., Mazumdar, S.: Phys. Rev. *B18*, 1991 (1978);
    Soos, Z. G., Bondeson, S. R., Mazumdar, S.: Chem. Phys. Lett. *65*, 331 (1979)
61. Mazumdar, S., Soos, Z. G.: Synth. Met. *1*, 77 (1979); Phys. Rev. *B23*, 2810 (1981)
62. Bondeson, S. R., Soos, Z. G.: Chem. Phys. *44*, 403 (1979); J. Chem. Phys. *71*, 380 (1979); *73*, 598 (1980)
63. Bondeson, S. R., Soos, Z. G.: Phys. Rev. *B22*, 1793 (1980); Solid State Commun. *35*, 11 (1980)
64. Ducasse, L. R., Miller, T. E., Soos, Z. G.: J. Chem. Phys. *76*, 4094 (1982); Ramasesha, S., Soos, Z. G.: ibid. *80*, 3278 (1984)

65. Soos, Z. G., Ducasse, L. R., Metzger, R. M.: ibid. *77*, 3036 (1982)
66. Mott, N. F.: Proc. Roy. Soc. *62*, 416 (1949); Rev. Mod. Phys. *40*, 677 (1968); Metal-Insulator Transitions, 1974, Taylor and Francis, London
67. Torrance, J. B., Mayerle, J. J., Bechgaard, K., Silverman, B. D., Tomkiewicz, Y.: Phys. Rev. *B24*, 4960 (1980)
68. Hawley, M. E., Poehler, T. O., Carruthers, T. F., Bloch, A. N., Cowan, D. O., Kistenmacher, T. J.: Bull. Am. Phys. Soc. *24*, 232 (1979);
    Bryden, W. A., Stokes, J. O., Cowan, D. O., Poehler, T. O., Bloch, A. N.: Mol. Cryst. Liq. Cryst. 1982, in press
69. Nordio, P. L., Soos, Z. G., McConnell, H. M.: Ann. Rev. Phys. Chem. *17*, 237 (1966)
70. Wheland, R. C., Gillson, J. L.: J. Amer. Chem. Soc. *98*, 3916 (1976)
71. Morosin, B., Plastas, H. J., Coleman, L. B., Stewart, J. M.: Acta Cryst. *B34*, 540 (1978)
72. Gundel, D., Sixl, H., Metzger, R. M., Heimer, N. E., Harms, R. H., Keller, H. J., Nöthe, D., Wehe, D.: J. Chem. Phys. *79*, 3678 (1983)
73. Metzger, R. M., Heimer, N. E., Gundel, D., Sixl, H., Harms, R. H., Keller, H. J., Nöthe, D., Wehe, D.: ibid. *77*, 6203 (1982)
74. Harms, R. H.: Dissertation, Univ. of Heidelberg 1980
75. Nöthe, D., Moroni, W., Keller, H. J., Soos, Z. G., Mazumdar, S.: Solid State Commun. *26*, 713 (1978)
76. Fritchie jr., C. J.: Acta Cryst. *20*, 892 (1966)
77. a) Bulaevskii, L. N., Zvarykina, A. V., Karimov, Y. S., Lyubovskii, R. B., Shchegolev, I. F.: Sov. Phys.-JETP *35*, 384 (1972)
    b) Shchegolev, I. F.: Phys. Stat. Solidi *12*, 9 (1972)
78. a) Sanny, J., Grüner, G., Clark, W. G.: Solid State Commun. *35*, 657 (1980)
    b) Clark, W. G.: in: Physics in One Dimension, Springer Series in Solid State Sciences, Vol. 23, 1981 (Bernasconi, J.; Schneider, T., eds.), Springer, Berlin, 289–301, and references therein
79. Epstein, A. J., Miller, J. S., Pouget, J.-P., Comès, R.: Phys. Rev. Letters *47*, 741 (1981)
80. Miller, J. S., Epstein, A. J.: J. Amer. Chem. Soc. *100*, 1639 (1978)
81. Epstein, A. J., Miller, J. S., Chaikin, P. M.: Phys. Rev. Letters *43*, 1178 (1979)
82. Epstein, A. J., Miller, J. S.: in The Physics and Chemistry of Low-Dimensional Solids (Alcácer, L., ed.), NATO-ASI Series, 1980, *C56*, Reidel Publ. Corp. Dordrecht, 333
83. Pouget, J.-P., Megtert, S., Cômes, R., Epstein, A. J.: Phys. Rev. *B21*, 486 (1980)
84. Endres, H., Keller, H. J., Moroni, W., Nöthe, D.: Acta Cryst. *B36*, 1435 (1980)
85. Flandrois, S., Keller, H. J., Ludolf, K., Nöthe, D., Bondeson, S. R., Soos, Z. G., Wehe, D.: Mol. Cryst. Liq. Cryst. *95*, 149 (1983)
86. McConnell, H. M., Hoffman, B. M., Metzger, R. M.: Proc. Natl. Acad. Sci. U.S. *53*, 46 (1965)
87. Soos, Z. G., Keller, H. J., Ludolf, K., Queckbörner, J., Wehe, D., Flandrois, S.: J. Chem. Phys. *74*, 5287 (1981)
88. Torrance, J. B., Silverman, B. D.: Phys. Rev. *B15*, 778 (1977)
89. Metzger, R. M., Bloch, A. N.: J. Chem. Phys. *63*, 5098 (1975)
90. Hubbard, J.: Phys. Rev. *B17*, 494 (1978)
91. Klymenko, V. E., Krivnov, V. Y., Ovchinnikov, A. A., Ukrainsky, I. I., Shretz, A.: Sov. Phys.-JETP *42*, 123 (1975)
92. Klymenko, V. E., Krivnov, V. Y., Orchinnikov, A. A., Ukrainsky, I. I.: J. Phys. Chem. Sol. *39*, 359 (1978)
93. Kamarás, K., Kertész, M.: Solid State Commun. *28*, 607 (1978)
94. Flandrois, S., Chasseau, D.: Acta Cryst. *B33*, 2744 (1977)
95. Shibaeva, R. P.: in Extended Linear Chain Compounds (Miller, J. S., ed.), Vol. 2, 1982, Plenum Press New York, 435–468
96. Bozio, R., Pecile, C.: in The Physics and Chemistry of Low-Dimensional Solids (Alcácer, L., ed.), NATO-ASI Series, 1980, *C56*, Reidel Publ. Corp. Dordrecht, 165–186
97. Suchanski, M. R.: PhD Thesis, Northwestern Univ., Evanston, 1977;
    Van Duyne, R. P.: private communication
98. Soos, Z. G., Mazumdar, S., Cheung, T. T. P.: Mol. Cryst. Liq. Cryst. *52*, 93 (1979)
99. Chappell, J. S., Bloch, A. N., Bryden, W. A., Maxfield, M., Poehler, T. O., Cowan, D. O.: J. Amer. Chem. Soc. *103*, 2442 (1981)

100. a) Megtert, S., Pouget, J. P., Comès, R.: in Molecular Metals (Hatfield, W. E., ed.), NATO-Conference Series VI, Vol. 1, 1979, Plenum Press, New York, 87

b) Pouget, J. P., Comès, R., Bechgaard, K.: in The Physics and Chemistry of Low-Dimensional Solids (Alcácer, L., ed.), NATO-ASI Series 1980, C56, 113, D. Reidel Publ. Corp. Dordrecht

101. Kobayashi, H., Kobayashi, A.: in Extended Linear Chain Compounds (Miller, J. S., ed.), Vol. 2, 1982, Plenum Press, New York, 259

102. a) Tomkiewicz, Y., Taranko, A. R., Torrance, J. B.: Phys. Rev. Letters 36, 751 (1976)

b) Tomkiewicz, Y., Taranko, A. R., Engler, E. M.: ibid. 37, 1705 (1975)

c) Tomkiewicz, Y.: in The Physics and Chemistry of Low-Dimensional Solids (Alcácer, L., ed.), NATO-ASI Series C, Vol. 56, 1980, Reidel Publ. Corp. Dordrecht, 187–196

103. a) Mulliken, R. S., Person, W. B.: Molecular Complexes: A Lecture and Reprint Volume 1969, Wiley, New York

b) Berlinsky, A. J.: Contemp. Phys. 17, 331 (1976)

104. Bonner, J. C., Fisher, M. E.: Phys. Rev. 135, 640 (1964);
a useful power-series fit for $\chi(T)$ is given by Torrance, J. B., Tomkiewicz, Y., Silverman, B. D.: Phys. Rev. B15, 4738, ref. 60 (1977)

105. a) Torrance, J. B., Girlando, A., Mayerle, J. J., Crowley, J. I., Lee, V. Y., Batail, P., LaPlaca, S. J.: Phys. Rev. Letters 47, 1747 (1981)

b) Hubbard, J., Torrance, J. B.: ibid. 47, 1750 (1981)

c) Jacobsen, C. S., Torrance, J. B.: J. Chem. Phys. 78, 112 (1983)

106. a) Mitani, T., Saito, G., Tokura, Y., Koda, T.: Phys. Rev. Letters, 53, 842 (1983)

b) Tokura, Y., Kaneko, Y., Okamoto, H., Tanuma, S., Koda, T., Mitani, T., Saito, G.: Mol. Cryst. Liq. Cryst. in press.

107. Torrance, J. B., Vazquez, J. E., Mayerle, J. J., Lee, V. Y.: Phys. Rev. Letters 46, 253 (1981)

108. Batail, B., LaPlaca, S. J., Mayerle, J. J., Torrance, J. B.: J. Amer. Chem. Soc. 103, 951 (1981)

109. Kobayashi, H.: Bull. Chem. Soc. Japan 48, 1373 (1975)

110. Ukei, K., Shirotani, I.: Commun. on Phys. 2, 159 (1977)

111. Sanz, F., Daly, J. J.: J. Chem. Soc. Perkin II, 1975, 1146

112. Harms, R. H., Keller, H. J., Nöthe, D., Werner, M., Gundel, D., Sixl, H., Soos, Z. G., Metzger, R. M.: Mol. Cryst. Liq. Cryst. 65, 179 (1981)

113. Harms, R. H., Keller, H. J., Nöthe, D., Wehe, D.: Acta Cryst. B38, 2838 (1982)

114. Morinaga, M., Nogami, T., Mikawa, H.: Bull. Chem. Soc. 52, 3739 (1979)

115. Matsunaga; Y.: Bull. Chem. Soc. Japan 51, 3071 (1978)

116. Harms, R. H., Keller, H. J., Nöthe, D., Wehe, D., Heimer, N., Metzger, R. M., Gundel, D., Sixl, H.: Mol. Cryst. Liq. Cryst. 85, 249 (1982)

117. a) Coleman, L. B., Khanna, S. K., Garito, A. F., Heeger, A. J., Morosin, B.: Phys. Lett. 42A, 15 (1972)

b) Morosin, B.: Acta Cryst. B32, 1176 (1976)

118. Morosin, B.: ibid. B34, 1905 (1978)

119. Sandman, D. J.: J. Amer. Chem. Soc. 100, 5230 (1978)

120. Sandman, D. J.: Mol. Cryst. Liq. Cryst. 50, 235 (1979)

121. Morosin, B.: Phys. Lett. A53, 455 (1975)

122. Yamaji, K., Pouget, J.-P., Comès, R., Epstein, A. J., Miller, J. S.: Mol. Cryst. Liq. Cryst. 85, 215 (1982)

123. a) Bloch, A. N., Weisman, R. B., Varma, C. M.: Phys. Rev. Lett. 28, 753 (1972)

b) Bloch, A. N., Varma, C. M.: J. Physics C6, 1849 (1973)

124. Schegolev, I. F.: Phys. Stat. Sol. (A), 12, 9 (1972)

125. Theodorou, G., Cohen, M. H.: Phys. Rev. Letters 37, 1014 (1976). The power-law distribution of exchanges is not necessary, as shown in Ref. 63 or 78b

126. Gruner, G., Jánossy, A., Holezer, K., Mihály, G.: in Lecture Notes in Physics, Vol. 96; Quasi One-Dimensional Conductors (Barisic, S., Bjelis, A., Cooper, J. R., Leontic, B., eds.), Springer, Berlin 1979, 246

127. Azevedo, L. J.: PhD Thesis, Univ. of California, Los Angeles 1977

128. Butler, M. A., Wudl, F., Soos, Z. G.: J. Phys. Chem. Sol. 37, 811 (1976)

129. Kwak, J. F., Beni, G., Chaikin, P. M.: Phys. Rev. B12, 4708 (1976)

130. Beni, G., Coll, C. F.: ibid. B11, 573 (1975)

131. a) Torrance, J. B., Scott, B. A., Kaufman, F. B.: Solid State Commun. *17*, 1369 (1975)
    b) Tanaka, J., Tanaka, M., Kawai, T., Takabe, T., Maki, O.: Bull. Chem. Soc. Japan *49*, 2358 (1976)
    c) Fujii, G., Shirotani, I., Nagano, H.: ibid. *50*, 1726 (1977)
132. Dong, Vu, Endres, H., Keller, H. J., Moroni, W., Nöthe, D.: Acta Cryst. *B33*, 2428 (1977)
133. Hoffmann, S. K., Corvan, P. J., Singh, P., Sethulekshmi, C. N., Metzger, R. M., Hatfield, W. E.: J. Amer. Chem. Soc. *105*, 4608 (1983)
134. Carrington, A., McLachlan, A. D.: Introduction to Magnetic Resonance, 1967, Ch. 8, Harper and Row, N.Y.
135. Silverstein, A. J., Soos, Z. G.: Chem. Phys. Letters *39*, 526 (1976)
136. Flandrois, S., Boissonade, J.: ibid. *58*, 596 (1978)
137. Leblanc, O. H. Jr.: J. Chem. Phys. *42*, 4307 (1965); in Physics and Chemistry of the Organic Solid State, Vol. III (eds. Fox, D., Labes, M. M., Weissberger, A.), Interscience, New York, 1967, 133–198
138. Moroni, W. F.: Dissertation, 1977, University of Heidelberg
139. Hughes, R. C.: private communication
140. Goldberg, I., Shmueli, U.: Acta Cryst. *B29*, 421 (1973)
141. Meneghetti, M., Girlando, A., Pecile, E.: Int. Conf. on Phys. and Chem. of Low-Dimensional Metals, Abano Terme, Italy, Mol. Cryst. Liq. Cryst. 1984, poster 2–35, in print
142. a) Dietz, K.: Dissertation, 1980, University of Heidelberg
    b) Dietz, K., Endres, H., Keller, H. J., Moroni, W. F., Wehe, D.: Z. Naturforsch. *37b*, 437 (1982)
143. Dietz, K., Endres, H., Keller, H. J., Moroni, W. F.: ibid. **36b**, 952 (1981)
144. Wehe, D.: Dissertation 1981, University of Heidelberg
145. Ludolf, K.: Dissertation, 1981, University of Heidelberg
146. Dietz, K., Keller, H. J., Nöthe, D., Wehe, D.: J. Amer. Chem. Soc. **000, 0000** (1982)
147. Fujita, I., Matsunaga, Y.: Bull. Chem. Soc. Japan *53*, 267 (1980)
148. Hanson, A. W.: Acta Cryst. *19*, 610 (1965)
149. Hughes, R. G., Hoffman, B. M.: J. Chem. Phys. *52*, 4011 (1970)
    Kinoshita, M., Akamatu, H.: Nature *207*, 291 (1965)
    Ohmasa, M., Kinoskita, M., Akamatu, H.: Bull. Chem. Soc. Japan *41*, 1998 (1968)
150. Girlando, A., Painelli, A., Pecile, C.: Mol. Cryst. Lq. Cryst. 1984, in press
151. Girlando, A., Painelli, A., Pecile, C.: Int. Conf. on Phys. and Chem. of Low-Dimensional Metals, Abano Terme, Italy 1984
152. Hanson, A. W.: Acta Cryst. *B24*, 768 (1968)
153. Dietz, K., Flandrois, S., Keller, H. J., Koch, P., Queckbörner, J., Schweitzer, D.: Chemica Scripta *17*, 93 (1981)
154. Cairns, T. L., Carboni, R. A., Coffman, D. D., Engelhardt, V. A., Heckert, R. E., Little, E. L., McGeer, E. G., McKusick, B. C., Middleton, W. J., Scribner, R. M., Theobald, C. W., Windberg, H. E.: J. Amer. Chem. Soc. *80*, 2775 (1958)
155. Williams, R. M., Wallwork, S. C.: Acta Cryst. *22*, 899 (1967)
156. Ikemoto, H., Kuroda, H.: ibid. *B24*, 383 (1968)
157. Ikemoto, I., Yakushi, K., Kuroda, H.: ibid. *B26*, 800 (1970)
158. Tanaka, M.: Bull. Chem. Soc. Japan *50*, 2881 (1977)
159. a) Khatkale, M. S., Devlin, J. P.: J. Phys. Chem. *83*, 1636 (1979)
    b) Hinkel, J. J., Devlin, J. P.: J. Chem. Phys. *58*, 4750 (1973)
160. Endres, H., Keller, H. J., Moroni, W. F., Nöthe, D.: Acta Cryst. *B35*, 353 (1979)
161. Dietz, K., Keller, H. J., Nöthe, D., Wehe, D.: Z. Naturforsch. *39b*, 452 (1984)
162. Keller, H. J., Nöthe, D., Pritzkow, H., Wehe, D., Werner, M., Koch, P., Schweitzer, D.: Mol. Cryst. Liq. Cryst. *62*, 181 (1980)
163. De Boer, J. A., Vos, A.: Acta Cryst. *B28*, 835 and 839 (1972)
164. Kehrmann, F.: Chem. Ber. *47*, 279 (1914)
165. a) Keller, H. J., Moroni, W. F., Nöthe, D., Scherz, M., Weiss, J.: Z. Naturforsch. *33b*, 838 (1978)
    b) Scherz, M.: Dissertation 1978, University of Heidelberg
166. Endres, H., Pouget, J. P., Comès, R.: J. Phys. Chem. Solids *43*, 739 (1982) and references cited therein

167. Endres, H., Keller, H. J., Moroni, W. F., Nöthe, D., Vartanian, H. M., Soos, Z. G.: J. Phys. Chem. Solids *40*, 591 (1979)
168. Chasseau, D., Filhol, A., Gaultier, J., Hauw, C., Steiger, W.: Chemica Scripta *17*, 97 (1981)
169. Endres, H., Pouget, J. P.: ibid. *17*, 137 (1981)
170. Steiger, W.: Dissertation 1981, University of Heidelberg
171. Martin, R.: private communication
172. Helberg, H. W.: private communication
173. Teitelbaum, R. C., Ruby, S. L., Marks, T. J.: J. Amer. Chem. Soc. *101*, 7568 (1979)
174. Teitelbaum, R. C., Ruby, S. L., Marks, T. J.: J. Amer. Chem. Soc. *102*, 174 (1980)
175. Phillips, T. E., Scaringe, R. P., Hoffman, B. M., Ibers, J. A.: ibid. *102*, 3435 (1980)
176. Noltemeyer, M., Saenger, W.: ibid. *102*, 2710 (1980)
177. Marks, T. J., Kalina, D. W.: in Extended Linear Chain Compounds (ed. Miller, J. S.), 1982, Vol. 1, Plenum Publ. Corp. N.Y., 197
178. Coppens, P.: in Extended Linear Chain Compounds (ed. Miller, J. S.), 1982, Vol. 1, Plenum Publ. Corp. N.Y., 333
179. Filhol, A., Rovira, M., Hauw, C., Gaultier, J., Chasseau, D., Dupuis, P.: Acta Cryst. *B35*, 1652 (1975)
180. Chesnut, D. B., Moseley, R. W.: Theor. Chim. Acta (Berlin) *13*, 230 (1969)
181. Williams, D. E.: in Crystal Cohesion and Conformational Energies, Topics in Current Physics, 1981, *26* (ed. Metzger, R. M.), Springer, Berlin
182. Govers, H. A. J.: Acta Cryst. *A34*, 960 (1978); *A37*, 529 (1981)
183. Hanna, M. W., Lippert, J. L.: in Molecular Complexes, 1973, *1* (ed. Foster, R., Elek, P.), London, Crane, Russak, and Co., New York, pp. 1–48
184. Kuroda, H., Amano, T., Ikemoto, I., Akamatu, H.: J. Amer. Chem. Soc. *89*, 6056 (1967)
185. Soos, Z. G., Strebel, P. J.: ibid. *93*, 3325 (1971)
186. Metzger, R. M.: J. Chem. Phys. *66*, 2525 (1972); Arafat, E. S., Metzger, R. M.: unpublished work
187. Jérome, D.: in The Physics and Chemistry of Low-Dimensional Solids, NATO ASI Series, 1980, *C56* (ed. Alcácer, L.), Reidel, Dordrecht, pp. 123–142
188. Hennessy, M. J., McElwee, C. D., Richards, P. M.: Phys. Rev. *B7*, 930 (1973)
189. Moriya, T.: ibid. *120*, 91 (1960);
     Abragam, A., Bleaney, B.: in Electron Paramagnetic Resonance of Transition Ions, 1970, Oxford Univ. Press, New York, Ch. 9
190. Willett, R. D., Wong, R. J.: Journ. Mag. Res. *42*, 446 (1981);
     Soos, Z. G., McGregor, K. T., Cheung, T. T. P., Silverstein, A. J.: Phys. Rev. *B16*, 3036 (1977);
     De Jongh, W. M., Rutten, W. L. C., Verstelle, J. C.: Physica *B82*, 288, 303 (1976)
191. Kittel, C.: Phys. Rev. *120*, 335 (1960);
     Harris, E. A.: Proc. Phys. Soc. Lond. *5*, 338 (1972);
     Lines, M. E.: Solid State Commun. *11*, 1615 (1975)
192. Soos, Z. G., McConnell, H. M.: J. Chem. Phys. *43*, 3780 (1965);
     see also Ref. 69
193. Bray, J. W., Interrante, L. V., Jacobs, I. S., Bonner, J. C.: in Extended Linear Chain Compounds, Vol. III, 1983 (ed. Miller, J. S.), Plenum, New York, 353–415
194. De Jongh, L. S., Miedema, A. R.: Adv. Phys. *23*, 1 (1974);
     Carlin, R. L., Van Duyneveldt, **000**.: Magnetic Properties of Transition-Metal Compounds, 1977, Springer-Verlag, New York;
     Hone, D. W., Richards, P. M.: Ann. Rev. Mater. Sci. *4*, 337 (1974)
195. Kokoszka, G. F., Gordon, G.: in Transition Metal Chemistry 1969, Vol. *5* (ed. Carlin, R. L.), Dekker, Marcel, New York, 181–277;
     Hatfield, W. E., Whyman, R.: ibid., 47–179
196. Torrance, J. B.: Ann. N.Y. Acad. Sci. *313*, 210 (1978); Torrance, J. B., Mayerle, J. J., Lee, V. Y.; Bozio, R., Pecile, C.: Solid State Commun. *38*, 1165 (1981)
197. Wudl, F., Schafer, D. E., Walsh, Jr., W. M., Rupp, L. W., Di Salvo, F. J., Waszczak, J. V., Kaplan, M. L., Thomas, G. A.: J. Chem. Phys. *66*, 377 (1977)
198. Zeller, H. R.: Adv. Solid State Phys. *13*, 31 (1973);
     Zeller, H. R.: in Low-Dimensional Cooperative Phenomena, NATO ASI Series *B7*, 1974 (ed. Keller, H. J.), Plenum New York, 215–234

Author Index Volumes 101–127

Contents of Vols. 50–100 see Vol. 100
Author and Subject Index Vols. 26–50 see Vol. 50

The volume numbers are printed in italics

Anders, A.: Laser Spectroscopy of Biomolecules, *126*, 23–49 (1984).

Asami, M., see Mukaiyama, T.: *127*, 133–167 (1985).

Ashe, III, A. J.: The Group 5 Heterobenzenes Arsabenzene, Stibabenzene and Bismabenzene. *105*, 125–156 (1982).

Austel, V.: Features and Problems of Practical Drug Design, *114*, 7–19 (1983).

Balaban, A. T., Motoc, I., Bonchev, D., and Mekenyan, O.: Topilogical Indices for Structure-Activity Correlations, *114*, 21–55 (1983).

Baldwin, J. E., and Perlmutter, P.: Bridged, Capped and Fenced Porphyrins. *121*, 181–220 (1984).

Barkhash, V. A.: Contemporary Problems in Carbonium Ion Chemistry I. *116/117*, 1–265 (1984).

Barthel, J., Gores, H.-J., Schmeer, G., and Wachter, R.: Non-Aqueous Electrolyte Solutions in Chemistry and Modern Technology. *111*, 33–144 (1983).

Barron, L. D., and Vrbancich, J.: Natural Vibrational Raman Optical Activity. *123*, 151–182 (1984)

Bestmann, H. J., Vostrowsky, O.: Selected Topics of the Wittig Reaction in the Synthesis of Natural Products. *109*, 85–163 (1983).

Beyer, A., Karpfen, A., and Schuster, P.: Energy Surfaces of Hydrogen-Bonded Complexes in the Vapor Phase. *120*, 1–40 (1984).

Böhrer, I. M.: Evaluation Systems in Quantitative Thin-Layer Chromatography, *126*, 95–118 (1984).

Boekelheide, V.: Syntheses and Properties of the [2$_n$] Cyclophanes, *113*, 87–143 (1983).

Bonchev, D., see Balaban, A. T., *114*, 21–55 (1983).

Bourdin, E., see Fauchais, P.: *107*, 59–183 (1983).

Charton, M., and Motoc, I.: Introduction, *114*, 1–6 (1983).

Charton, M.: The Upsilon Steric Parameter Definition and Determination, *114*, 57–91 (1983).

Charton, M.: Volume and Bulk Parameters, *114*, 107–118 (1983).

Chivers, T., and Oakley, R. T.: Sulfur-Nitrogen Anions and Related Compounds. *102*, 117–147 (1982).

Consiglio, G., and Pino, P.: Asymmetrie Hydroformylation. *105*, 77–124 (1982).

Coudert, J. F., see Fauchais, P.: *107*, 59–183 (1983).

Dyke, Th. R.: Microwave and Radiofrequency Spectra of Hydrogen Bonded Complexes in the Vapor Phase. *120*, 85–113 (1984).

Ebel, S.: Evaluation and Calibration in Quantitative Thin-Layer Chromatography, *126*, 71–94 (1984).

Edmondson, D. E., and Tollin, G.: Semiquinone Formation in Flavo- and Metalloflavoproteins. *108*, 109–138 (1983).

Eliel, E. L.: Prostereoisomerism (Prochirality). *105*, 1–76 (1982).

Fauchais, P., Bordin, E., Coudert, F., and MacPherson, R.: High Pressure Plasmas and Their Application to Ceramic Technology. *107*, 59–183 (1983).

H. J. Fischbeck, K. H. Fischbeck

Formulas, Facts and Constants

for Students and Professionals in Engineering, Chemistry and Physics

1982. XII, 251 pages. ISBN 3-540-11315-0

Contents: Basic mathematical facts and figures. – Units, conversion factors and constants. – Spectroscopy and atomic structure. – Basic wave mechanics. – Facts, figures and data useful in the laboratory.

This book provides a handy and convenient source of formulas, conversion factors and constants for students and professionals in engineering, chemistry, mathematics and physics. Section 1 covers the fundamental tools of mathematics needed in all areas of the physical sciences. Section 2 summarizes the SI system (International System of Units of measurement), lists conversion factors and gives precise values of fundamental constants. Section 3 and 4 review the basic terms of spectroscopy, atomic structure and wave mechanics. These sections serve as a guide to the interpretation of modern literature. Section 5 is a resource for work in the laboratory, listing data and formulas needed in connection with frequently used equipment such as vacuum systems and electronic devices. Material constants and other data are listed for information and as an aid for estimates or problem solving.

Formulas and tables are accompanied by examples in all those cases where their use might not be self-explanatory.

Springer-Verlag
Berlin
Heidelberg
New York
Tokyo

Reactivity and Structure

Concepts in Organic Chemistry

Editors: K. Hafner, J.-M. Lehn, C. W. Rees, P. v. R. Schleyer, B. M. Trost, R. Zahradník

Springer-Verlag
Berlin
Heidelberg
New York
Tokyo